Beyond Neo-Darwinism

Beyond Neo-Darwinism

An introduction to the new evolutionary paradigm

edited by

MAE-WAN HO

Biology Discipline
The Open University
Milton Keynes
Buckinghamshire

PETER T. SAUNDERS

Department of Mathematics
Queen Elizabeth College
University of London
London

ACADEMIC PRESS, INC.

(Harcourt Brace Jovanovich, Publishers)

London   Orlando   San Diego   New York
Toronto   Montreal   Sydney   Tokyo

ACADEMIC PRESS INC. (LONDON) LTD
24/28 Oval Road, London NW1 7DX

United States Edition published by
ACADEMIC PRESS INC.
(Harcourt Brace Jovanovich, Inc.)
Orlando, Florida 32887

British Library Cataloguing in Publication Data
Beyond neo-Darwinism.
 1. Human evolution
 I. Ho, Mae-Wan   II. Saunders, Peter T.
 573.2′01   GN281

ISBN 0-12-350080-X

PRINTED IN THE UNITED STATES OF AMERICA

85 86 87 88   9 8 7 6 5 4 3 2

Contributors

Margaret A. Boden F.B.A.

Department of Philosophy
School of Social Sciences
University of Sussex
Falmer, Brighton
Sussex BN1 9QN, UK

Sidney W. Fox

Institute for Molecular and Cellular
  Evolution
521 Anastasia
University of Miami
Coral Gables
Florida 33134, USA

Brian C. Goodwin

Biology Discipline
The Open University
Walton Hall
Milton Keynes MK7 6AA
UK

Mae-Wan Ho

Biology Discipline
The Open University
Walton Hall
Milton Keynes MK7 6AA, UK

Søren Løvtrup

Department of Zoophysiology
University of Umea
S-901 87 Umea
Sweden

Koichiro Matsuno

Technological University of Nagaoka
1603-1 Nagamine
Kamitomioka
Nagaoka 949-54
Japan

Gareth J. Nelson

Department of Entomology
American Museum of Natural History
Central Park West at 79th Street
New York, New York 10024, USA

Norman Platnick

Department of Ichthyology
American Museum of Natural History
Central Park West at 79th Street
New York, New York 10024, USA

Jeffrey W. Pollard

Department of Biochemistry
Queen Elizabeth College
University of London
Campden Hill
London W8 7AH, UK

Peter T. Saunders

Department of Mathematics
Queen Elizabeth College
University of London
Campden Hill
London W8 7AH, UK

Chris Sinha

Avery Hill College
Bexley Road
Eltham
London SE9 2PQ, UK

Elizabeth S. Vrba

Department of Palaeontology
Transvaal Museum
Paul Kruger Street
PO Box 413
Pretoria
South Africa

Gerry Webster

Development Biology Group
School of Biological Sciences
University of Sussex
Falmer, Brighton
Sussex BN1 9QN, UK

Jeffrey S. Wicken

Division of Natural Science and
   Engineering
Behrend College
Pennsylvania State University
Erie, Pennsylvania 16563, USA

Foreword

It was a great honour for me to have been asked by Ho Mae-Wan and Peter Saunders to write an introductory note to this exciting book. It needs none, of course, but presumably they felt that a few words by 'the oldest inhabitant' might be appropriate at the outset. Even more impressive for me was their decision to dedicate it to the memory of the Theoretical Biology Club.

This group, led by J. H. Woodger, met annually between 1932 and 1938 to discuss just the kind of issues raised by the present essays. Woodger ('Socrates') was a philosopher of biology interested in the applications to it of mathematical logic; Dorothy Wrinch was a mathematician interested in protein chemistry; C. H. Waddington, my particular friend and collaborator, was an experimental morphologist and embryologist; while J. D. Bernal was a crystal physicist interested in biological materials. He was also a notable philosopher of dialectical materialism and a historian of science. As for me, I was primarily a biochemist but specialized in embryology and experimental morphology. People who came less often included Max Black the philosopher, B. P. Wiesner the endocrinologist, and my wife Dorothy Needham, whose line was muscle biochemistry.

We were all devoted to a non-reductionist biological organicism which aimed at being at the same time non-obscurantist; and of course we discussed evolution-theory now and then. One of the products of the group's thinking was the book 'Order and Life', my Terry Lectures at Yale University in 1935. The three chapters dealt successively with the Nature, the Deployment, and the Hierarchical Continuity of Biological Order; and among the subjects taken up were the existence of morphogenetic hormones, the organization and 'anatomy' of the cell, the role of fibre-molecules, and the possible importance of liquid crystals, together with the field concept, and the notion of 'envelopes' in space and time. One could say that in the end this book became a minor classic, as it was reissued in paperback in 1968.

Eventually we all proposed an Institute of Mathematico-Physico-Chemical Embryology which should combine these studies at Cambridge. The Rockefeller Foundation toyed with the idea of financing this for some time, but was eventually frightened off by the hostility of the British, and the Cambridge, establishment. There are those who maintain that if the idea had been proceeded with, molecular biology might well have come into being 20 years or more before it actually did. This is as may be, but for my own part the failure of the plan enabled me to be released into Chinese studies, as could never have been the case otherwise, a circumstance which has long led me to remark with oracular brevity that the Tao of all things probably knows what it is doing.

Waddington always used to say that molecular biology and the study of the properties of DNA and RNA was all very well, but that people would have to come back to embryology in the end, if they wanted really to understand the properties of the higher living systems. As Ho Mae-Wan says in the present book, 'evolution-theory' based solely on natural selection must reckon with developmental physiology'. Another great influence on me was Teilhard de Chardin, with his interpretation of cosmic, inorganic, biological and social evolution as a single process; and I had been brought up on Lawrence J. Henderson's 'Fitness of the Environment', which defined the conditions that had to pertain before the forms of life could develop. D'Arcy Thompson was another powerful influence, with his lore of how related species could be distorted mathematically, with the essential pattern still remaining the same. And finally there was A. N. Whitehead, who showed how organicism could be non-obscurantist, and prepared me for the overwhelmingly organic quality of mediaeval Chinese philosophy.

With all this background one can see how the essays in the present volume reinforce and carry forward all these ideas. For example, Sidney Fox tells how 'holistic determinism is rooted in non-randomness at each stage', and how 'natural selection preserves non-random variants'. The idea of self-assembly and self-organization is pursued by Matsuno san who studies the increase in complexity during evolution in the light of organizing pattern-principles. Might he not have been influenced by the Neo-Confucian ideas of East Asia? With Jeffrey Wicken we look again at the increase of complexity during evolution, a writer who audaciously seeks to reconcile Herbert Spencer with Willard Gibbs. As he well says, the 'systematically directional parameters are not catered for in Neo-Darwinism'. Then the field concept is pursued by Brian Goodwin. Finally, Margaret Boden considers the ability of cells to react to positional information, and shows how this affects such problems as those of determination and fate-fixation in embryonic development. All through this book the dissatisfaction of modern biologists with reductionism is apparent.

I have mentioned only a few of the contributors, but I do commend the book as a whole to candid and judicious readers, especially those not *a priori* enamoured of the paradigms of mechanistic causality and physico-chemical reductionism. I have often thought that the reductionist concept is perhaps a *Scheinproblem*, for if you knew *all* the properties of the sub-atomic particles you would of course know how the quarks combine to form atoms, molecules, organelles, cells, organs and organisms. But only a divine intelligence (if such there is) could be capable of this before the event. And still you would have to know their properties at the organismic level, which never arise at all where lower levels are concerned. An old science like genetics could enunciate perfectly valid laws, which are always adhered to, without knowing anything definite about the chromosomes and the genes.

So, in sum, I have the greatest pleasure in introducing this book of essays, edited by my friends Peter Saunders and Ho Mae-Wan. I can only hope it will be widely read, and in due course become a classic in its own right.

Joseph Needham F.R.S., F.B.A.

Preface

Until only a few years ago, the 'synthetic' or 'neo-Darwinist' theory of evolution stood virtually unchallenged as the basis of our understanding of the organic world. There were, to be sure, a few who held out against the consensus, but they had very little influence on the majority of biologists. Almost all the research that was undertaken in evolution was designed to investigate the operation of natural selection, and was seen as confirming the theory.

Today, however, the picture is entirely different. More and more workers are showing signs of dissatisfaction with the synthetic theory. Some are attacking its philosophical foundations, arguing that the reason that it has been so amply confirmed is simply that it is unfalsifiable: with a little ingenuity any observation can be made to appear consistent with it. Others have been deliberately setting out to work in just those areas in which neo-Darwinism is least comfortable, like the problem of the gaps in the fossil record or the mechanisms of non-Mendelian inheritance. Still others, notably some systematists, have decided to ignore the theory altogether, and to carry on their research without any *a priori* assumption about how evolution has occurred. Perhaps most significantly of all, there is now appearing a stream of articles and books defending the synthetic theory. It is not so long ago that hardly anyone thought this was necessary.

All the signs are that evolution theory is in crisis, and that a change is on the way. Moreover, it is already possible to see what sort of change this is going to be. For while those whose research does not conform to the orthodox pattern represent many different disciplines, and while they appear for the most part to have arrived at their points of view independently, they still have enough in common that a careful reading of their work gives a good indication of the way in which the study of evolution is going to develop in the future.

Unfortunately, the relevant work is rather widely dispersed throughout the literature, and much of it is not immediately accessible. The authors have naturally chosen to submit their papers to journals specializing in their own subject areas, and to write them for colleagues already familiar with the field. And indeed, even if they had wanted to reach a wider audience, they would have discovered that those journals that customarily publish articles on evolution generally interpret this to mean only neo-Darwinist research. Our aim in this volume has been to gather together in one place, and in a form suitable for the non-specialist, a sufficiently broad sample of this work to provide an introduction to the emerging alternative to neo-Darwinism.

The most important thing to recognize about the alternative is that it is not a theory, as neo-Darwinism claims to be, but a paradigm. It is a way of approaching problems, rather than a solution. It accepts that different aspects of evolution may require different sorts of explanations, and in particular it rejects the fundamental tenet of neo-Darwinism that the natural selection of random variations is both necessary and sufficient to account for all of evolution. Put at its simplest, the new paradigm is an insistence on pluralism in evolutionary studies, a pluralism which is central, not something to be paid lip-service to or possibly tacked on at the end. This leads to a fundamentally different approach.

The synthetic theory is based on a definite assumption about how evolution occurs. The evolutionist is supposed to accept this as given and simply to show how the process may have operated in various different cases. The theory also has a profound influence on how organisms themselves are studied. The most important effect is the tendency to regard organisms and even societies as being decomposable into distinct traits, each of which must necessarily confer some selective advantage.

Within the new paradigm, in contrast, we begin by studying organisms and societies as integral wholes. We then try to use what we have learned about them to help us discover how they have evolved. Now there are many similarities among organisms, and so there are likely to be similarities in the ways in which they came to exist. Consequently we expect to be able to derive general statements about evolution, or at least about some aspects of it. But our account of evolution follows our understanding of that which has evolved, not the other way round.

How the organisms that we see about us came to be, and why they are the way they are, are questions that have engaged the human mind for as long as we have records, and probably even longer. For much of this time the causes were believed to be supernatural. Eventually there arose a movement whose aim was to find a scientific explanation, and many consider that this culminated with the theory of natural selection. Yet on reflection we can see that this was not quite the end of the matter, because while the subject was now a part of science it was significantly different from the rest of science. The properties of organisms were seen as being determined by natural selection, not by the same forces that mould inanimate objects. And as Darwinists and neo-Darwinists have become ever more adept at finding possible selective advantages for any trait one cares to mention, explanation in terms of the all-powerful force of natural selection has come more and more to resemble explanation in terms of the conscious design of the omnipotent Creator.

Those who work within the new paradigm seek to explain as much as possible by the same means that other scientists employ, appealing to natural selection only at the end of their study, rather than at the beginning. Thus the principle of natural selection, while it continues to impart a distinctive character to evolutionary studies, no longer sharply divides them from other disciplines. In bringing evolution theory into the mainstream of modern science, the new paradigm aims to complete the process which Lamarck and Darwin and their contemporaries began.

Milton Keynes and London   M.-W. Ho
November 1983   P. T. Saunders

*To the memory of the Theoretical Biology Club,
whose members helped to keep these ideas alive.*

Contents

Section I

NECESSITY THE MOTHER OF INVENTION

1

Pluralism and Convergence in Evolutionary Theory

MAE-WAN HO and PETER T. SAUNDERS

INTRODUCTION

It has been more than 50 years since the neo-Darwinian synthesis began with Fisher's *Genetical Theory of Natural Selection*. In that time, it has generated a veritable industry of theoretical and experimental research devoted to explicating and demonstrating the operation of the principle of natural selection. Of late, the extension of that principle to social behaviour has given birth to the 'new science' of sociobiology which, Gargantua-like, immediately threatens to subsume both anthropology and sociology. Yet, paradoxically, just as the grand synthesis seems on the verge of being complete, it has come under scientific attack simultaneously from a number of different directions. Why should this be? Perhaps it is an expression of the *Zeitgeist* which has mysteriously captured the period; or perhaps it is simply that the time is ripe for a post-Darwinian revolution.

There have always been critics of the neo-Darwinian synthesis: independent thinkers who steadfastly refused to lose sight of the fundamental problems of evolution which the theory does not address. The most influential among the critics were undoubtedly Waddington, Goldschmidt, and D'Arcy Thompson, who did much to keep the real issues alive throughout the period of the grand synthesis. Their writings continue to inform the present evolutionary controversy, though they were not sufficient by themselves to bring that about. The precipitating factors for

BEYOND NEO-DARWINISM
ISBN: 0-12-350080-X

the controversy were, we suggest, important discoveries made in molecular evolution and paleontology. These have led to major conceptual crises within the orthodox theory, compelling many workers to reassess its basic tenets.

The discovery that the rates of amino acid substitution in proteins are nearly constant during evolution gave rise to the hypothesis that many mutations may be 'neutral' with regard to natural selection (Kimura, 1968). This hypothesis was corroborated by the observation that abundant electrophoretic variations exist in natural populations (Lewontin and Hubby, 1966; Harris, 1966)—too abundant in fact to be accounted for by natural selection. The conclusion seems inescapable that the majority of genetic variations are of no physiological or phenotypic consequence. Indeed, the general disparity between genic and organismic evolution which soon emerged suggested an effective decoupling between the two. The best known example is provided by the finding that humans and chimpanzees, which are biologically so distinct as to be classified in different families, nevertheless are 99 per cent identical in their genes (King and Wilson, 1975). Such observations prompted us and others to question the validity of a theory of evolution that is essentially based solely on genes. The neutral mutation concept represents an important turning point in the history of ideas. It presaged the fall from dominance of the genetic theory of natural selection—and the concomitant return of theories on organismic structure and form.

Concurrent with the developments in molecular evolution, the Darwinian picture of gradual phyletic transformation of species was being slowly and surely eroded by improved stratigraphic analyses of the fossil record. The latter soon appeared to confirm earlier speculation of paleontologists such as Schindewolf that the postulated continuous series of intermediates during species transformations simply do not exist. Hence the origin of species may require other explanations than those offered within neo-Darwinian orthodoxy (Eldredge and Gould, 1972).

While the above discoveries and their consequences were crucial, they were precipitating factors only. For the evolutionary controversy which ensued is wide-ranging in scope. Development and the nature of the organism rapidly come to the fore as problems unintelligible within neo-Darwinism. Long cherished dogmas of heredity on which neo-Darwinism is firmly based, are being challenged from within molecular biology itself. The science of systematics and rational morphology, after a century of eclipse under the Darwinian dictum of 'descent with modification' has finally come into its own again, overturning deeply held assumptions in phylogenetic classification.

The present intellectual 'uprising'—a cross-section of which is represented in this volume—is unique. It springs simultaneously and independently out of diverse disciplines: from chemists contemplating the origin of life to developmental psychologists preoccupied with the origin of human nature. At first glance, they seem to have little in common save a dissatisfaction with the neo-Darwinian framework. A closer inspection reveals the connecting threads which converge on those fundamental issues of evolution left largely untouched by Darwin and his followers. Problems such as determinism and direction in evolution, global patterns of speciation and extinction, and the origin of biological form and function, have been regarded by neo-Darwinists either as irrelevant—because they are outside the scope of the theory—or as explicable by natural selection in combination with different *ad hoc* assumptions. To us, on the other hand, these problems are primary to evolution, and hence epistemologically prior to natural selection. Our common goal is to explain evolution everywhere by necessity and mechanism with the least possible appeal to the contingent and teleological. Accidental variation and selective advantage—the foremost categories of explanation in neo-Darwinism—are thus relegated here to the last resort.

It must not be supposed however, that there is anything approaching the 'consensus' which is often claimed for the neo-Darwinian synthesis. Pluralism is a predominant feature of the emerging paradigm of evolution. Not only is there a genuine (and in our view, healthy) diversity of opinion and emphasis, but evolution is a complex phenomenon and it is to be expected that different kinds of explanations will be appropriate to different aspects. In particular, higher level explanations cannot always be collapsed or reduced into lower level ones. Thus, pluralism ought in principle to be a permanent feature of evolutionary studies.

Above all, however, our emphasis is on process—surprisingly rich and varied in texture seen from different vantage points. The result is a transcendence of the predominantly Aristotelian framework of neo-Darwinism—in which organisms are explained in terms of essences or genes—to the post-Galilean world view in which relation and process are primary. Perhaps this is but part of the general scientific revolution that has been long overdue in biology.

RANDOMNESS AND DETERMINISM

The neo-Darwinian concept of random variation carries with it the major fallacy that everything conceivable is possible. This is made most explicit in discussions on the origin of life, when it is claimed that life is utterly improbable without natural selection. The argument is usually couched in terms of the probability that a functional polypeptide of a *specific* amino acid sequence could arise by chance, which is vanishingly small, if one supposes that all conceivable random sequences of 20 amino acids are equally likely to occur. Now, the latter supposition contains the false assumption that we have a universe of pure numbers devoid of physics or chemistry. In fact, experiments under simulated prebiotic conditions consistently tell us that the probability space of prebiotic proteins is much more restricted (Fox, Chapter 2).

The same experiments show that the other assumption implicit in such formulations—that *function* is a very rare and special quality created only as the result of natural selection—is also false. Proteinoids synthesized in the laboratory already possess a variety of catalytic functions. This is indeed a vindication of Henderson's (1913) thesis of the 'fitness of the environment'. The Darwinian notion that organisms are fitted (or adapted) to the environment is incomplete, for it misses the reciprocal in the relationship: the environment is fit for the origination and evolution of organisms. Ultimately, mechanism and teleology may be one and the same, being part and parcel of the properties of matter of which living things are composed.

Inherent in the idea of non-randomness and material self-organization (Fox; Matsuno, Chapter 3) is the concept of determinism in evolution. This is best expressed by a quotation from T. H. Morgan (1929), founder of chromosomal genetics:

> When, if ever, the whole story can be told, the problem of adaptation of the organism to its environment and the coordination of its parts may appear to be a self-contained progressive elaboration of chemical compounds–a process no more fortuitous than the constitution of the earth or its revolution about the sun. The outcome would be as determined as any natural event, subject always to the principle of survival . . .   (p. 77)

Determinism in turn implies a general direction or 'time's arrow' in evolution. Many recent attempts have been made to identify a relatively objective quality if not quantity which increases in evolution; one suggestion is 'complexity' (*see* Saunders and Ho, 1976). Wicken (Chapter 4) attempts to account for the increase in complexity in evolution from a thermodynamic perspective: the second law promotes chemical structuring through the randomization of potential energy and material configuration. This is reminiscent of Joseph Needham's (1943) earlier attempt to reconcile the apparent contradiction between the increase in entropy predicted by the second law and the increase in biological organization in evolution. Wicken and Matsuno both envisage an inexorable impetus for evolutionary change: the former in the flow of energy along ordered pathways of entropy production, the latter in the flow of material through self-organizing systems. For Wicken, the evolutionary 'motor' is cosmic energy external to the biosphere, whereas for Matsuno, it is the process of equilibration endogenous to self-organizing systems which incessantly gives rise to further evolution autonomously.

PATTERNS OF ORGANISMIC EVOLUTION

If deterministic constraints exist, then certain regularities or trends in the large scale pattern of evolution should be evident. Yet very few studies have addressed this problem. One main reason is that natural selection is strictly a local mechanism (Saunders and Ho, 1976) and hence inherently unable to account for any global trend or pattern (Vrba, Chapter 5). Another reason is that evolutionary pattern itself is the product of inference from available data. Where inference is habitually made under certain presumptions, the resulting pattern becomes correspondingly biased. A case in point is the phylogenetic classification of organisms.

Organisms exist in a naturally ordered hierarchy, and it is that which makes a hierarchical system of classification possible. This is independent of the Darwinian notion of descent with modification, as the success of pre-Darwinian systematics amply demonstrates. But when phylogenetic classification is performed under presumed adaptive pathways, as implied by Darwinian natural selection, pattern becomes confounded with hypothetical process. Cladistic analysis is an attempt to re-dissociate inferences of organismic relationship and evolutionary pattern from assumptions concerning process (Nelson and Platnick, Chapter 6). The first results of its application already present views of evolution very much at odds with the orthodoxy.

The predominant pattern of the fossil record is punctuated equilibria: long periods of relative stasis interrupted by bursts of speciation events at which almost all morphological divergences take place. Many interpret this to be an indication that the process occurring within species (neo-Darwinian micro-evolution) differs from that which establishes species (macro-evolution, Vrba). Biogeographical data begin to yield families of species relationships which are congruent for widely different groups inhabiting the same areas. This implies the existence of historical geological events which subdivide geographic areas, resulting in parallel paths of divergence for all included groups of species (Nelson and Platnick). Orthodox explanations based on centres of dispersal of ancestral taxons are thus called into serious question.

The significance of the above observations lies not so much in their obvious challenge to established ideas as in their demonstration of how the arbitrariness of description of data can be reduced. This allows the proper integration of phylogenetic data, paving the way to more rigorous formulation and testing of evolutionary hypotheses concerning process.

THE PROBLEM OF FORM

The abstraction of empirical regularities or patterns is not enough, however. The regularities of biological form have been extensively commented on by students of morphology since Goethe's hypothesis of the 'unity of type'. Yet only relatively recently have there been attempts to look beneath the regularities for the *generative* germs of form (Webster, Chapter 8).

Pre-Darwinian morphologists like von Baer, Geoffroy St. Hilaire and Richard Owen clearly perceived a deep connection between the systematics of natural forms and the organization of developmental processes by which forms are realized. Since Darwin, this connection has been interpreted purely in terms of descent with modification. Organisms are seen to be the sum of past random variations accumulated by natural selection and preserved by heredity. Form therefore requires no other explanation than a combination of utility and historical contingency. Similarly, development (or ontogeny) is nothing more than the recapitulation of evolutionary history (or phylogeny). Such was the basis of Haeckel's famous biogenetic law.

But organisms are not preformed in the germ, they take shape epigenetically, in the course of development. As von Baer noted long ago, the regularities of ontogeny point to the existence of an underlying universality of process independent of recapitulation. Thus, the cause of phylogency may be sought within the epigenetic processes involved in ontogeny itself (Løvtrup, Chapter 7). Løvtrup indeed interprets cladistic phylogenetic trees literally in the sense that the evolutionary pathways — or pathways of changes in ontogeny—can be read off directly from them. The corollary of this is that ontogenetic sequences can be used to test phylogenetic hypotheses (Nelson, 1978). At any level of a phylogenetic hierarchy, primitive (plesiomorphic) characters should be common to the ontogenies of two or more groups, and derived (apomorphic) characters should be group specific.

The thesis that development exerts deterministic influence on evolution is assumed by many authors: it is certainly a strong underpinning of contemporary evolutionary thinking. However, there is a clear divergence in emphasis as to the precise nature of the developmental 'constraints' on evolution.

The major divide is a diachronic–synchronic one. At the extreme synchronic level, catastrophe theory defines a taxonomy of pure forms which is independent of the substrate (Thom, 1972). This has the important implication that all, or nearly all dynamical systems may be reducible to a limited number of canonical classes. At once, it becomes tempting to consider if the multifarious array of biological forms could be described in terms of combinations of the elementary catastrophes in such a way as to constitute a 'theory of archetypes'. This would be an extremely ambitious undertaking even if it were feasible in principle, as there are layers of analysis yet to be performed before empirical observations and abstract theory could be bridged. Granted that there might be a structuralist core of 'invariants' to biological forms, would that adequately describe much that is of interest and relevance in evolution? It might well be that the historical and functional contributions to biological form are as considerable as the purely structural. To paraphrase Polanyi's aphorism, 'mathematics is dumb without the gift of boundary conditions'.

In the intermediary realm between empirical observations and abstract forms we find several levels of theorization concerning the origin of spatial organization (Saunders, Chapter 10)—the fundamental problem in development. The aim is to generate spatial heterogeneity from a relatively homogeneous state. A variety of models demonstrate the extent to which mathematics, physics and chemistry can contribute to the genesis of specific forms—a project initiated by D'Arcy Thompson (1917). Constraints to form thus arise as natural necessity rather than from natural selection. The major evolutionary consequence is that it is dynamic expediency and not selective advantage which may largely determine evolutionary pathways, and hence phylogeny.

At a higher level of analysis we find Goodwin's field theory description of development (Chapter 9). The concept of fields dates from the great embryologist Hans Driesch (1892). It derives from the phenomenon of regulation—the reconstitution of developmental fields of whole organisms from material parts of early embryos—which has consistently resisted assimilation into the dominant genetic theory of development. (Ho, Chapter 11, argues for the necessary independent localization of field or field-like properties in the egg cytoplasm as the means of 'translating' one-dimensional 'instructions' in the genes to the 3-dimensional spatial domain of the organism.) Goodwin identifies general processes within which detailed mechanisms could be coordinated. The articulation of process is independent of the description of underlying mechanisms, however (*see*

Boden, Chapter 13); in particular, the latter does not translate directly into the former. Thus, the overwhelming diversity of molecular mechanisms recently uncovered by recombinant DNA technology has contributed little apart from puzzlement to our understanding of the process of development (Dawid, 1982). This speaks most eloquently against any claim that organisms, and perforce, societies and cultures, can be reduced to molecules and genes. In order to flesh out our understanding of nature, we must apprehend the unity of process at the intermediate levels of generality between mechanisms and pure form. The field theory description can be seen as an attempt to fulfil that role. At the same time, it is also capable of addressing the diachronic aspect of evolution in the 'internalization of constraints' of which more will be said below.

DESIGN AND FUNCTION

It is perhaps significant that the inventor of catastrophe theory himself is not a pure structuralist. He has this to say about evolution (Thom, 1968): '. . . I believe that in biology there exist formal structures . . . which prescribe the only forms which a dynamic system of autoreproduction can present *in a given environment.*' (italics ours)

Thus, an organism is not uniquely defined unless its environment is specified. The relationship between an organism and its environment constitutes its functions or adaptations. Ho (Chapter 11) argues that organism–environmental interactions contribute to the genesis of form via an essentially Lamarckian process of canalization and genetic assimilation (or 'internalization') of experienced environments. In a similar vein, Pollard (Chapter 12) reviews the extraordinary diversity of recently discovered molecular mechanisms which can render Weismann's barrier permeable to environmental influences, especially of the kind involving the transfer of nucleic acid coded information. Preoccupations such as the above with the origins of adaptive molecular functions or morphology, as well as behaviour (Sinha, Chapter 14) give emphasis inevitably to the diachronic aspect of evolution.

It is impossible—certainly at present—to state how much of biological form is due to structural considerations and how much to functional considerations. Perhaps it would not be out of place to suggest that from the functionalist perspective, evolution is emergent or creative. This is not to say that evolution is indeterminate, only that it is not *finalistic*. In fact, it is only by virtue of the structural (deterministic) core that organic systems can exist which can in turn assimilate and evolve to further novelty and complexity. Thus, epigenetic systems, like representational systems in psychology (Boden; Sinha) are neither synchronic nor diachronic, for structural and functional constraints are inextricably interwoven. (This view is broadly compatible with those of Fox, Wicken, and Matsuno for the evolution of the biosphere as a whole.)

THE UNIQUENESS OF THE HUMAN SPECIES

The applicability of Darwinian natural selection to the evolution of the human species has been strenuously debated since Darwin's days. The usual answer to Social Darwinism is to appeal to the uniqueness of our species—our social, cultural and mental attributes—which somehow override our genetic biological heritage. Such arguments are without force, for it is precisely those social, cultural and mental attributes that the present day counterpart of the Social Darwinists—the sociobiologists—claim to explain in terms of the genetic and biological.

Both sides in the debate suffer from the inherently fallacious, but unquestioned assumption that all or almost all of evolution can be explained by the natural selection of random mutations. Once we go beyond the restricted framework of neo-Darwinism there is indeed no need to deny our biological heritage, nor need we shun the study of such characteristically human attributes as the mind (Boden). Artificial intelligence allows the exploration and illumination of natural intelligence. It provides a description of the mental process which is irreducible to molecules and detailed mechanisms; thus emphasizing the richness and subtlety of that process.

But mind, like organism, arises epigenetically (Sinha). The social environment enters as formative influence into development. Perhaps human freedom and uniqueness lie precisely in that responsibility for an environment in which our species can develop and evolve.

REFERENCES

Dawid, I. (1982). Genomic change and morphological evolution. Group report. *In* 'Evolution and Development' (J. T. Bonner, *Ed.*), pp.19–39. Springer-Verlag, Berlin, West Germany.

Driesch, H. (1982). Entwicklungsmechanische Studien I. Der Werth der beiden ersten Furchungszellen in der Echinodermenentwicklung. Experimentelle Erzeugung von Theil und Doppelbildungen. II. Uber die Beziehungen des

Lichtes zur ersten Etappe der thierischen Formbildung. *Z. Wiss. Zool.* **53**, 160–184.

Eldredge, N. and Gould, S. J. (1972). Punctuated equilibria: an alternative to phyletic gradualism. *In* 'Models in Paleobiology' (T. J. M. Schopf, *Ed.*), pp.82–115. Freeman, San Francisco, USA.

Harris, H. (1966). Enzyme polymorphisms in man. *Proc. R. Soc. B* **164**, 298–310.

Henderson, L. J. (1913). 'The Fitness of the Environment'. The MacMillan Company, New York, USA.

Kimura, M. (1968). Evolutionary rate at the molecular level. *Nature (London)* **217**, 624–626.

King, M. C. and Wilson, A. C. (1975). Evolution at two levels in humans and chimpanzees. *Science* **188**, 107–116.

Lewontin, R. C. and Hubby, J. L. (1966). A molecular approach to the study of genic heterozygosity in natural populations. II. Amount of variation and degree of heterozygosity in natural populations of *Drosophila pseudoobscura*. *Genetics* **54**, 595–609.

Morgan, T. H. (1929). 'What is Darwinism?' W. W. Norton, New York, USA.

Needham, J. (1943). 'Time: the Refreshing River'. Allen and Unwin, London, UK.

Nelson, G. (1978). Ontogeny, phylogeny, paleontology and the biogenetic law. *Syst. Zool.* **27**, 324–345.

Saunders, P. T. and Ho, M. W. (1976). On the increase in complexity in evolution. *J. Theor. Biol.* **63**, 375–384.

Thom, R. (1968). Comments by René Thom. *In* 'Towards a Theoretical Biology' (C. H. Waddington, *Ed.*), Vol 1, pp.32–41. Edinburgh University Press, Edinburgh, UK.

Thom, R. (1972). 'Stabilité Structurelle et Morphogénèse'. Benjamin, Reading, Massachusetts, USA.

Thompson, D'A. W. (1917). 'On Growth and Form'. Cambridge University Press, Cambridge, UK.

Section II

ORIGINS AND DIRECTION

2

Proteinoid Experiments and Evolutionary Theory

SIDNEY W. FOX

Abstract

The theory of the origin of life that has emerged from experimental reconstruction of evolution to and beyond the first cells on Earth has been collated with aspects of standard Darwinian theory. While a unified evolutionary continuum is suggested by conceptual interdigitation of origins with evolution, the premise of randomness that is central to neo-Darwinism has been found to be incompatible with the experiments.

The proteinoid theory emphasizes in an evolutionary context: non-random reactions of monomers, self-organization of polymers, endogenous control of processes, stepwiseness, cells-first, frequent events of spontaneous generation on the primitive Earth, and discontinuous evolutionary events emerging from synthesis and assembly. Some criticisms of these views are discussed. Some implications of these foci of thought for established evolutionary theory are examined.

The relationships are further examined in a context of cosmic evolution and thermodynamic principles. Deterministic influences are seen to entwine the evolutionary sequence. Definitions and connotations of randomness and non-randomness are reviewed in the light of developing evolutionary theory.

HISTORICAL INTRODUCTION

Illumination of the theory of protobiogenesis by experimental retracement of steps leading to and beyond the first cells on Earth has reflected light on the Darwinian theory of evolution. A guiding principle of non-randomness has proved to be essential to understanding origins; an explicit expression of this tenet supplies a long needed perspective, at least by way of emphasis, for organismal evolution. As a result of the new protobiological theory the neo-Darwinian formulation of evolution as the natural selection of random variations should be modified (Fox, 1980c) to the *natural selection of non-random variants resulting from synthesis of proteins and assemblies thereof*.

Discussions on evolutionary theory tend to become diffusely wordy because the terms used carry many connotations that differ from one evolutionist to another. The word 'random' is an example. The 'randomists' apply the term to the evolutionary matrix, to polymerization, to mutation, and other processes. Clarification of the terminology is now needed because students originally educated in such diverse subjects as biophysics, molecular biology, population genetics, origin-of-life, and electrical engineering are now coming together in a largely unorganized effort to understand evolution.

The present chapter is an attempt to present an overview of protobiological events inferred from laboratory experiments and their interpretation in the context of evolution. Data and mechanisms at the strictly chemical or physical level are presented in the relevant papers cited. Other discussions at that level are treated also in omnibus books (Fox and Dose, 1977; Matsuno, 1983). In this chapter, an attempt is made to explain relationships mainly at the evolutionary level.

Since the interpretations of the experiments and their extrapolation to evolution depend upon how randomness is conceived, or preconceived, by physical and biological scientists, some examples follow.

Monod (1971, p.98), in *Chance and Necessity* ascribed randomness to amino acid sequence in modern proteins. This is inferred from data which is consistent with similar interpretations of data collected by others (Gamow *et al.*, 1956; Vegotsky and Fox, 1962). From his factually based observation, however, Monod evidently back-extrapolated to the view that the original protein molecules were disordered. Monod expressed this relationship of primordial disorder to modern disorder by characterizing proteins as follows:

in its basic make-up it discloses nothing other than the pure randomness of its origin.

Monod recognized that the relationship *between* (rather than within) protein molecules is non-random. For this he invoked, as have many others, the nucleic acids as the sole directing influence for a given sequence of amino acids in each protein molecule.

In at least two of his papers describing a physical approach to understanding evolution, Eigen (1971 a,b) also assumed a random matrix by declaring that 'evolution must start from random events'. In one, Eigen (1971b) wrote,

> At the 'beginning' there must have been a molecular chaos, without any functional organization among the immense variety of chemical species. Thus, the self-organization of matter we associate with the 'origin of life' must have *started* from random events.

The randomness imputed to the matrix, or what Luria (1973, p.117) has called the 'prebiotic chaos', interfaces conceptually with protein synthesis. This is evident from Crick's writing (Crick *et al.*, 1976) in which he commented on the origin of protein synthesis as a notoriously difficult problem, and excluded from consideration the formation of what he assumed to be random polypeptides.

The concept of random synthesis is repeated elsewhere; a well known book on the origin of life (Miller and Orgel, 1974), for example, assumes 'random polymers' for the early stages of molecular evolution. These various references, and others, envisage a random matrix to and including random polypeptides, which were however rescued from chaos by a nucleic acid templating mechanism. No explanation, however, is given for the origin of the templating mechanism, which must itself have been ordered, according to molecular logic.

This dominant picture of random matrix and random polymerization has led easily into the concept of random mutation which is a central tenet of neo-Darwinism. As one example, Luria (1973, p.20) states, 'the process of genetic mutation is strictly random'. The word strictly, which accurately describes much of the thinking by biologists, leaves little room for qualification of the word *random*.

However, when we turn to population geneticists and other biologists we *sometimes* read of qualifications to the term randomness. Wright (1967) for example, stated,

> The meaning of 'random' is that the variations are, as a group, not correlated with the course subsequently taken by evolution (which is determined by selection). The variations are, of course, severely limited in kind by the accumulated results of past evolution.

Gould (1982) states:

> By 'random' in this context, evolutionists mean only that variation is not

inherently directed toward adaptation, not that all mutational changes are equally likely.

Accordingly, random does not mean random for some, whereas for the biologist Luria, for example, and the physicists it does mean random in the pure statistical sense.

Gould and other neo-Darwinists ascribe creativity to natural selection operating from without. The alternative that natural selection might operate on variants that arise from within, i.e. at the molecular level (Morgan, 1919, 1932; Haldane, 1966; Lima-de-Faria, 1962; Whyte, 1965), is not discussed. Gould's arguments are in line with Darwin's first title *The Origin of Species by Means of Natural Selection* but do not honor Darwin's later title: *The Preservation of Favored Races in the Struggle for Life*, one which suggested that Darwin himself later understood that selection of itself is not, or may not be, a creative process.

Darwin's total theory is one in which he blended the facts of variation, reproduction, and natural selection to give his contemporaries an understanding of the profusion of living things on a natural basis. By this integration he changed the course of civilization, and in fact he can be said to have introduced along with his contemporary, Gregor Mendel, a biological civilization. But his ability to process criticism by others, and to be 'self-critical', as in the amended title of his opus places him, already in his time, beyond neo-Darwinism. Had molecular science been sufficiently advanced in Darwin's day, we might now anticipate that he would also have looked at evolution from the inside out, not just the outside in.

Darwin did not use the words *random* nor *non-random* (Barrett *et al.*, 1982). Although one may impute randomistic meanings to some of his statements, it is fairer to point out that in his time related concepts were undeveloped. Thus, he stated (Darwin, undated, p.372) 'A grand and almost untrodden field of inquiry will be opened, on the causes and laws of variation . . .'

Darwin's thinking on random variation was probably closer to Luria's than to Gould's (Darwin, undated, p.16),

We see indefinite variability in the endless slight peculiarities which distinguish the individuals of the same species, and which cannot be accounted for by inheritance from either parent or from some more remote ancestor.

And in a section on fossils (Darwin, undated, p.128),

If my theory be true, numberless intermediate varieties, linking closely together all the species of the same group, must assuredly have existed.

Thus Darwin came close to conveying the meaning of the word *random* with his remarks on 'numberless' and 'endless' variations.

We accordingly have two principal definitions of random, as well as shaded definitions in between. In one definition, random means random, as now statistically defined. In the other, it means essentially undirected. A principal thesis of this chapter is that variations are directed, and that they are directed from the molecular level within an hierarchical organization. Either of these two definitions of *random* thus fails to fit a newer description of evolution in which self-ordered synthesis connotes much direction from the molecular level (for which explanation will be extended in this chapter). The new perspective is in disagreement with *either* random or 'undirected' variation.

An especially perceptive earlier criticism of randomness in evolution was that of Eden (1967), who stated,

Any principal criticism of current thoughts on evolutionary theory is directed to the strong use of the notion of 'randomness' in selection. The process of speciation by a mechanism of random variation of properties in offspring is usually too imprecisely defined to be tested. When it is precisely defined, it is highly implausible.

It is the discovery that amino acids are self-sequencing that is responsible for our emphasis on non-random variation and for the view that evolution is self-limiting, in contrast to the widely prevalent assumption of random variation. When the inference of non-randomness was bolstered by a newer perspective of cosmic order derived from the Big Bang (Fox, 1980b), the possibilities for quantitative testability of evolutionary sequences and theoretical testing of modelled processes (Matsuno, 1982) emerged.

The phrase 'natural selection of random variants' is a compressed statement of Darwin's theory. It fits the original Darwinism approximately as well as it fits neo-Darwinism. As had to be learned after Darwin, causes are fundamentally genetic and even more fundamentally molecular.

In this respect, it is of historical interest to examine the views expressed in T. H. Morgan's (1932) writing on *The Scientific Basis of Evolution*, published after Morgan had consolidated his theory of gene linkage in the chromosome.

In that book, Morgan wrote, 'Neither the genetic factors responsible for a part of the initial variability, nor the environmental factors can bring about such an advance'. Morgan was here stating that the process of natural selection is incapable of moving variability beyond determinate limits. This view has been disputed (Mayr and Provine, 1980).

Some modern geneticists believe that 'modern genetics' has explained variation adequately. They point to recombination, the founder effect, genetic drift, and punctuated equilibria; these have even been summarized in a legal opinion (Overton, 1981).

None of these processes was totally unfamiliar to Morgan; there is no reason to believe that they would have led him to change his evaluation, especially when the processes are rationalized as having arisen from a random matrix, i.e. by chance. We may look elsewhere for causes of variation, especially in the internal non-randomness that Berg (Haldane, 1966, p.12) and Morgan[1] favored (cf. Mayr, 1980). A fresh source of such perspective is the sharpening picture of protobiology.

Accordingly, we shall examine the model for protobiology in some detail and then analyze its relationship to the Darwinian concepts.

HOW DO WE LEARN HOW LIFE BEGAN?

The problem of how life began has long been considered to be imponderable. Much of the reason for this is that modern science is predominantly reductionistically analytical. For many modern scientists, an adequate feeling of assurance can be obtained only by analyzing what is here in hand.

We do not however have any primordial organisms to analyze. All we have are distant descendants of primordial organisms. Even if we had on hand a primordial organism, a complete analysis would be a long way from answering how the components identified by that analysis were assembled into that first organism. What could be learned we would learn by 'taking apart'; what we want to know requires 'putting together'. The latter is synthesis and assembly, or *constructionism* (Fox, 1977).

However, the starting point for constructionistic studies was analytical. The basic enabling principle is the 'unity of biochemistry' (Florkin and Mason, 1960). The labours of thousands of analytical biochemists have taught us that all life, all living units, are constructed in an amazingly ramified but highly limited variety from a remarkably small number of molecular units: nucleic acids from five main nitrogen bases; proteins from twenty amino acids; lipids from three components of which the fatty acids are fewer than ten kinds; and carbohydrates from a handful of monosaccharide types (Lehninger, 1975; Needham, 1965). The principle of the unity of biochemistry is perhaps the most pervasive evidence that evolution has occurred by descent with modification at the molecular level.

Had there not been a discernible unity underlying biochemical variety, living things would have seemed chaotic and perplexing, and in such a situation, the prospects for selecting a protobiochemical model would have indeed been hopeless. Since organisms are so much alike chemically the possibility for experimenting with protobiochemical recipes was encouraging, especially when aided by some inferred geochemical guidelines.

The other potential approach is that of back-extrapolation from what is known of modern organisms, a large catalog of knowledge indeed. Crick (1981, p.37f) has analyzed the problem similarly, but he does not bother to consider the investigative approach of back-extrapolation, which obviously cannot identify an initial assembly mechanism (Fox, 1977). He does, however, seem to be sufficiently dismayed by the difficulties of a constructionistic approach that he prefers to devote his efforts to explaining how life arrived here from elsewhere. Perhaps this is not so much dismay as the exercise of a characteristically theoretical, rather than an experimental, approach to problems. Crick (1981, p.148) acknowledges that many scientists plus his own wife regard 'directed panspermia' as 'science fiction'.

The panspermia concept stems from Svante Arrhenius (Oparin, 1957); it received much attention in an era which could look back upon no experimental investigation other than the discredited nineteenth-century experiments of 'spontaneous generation' (Oparin, 1957). Since a modern kind of experiment on chemical spontaneous generation began in the middle of this century (Fox, 1957), numerous eminent scientists have turned their attention to the problems of the origins of life and of the genetic code (e.g. Crick, 1968, 1970; Calvin, 1969; Monod, 1971; Eigen, 1971a,b; Luria, 1973; Nicolis and Prigogine, 1977). The majority of the authors cited used the Aristotelian approach of developing a picture by purely theoretical means from data obtained from modern organisms. Their assumptions about the original matrix are neo-Darwinian. The *experimental* approach has however highlighted the inadequacy of neo-Darwinian principles and has provided a theory of a markedly different kind. The differences will be analyzed here after the comprehensive (Lehninger, 1975; Florkin, 1975) experimentally derived theory is itself described.

MOLECULAR EVOLUTION TO LIFE

The flowsheet of primitive molecules to protocells is seen in fig. 2·1. This flowsheet was established by over 200 man-years of experimental exploration.

The evidence for the existence of sets of amino acids is from returned samples from the lunar surfaces and meteorites, and from numerous simulation syntheses (Fox *et al.*, 1981).

The fact that sets of amino acids containing trifunctional types (aspartic acid) undergo copolymerization in a non-random manner is crucial to molecular evolution and to all of evolution; the evidence for such self-ordering is firmly established from experiments in numerous laboratories (Fox, 1980b). Numerous other studies have also shown that the highly specific polymers obtained have 'protoenzymic' activities (Fox, 1980a). The

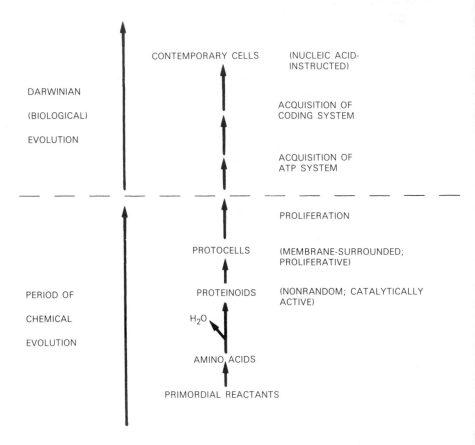

FIG. 2·1 Flowsheet of molecular and protocellular evolution (Fox, 1981c) from primordial matter.

polymers or proteinoids (thermal proteins), are catalytically active in various specific ways due to the interactions of amino acid sidechains, much as in modern proteins (with amplification of functional diversity by metal ions and other prosthetic groups).

The polymerization of amino acids is operationally simple enough, but the self-organization of proteinoids to simulated protocells in the laboratory is simpler yet. Mere contact with free water is all that is necessary (Fox, 1976). The tendency to reproduction involving heterotrophic growth has been demonstrated by binary fission (Fox and Yuyama, 1964), budding (Fox, McCauley, and Wood, 1967), and other means (Hsu et al., 1971; Brooke and Fox, 1977).

One remarkable aspect of the simulated protocells is their simultaneous content of varied activities. This perspective is a revelation; it contrasts with the assumption of Oparin (1957) that molecular order and function would have had to evolve in the protobionts he modelled. It must be remembered, however, that Oparin's coacervate droplets could not assume the role of protocells since they were produced from evolved biopolymers instead of from the equivalent of protobiopolymers.

In modern cells the barrier function of membranes depends upon phospholipid and the order of amino acids in proteins results from the functions of nucleic acids. In the modeled proteinoid protocell, however, the order of monomers is due to the self-sequencing of the amino acids (Fox, 1978), and the membrane of the microsphere is proteinoid (Fox et al., 1982; Przybylski et al., 1982). Accordingly, a functional proteinoid protocell required neither nuclcic acid nor phospholipid. The protocell had, moreover, according to the experiments, an ability for (heterotrophic) growth (Fox et al., 1967; Fox, 1973a).

In the light of this analysis, the probability that the proteinoid microsphere represents the original terrestrial protocells, and that no other entity could fill that role, is much greater than if primitive types of protein, nucleic acid, and lipid were each to have arisen separately and then to have coassembled.

Once the functions of the laboratory protocell had been catalogued sufficiently, it became clear which functions were required in order to bridge the gap from modeled protocell to modern cell. Outstanding among these requirements were the abilities to synthesize proteins and to synthesize nucleic acids. For example, a principal difference between the reproductive mode of the proteinoid microsphere and that of the modern cell is that the latter synthesizes its own proteins (Weissbach and Pestka, 1977), whereas the former obtains its proteins in a preformed state from the environment.

What has been learned recently is that proteinoids rich in basic amino acids are capable of catalyzing the synthesis of peptides and of enlarging the peptidic proteinoids themselves by conjugation with amino acids or peptides; ATP or pyrophosphate (Baltscheffsky, 1981) must be supplied as energy source. They are capable also of catalyzing the synthesis of polynucleotides (Fox et al., 1974). Since the processes of synthesis of protein and of polynucleotides are catalyzed by the same basic polyamino acid and the catalysts are effective within microspheres (and within complexes with polynucleotides) the proteinoid microspheres on the early Earth could have served as effective locales for the evolution of the genetic mechanism and code. This sequence is represented in fig. 2·2.

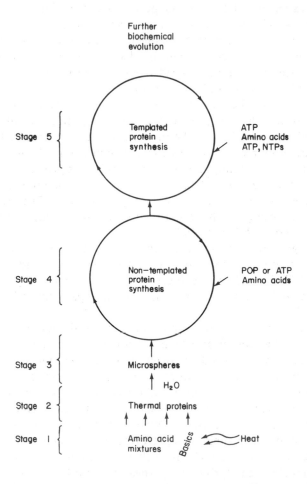

FIG. 2·2 Flowsheet of experimentally suggested evolution from reproductive protocells to modern cells. Thermal peptides were synthesized in geological stage 1. Some of the mixtures contained sufficient basic amino acid for the early part of stage 4 and stages 2 and 3. In stage 5 some of the ATP, and other nucleoside triphosphates, polymerized to polynucleotides in microspheres. In stage 4, large peptides would presumably yield offspring microspheres Fox, 1981).

LESSONS FROM PROTEINOID EXPERIMENTS

Ease and speed of formation of protocells

An outstanding feature of the experimental findings is the ease with which proteinoid microspheres possessing remarkable properties can come into existence. The expectations were quite otherwise; the view was that anything as complex as a cell could have arisen only after a prolonged and intricate evolution. An example of this premise was a detailed symposium in 1964, titled *Evolving Genes and Proteins* (Bryson and Vogel, 1965), at which it was possible to point out that genic and enzymic evolution must occur within *cells* (Fox, 1965; cf. Prosser, 1965)—entities omitted from the title. If one focuses attention on a modern, evolved cell, the need for prolonged, intricate evolution appears to be justified. That the evolution could occur from a less developed cell is the saving concept.

The expectation that the origins of cells would be rare and would require a long evolution (Crick, 1981; Eigen *et al.*, 1981) is a plausible one. Eigen *et al.* (1981), for example, state, 'Organization into cells was surely postponed as long as possible'. The arbiter for the correct point of view must however be the results of experiments, and experiments only. Although it is not always possible to determine where premises end and experimental results begin, the ease of formation of *proto*cell early in molecular evolution must stand out as a view firmly supported by experiment.

Neither scientific logic nor data permit the instantaneous origin of a fully evolved modern cell; that would have been instant creation. The modern cell accordingly evolved from the protocell by a series of *self-organizing steps* (Fox, 1960). The protocell can thus be thought of as discernibly related to the modern cell, albeit deficient in some functions, either quantitatively or qualitatively.

Number of protocells

Another aspect revealed by experiment is that protocells would have arisen in astronomical numbers. Prior to the first experiment, this is not what was expected (Fox, 1983a). However, 1 gm of proteinoid can yield 10^{10} microspheres.

Sociality of protocells

Because of large numbers in a dense population, the kind of sociality described by the psychologists Tobach and Schneirla (1968) would have been present in the first organisms. In view of the obvious tendency of all living forms to join into societies large or small, the observation of this phenomenon in laboratory protocells, as representative of initial behavioral units, was arresting. Tobach and Schneirla (1968) stated:

> Until forms intermediate between organic and inorganic matter are better known, the fact that all cells issue directly from other cells may be taken to indicate that no existing form of life is truly solitary, and no organism is completely independent of others at all times in its history. This dependence of every individual on others is the prerequisite to social behavior.

With the development of protobiology and the opportunity to examine the behavior of laboratory protocells 'intermediate between organic and inorganic matter', this intimation of Tobach and Schneirla was borne out. Laboratory protocells display strong tendencies to associate and to communicate (Hsu *et al.*, 1971). The individual microspheres not only influence each other, they associate in ways that constitute behavior that could not be predicted from examining individuals alone (Fox, 1983a). The experiments even suggest parent–offspring relationships that have not earlier received attention — that many of the first or originally predominant parent–offspring relationships were those of foster parent-children.

Physical properties, general

The properties that have been described in previous paragraphs have been selected for their relevance to evolutionary theory. The other properties that have been characterized are numerous, and are available for perusal in monographic treatments (Fox and Dose, 1977; Fox, 1978).

It will be of value however in a chapter to be read by evolutionists again to correct some misleading statements, exemplified by those that appeared in a recent book on evolution (Stebbins, 1982). One statement claims that 'the majority of protenoid (sic) spheres show no chemical activity at all, and only a few have a weak enzymic activity'. Another implies the lack of ordered sequence in proteinoids and a third assigns a short life to microspheres.

Chemical activity is however found in all microspheres (Fox, 1980a) in which it has been sought. To assert the opposite, as Stebbins has done, is merely to indulge in negative armchair ratiocination.

The assertion that proteinoid microspheres are short-lived is also false. Aseptically prepared microspheres (bacteria from the air can feed on proteinoids) have been observed to last for six years before they were discarded (Hsu and Fox, 1976). The statement that proteinoids will thermodynamically tend toward hydrolysis (Miller and Orgel, 1974) may be 'theoretically' justified, but this ignores the realities of kinetics and of stabilization by intramolecular interactions in proteinoids, not to mention the published experimental evidence.

The mis-statement about amino acid sequence ignores the self-ordering power of amino acids, which is a thesis of this chapter and is now seen as a cornerstone of any valid theory of evolution.

These errors are illustrative of the kinds of difficulty that non-chemists sometimes encounter in incorporating chemical or biochemical facts into the generally sought evolutionary synthesis (Mayr and Provine, 1980). On the other hand, chemists and biochemists have themselves not aided the enterprise, since nearly all of them have avoided evolutionary theory until the 1950s. Even then, many tended to look on evolution as a term to be uttered like a password that would let them enter through the dark green door of a 1930s speakeasy. Some of the shining exceptions to this last generalization were Henderson (1913) and Joseph Needham (1935); and later on Florkin and Mason (1960) and Dayhoff (1972) and associates.

Other objections have been systematically discussed elsewhere (Fox, 1983b).

Main theoretical advances

The principal advances that can be credited to the proteinoid experiments are perspectives that were attained either earlier than by any other approach or *only* because of the exploratory experiments. The main advances listed below are all conceptual, but are derived from the physical advance of the production in the laboratory of an evolvable organized cell-like structure. This structure is composed of internally ordered macromolecules (thermal copolyamino acids), which support in the structure primitive aspects of metabolism. The microstructures have double-layered boundaries, they can participate in reproduction, and they have many, if not all, of the properties found in modern cells in some primordial form. The physical advance occurred more than twenty years ago (Fox, 1981a), when its general or detailed nature was barely understood. The conceptual advances resulting therefrom are much of what will be discussed here.

The self-sequencing of amino acids
The fact that amino acids are self-sequencing is the essence of an evolutionary theory that takes us beyond, and out of, neo-Darwinism. It is also, in hindsight, what made possible the production of a laboratory protocell and the related idea of the spontaneous generation of cells on the primitive Earth 3–4 billion years earlier. This is the crucial phenomenon in

the understanding of protobiological evolution (Fox, 1978). It is the products of self-sequencing, the thermal proteinoids, that explain the origin of enzymes, the origin of cells, and the ability of such cells to make the materials, with controlled help from the environment, that could yield a continuing stepwise generation of evolving cellular organisms.

The initial experiments that explained self-sequencing, or self-ordering (Fox, 1960), were based on theoretical inferences derived from an earlier kind of experiment in which self-ordering was observed. The impetus for continuation throughout was a stream of new suggestions from experimental results.

The first experiments had objectives in protobiology. They were experiments on the synthesis of substituted peptide derivatives in aqueous solution in the presence of proteolytic enzymes (Fruton, 1982). Some of the experimental results suggested that the identity of the substituted peptide products was a function not only of the enzyme, but of the small molecules that constituted substrates (Fox et al., 1953). It has been fashionable to speak of the specificity of proteolytic enzymes, a phrase that immediately tends to suggest that no other component of the reaction system contributes to the specificities in synthesis or degradation by enzymes. With the experimentally derived inference that (substituted) amino acids could, along with enzymes, contribute to the specificity of reactions, the way was cleared for testing the interreaction effects of amino acids alone, *without enzymes*, in the formation of specific acellular proteins (Fox, 1956). The self-ordering was then observed (Fox et al., 1953); like the polymerization, such specific behavior did not require enzymes. Moreover, the origin of specific enzymic activity in the polymeric products could be visualized and tested from this enzyme-free matrix. Activities, albeit weak ones by standards of evolved modern enzymes, were found in abundance (Rohlfing and Fox, 1969; Oshima, 1971; Dose, 1983; Fox et al., 1974).

Most of the first indications that the polyamino acids are ordered emerged in several laboratories from fractionation of proteinoids (Fox, 1981b). Subsequently most of the evidence that self-ordering is a general process including self-sequencing of the amino acids (Nakashima et al., 1977; Hartmann et al., 1981; Melius, 1977) was reported (Table 2·1).

While the sequences reported as dominant are two tripeptides, peptides of 10 000 molecular weight or more are known to arise from thermal condensation of amino acids (Fox and Harada, 1960; Melius and Sheng, 1975). The explanation for the recovery of molecules as small as tripeptides seems to be that the polyamino acid chains are punctuated by molecules like flavins and pteridines (Heinz and Reid, 1981); it is to be expected that the linkages between amino acid residues and the heterocycles would split most easily.

TABLE 2·1 Tyrosine-containing tripeptides found versus those expected on the basis of the random hypothesis

Expected from random polymerization		Found from non-random polymerization
αUαUY	YαUU	
αUγUY	YγUU	
γUαUY	YαUG	
γUγUY	UYγUG	
αUGY	YαUY	
γUGY	YγUY	
αUYU	PαUY	
γUYU	PγUY	
αUYG	PGY	PGY
γUYG	PYU	
αUYY	PYG	PYG
γUYY	PYY	
GαUY	YGU	
GγUY	YGG	
GGY	YGY	
GYU	YYU	
GYG	YYG	
GYY	YYY	

The dominant fraction obtained from the thermal copolymerization of glutamic acid, glycine, and tryosine proved to be an equimolar complex of pyroglutamylglycyltyrosine and pyroglutamyltyrosylglycine (Nakashima et al., 1977; Hartmann et al., 1981). U = glutamic acid residue, Y = tyrosine residue, G = glycine residue, P = N-pyroglutamyl.

The results of the sequence study reported in Table 2.1 are not the only ones derived from analysis of terminal residues. Striking examples of non-randomness are found in the studies of Melius and Sheng (1975), and in other analyses (Fox, 1980c).

Moreover, numerous other kinds of study have verified the basic processes of self-ordering (Fox, 1980b; Fox, 1981b).

The fact that the ordering could occur without nucleic acids to guide it resolved a principal problem of origins, the so-called chicken–egg question (Lehninger, 1975; Eigen, 1971a; Florkin, 1975).

Early formation of a cell

Viewed as a problem in pure Aristotelian logic, the emergence of the cell should have followed that of the cell's components. This appears to be logical on the basis that the totally assembled cell is a final packaging of all its component molecules. Luria (1973, p.115) has, for example, explained the difficulty of direct genesis of a modern cell in that cells as we know them are enormously complicated objects.

Our experiments, however, suggested another sequence. Thermal proteins alone aggregate on contact with water with consummate ease. Upon experimental extension of this unexpected revelation, a new picture formed of an early cell that could then evolve to a modern cell. During that evolution, and probably early in that evolution, the evolving cell learned to make both its own protein and its own nucleic acids and other components (Fox, 1981c). The evolving cell then passed through a stage of non-templated protein synthesis, into templated protein synthesis, as explained in detail elsewhere (Fox, 1981c).

A principal rethinking required by the recognition that the cell emerged early is that required for evolutionary processes in general — that the primordial processes must be understood and mimicked in a manner that is partly detached from modern circumstances (Fox, 1981a). The protocell, a first cell, could have emerged without having had all of the properties of the modern cell, or without having any or all of those functions in as full measure or as specialized as they are found to be in the modern cell.

Also, one cannot extract any protein from modern organisms and cause it to form cell-like structures in the simple, efficient, largely encompassing way that thermal proteins perform this act. As proteins became more specialized in evolution, many evidently lost this propensity.

Numerous associated properties in the protocell

The lengthy roster of properties and functions that the proteinoid microspheres are found to have, include manifold protoenzymic, protophysiological, and protobehavioral types (Table 2·2). The evidence in the literature is keyed in reviews (Fox and Dose, 1977; Fox, 1978; Fox et al., 1978; Fox and Nakashima, 1980).

The accumulated knowledge has had to contend with assumptions constituting a belief that the first cells were inert, and that the typical properties of cells as we know them were developed during the evolutionary steps. In other words, not only was the cell assumed to have arisen late in molecular evolution, the functions of the evolved cell were assumed to have been acquired entirely during evolution.

This latter view was due largely to Oparin's studies. It is true that the role of Oparin's coacervate droplets as models for protocells, his 'protobionts', has often been regarded as equivalent to that of proteinoid microspheres; the two models are treated side-by-side in textbooks (Lehninger, 1975; Korn and Korn, 1971; Scott, Foresman and Co., 1980). Some authors however recognize that some of the differences between standard coacervate droplets and proteinoid microspheres are crucial.

Coacervate droplets have typically been made from materials like gum arabic and gelatin, oppositely charged colloids (Oparin, 1957, p.301) obtained by extraction from highly evolved organisms. Their use fails to

TABLE 2·2 Associated properties in the laboratory protocell

Enzymic	
In proteinoids	*In microspheres*
Esterolytic	Esterolytic
Phosphatatic	Phosphatatic
Decarboxylatic	Decarboxylatic
Aminatic	
Deaminatic	
Peroxidatic	Peroxidatic
Peptide-synthetic	Peptide-synethetic
Internucleotide-synthetic	Internucleotide-synthetic

Photochemical	
In proteinoids	
Decarboxylatic	

Hormonal	
In protenoids	
MSH	

Behavioural	
In proteinoids	*In microspheres*
Electrophoretic	Electrotactic
Enzymic	Enzymic, metabolic
Aggregative	Aggregative, thigmotactic
	Motile
	Osmotic
	Selectively diffusive
	Fissive
	Reproductive
	Conjugative
	Communicative
	Protective
	Excitable

[a] A review of the protobehavioral properties is found in Fox and Nakashima (1980).

illuminate how cells came into existence when there were in existence neither cells nor cellular polymers. Moreover, to perform some highly interesting experiments, Oparin (1971) incorporated enzymes from modern cells.

Against this background it is easier to understand why Oparin (1957, p.290) believed,

All that we can expect . . . is (to) . . . explain the formation of organic polymers in the shape of polypeptides and polynucleotides, assemblages having, as yet, no orderly arrangement of amino acid and nucleotide residues adapted to the performance of particular functions.
These polymers were, nevertheless, able to form multimolecular systems. . . . It is only by the prolonged evolution of these . . . that there

developed . . .: metabolism, proteins, nucleic acids and other substances with
complicated and 'purposeful' structures. . . .

This view is the logical inference from data on coacervate droplets.

The proteinoid microspheres, which are historically more recent, represent
the emergence of protocells from precursors formed in the geological realm,
proteinoids. These proteinoids and the aggregation products thereof (Fox,
1980a), are crowded with functions.

Salient functions within a protocell

The association of functions in the proteinoid laboratory protocell is best
understood by focusing on those functions which are considered in the
modern view to be essential to the living system. These are (a) ordered
macromolecules, as in the coded genetic system, (b) metabolism from the
actions of specific enzymes, and (c) membranes of cells, all in association.
All of these are present simultaneously in proteinoid microspheres with no
other substances included, which is not true in evolved specialized proteins.
This strengthens the view that proteinoid microspheres could have evolved
to modern cells with the same functions provided by three types of modern
substance: protein, nucleic acids, and lipids.

TABLE 2·3 Evidence for the membrane nature of the boundary of
proteinoid microspheres

Electron micrographs (Fox and Dose, 1977)

Selective permeability (Fox *et al.*, 1969)

Osmotic properties (Fox *et al.*, 1969)

Black films (Dr. Gilbert Baumann in Fox *et al.*, 1978)

Polarization in microsphere boundary (Ishima *et al.*, 1981)

Electrical discharge from microsphere boundary (Przybylski *et al.*,
   1982)

That the protein functions can be provided by proteinoids was the easiest
to grasp and the first to be established (Rohlfing and Fox, 1969). That
the proteinoids could be ordered macromolecules having no DNA nor
RNA in their evolutionary history required the solution that the first
information came from reactions of simpler substances, the amino acids.
The surprising aspect of this was the high degree of precision in self-
ordering. That the proteinoid microspheres had at least moderately efficient

membranes without the benefit of modern phospholipid was established by several lines of evidence, not the least of which was the strong electrical behavior observed (Table 2·3).

TESTS OF VALIDITY OF THE THEORY

Science proceeds from initial periods of direct observation to admixtures of observations and inferential interpretation. The substance of protobiology represents a discipline that has appeared within science as a whole when the latter was already well stocked with data from geology and biochemistry. Atop this base were introduced corrective inferences derived from building models of the origin of life. The corrections ensued because the knowledge gained by analysis of the objects of geology, biochemistry, and physiology could not explain how these objects came into existence by what was, in a sense, a process that was the inverse of analysis, i.e. synthesis and assembly.

Some tests for validity have however been available.

Microfossils

One may ask whether any objects resembling the proteinoid microsphere (interpreted as laboratory protocells = units of protolife) are found in ancient strata. This question is thus phrased in a proper historical sense, since the proteinoid microspheres were reported in 1959 (Fox et al., 1959) whereas the spheroidal microfossils were first reported in 1962 (Barghoorn and Tyler, 1963). The possibility that the microfossils are lithified relics of spontaneously originating proteinoid microspheres was suggested at the same conference in which the microfossils were first reported (Fox and Yuyama, 1963a).

To at least one physiologist (Keosian, 1974), this explanation (Fig. 2·3) appeared likely. Following the application of a new method of artificial fossilization (Leo and Barghoorn, 1976) to both algae and proteinoid microspheres, a group of micropaleontologists (Francis et al., 1978) reported that artificially fossilized proteinoid microspheres and artificially fossilized algae resembled each other. The two sets of objects could not be distinguished. Other micropaleontologists (Cloud and Morrison, 1979; Pflug and Jaeschke-Boyer, 1979) have recorded similar judgments.

In addition to the spheroids, microfossils caught in the act of binary fission (Knoll and Barghoorn, 1977) were imitated, in advance of their finding, by proteinoid microspheres. The proteinoid microspheres can be studied dynamically, which is of course not true for fossils. Such study has

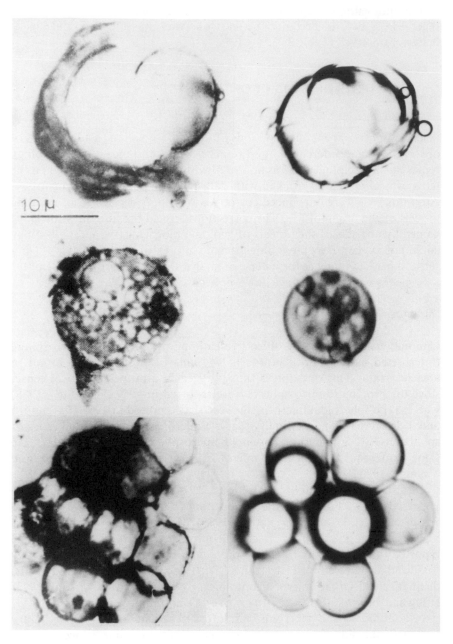

FIG. 2·3 Comparison of ancient microfossils (on the left) with proteinoid microspheres (on the right).

revealed that some microspheres that appear to be dividing are actually two microspheres undergoing coalescence. Although that was not offered as an alternative by Knoll and Barghoorn, it cannot be ruled out by morphological examination of the twinned microfossils.

De novo organisms from hydrothermal vents

The micro-organisms that appear in events from hydrothermal vents have been proposed as being closely related to proteinoid microspheres (Corliss *et al.*, 1981). Part of the reason for this proposal is that such events have a thermal beginning followed by processes that occur in less time than a diurnal cycle, and which can occur in the dark. These conditions are met by the proteinoid microspheres (Fox and Dose, 1977). The field occurrences have been recorded or suggested for the Galapagos Rift, the East Pacific Rise, Spirit Lake near Mt. St. Helens (Corliss *et al.*, 1981), and earlier in a similar geological context for Yellowstone Hot Springs (Copeland, 1936). The function of a submarine volcano or hydrothermal vent for protobiogenic occurrences had been predicted earlier from observations in the laboratory (Fox, 1957; Mueller, 1972).

Prions

The prions are pathogenic organisms responsible for diseases such as scrapie (Griffith, 1967; Prusiner, 1982). Although purified only with great difficulty, the prions have many properties in common with proteinoid microspheres as indicated in Table 2·4. The scrapie organisms, however, are much smaller in size than the proteinoid microspheres.

TABLE 2·4 Properties common to prions and to proteinoids and microspheres

---

Stickiness [aggregation and thigmotaxis in microspheres (Kimberlin, 1982; Fox and Nakashima, 1980)]

Molecular weights below 50 000 (Prusiner, 1982; Fox and Dose, 1977)

Ability to reproduce (Prusiner, 1982; Fox, 1974)

Physical stability (Kimberlin, 1982; Fox and Dose, 1977)

Nonantigenicity (Prusiner, 1982; Fox and Dose, 1977)

Although neither Prusiner (1982), nor a commentator on his studies (Kimberlin, 1982), has published any statement about such a relationship, the possibility of a close relationship between prions and proteinoids and microspheres has suggested itself. This similarity in unique properties is seen in Table 2·4. Such similarity does not constitute evidence in the way that a comparison of artificially fossilized algae and artificially fossilized microspheres does. However, it should be considered with that evidence and with the suggestions made for products from hydrothermal vents. What also should be considered is the considerable disparity in size between scrapie units and proteinoid microspheres.

The significance of heritable or endogenously similar protein production to prion replication has been explained by Melius (1982) as due to self-ordering of amino acids, and by Root-Bernstein (1983) as due to amino acid complementarity. Each mechanism operates without nucleic acids. Melius reviews other known mechanisms for non-ribosomal synthesis of peptides in organisms.

COMPATIBILITY OF PROTOBIOLOGY WITH BIOLOGY

One fundamental general test for validity that must be passed is whether the principles and phenomena of protobiology bridge to the principles and phenomena of biology.

A number of such relationships are documented here. In each case, the protobiological tenet was proposed, or largely emphasized, before the counterpart biological tenet received widespread attention.

Microfossils

The resemblances between microfossils and lithified proteinoid protocells has been pointed out above. The proteinoid microsphere as a protocellular bridge between protobiology and biology is thus supported by circumstantial evidence.

The concept of proteins-first

The two great classes of informational biomacromolecule are the nucleic acids and the proteins. The question that has long evaded resolution is which of these is primary in the biochemical economy of the cell and which was primary in evolution. The general consensus of understanding in biochemistry is recapitulationist, i.e. is such that biochemical sequences are expected to have arisen from a similar sequence of earlier evolutionary events.

The idea that protein and nucleic acids are both primary is not new. Commoner has been a leading proponent of joint primacy. In a paper in which he first pointed out that inheritance is uniquely associated with life (life is much more than inheritance!), Commoner (1968) said, 'the biology of inheritance is embodied not only in nucleic acids but in proteins as well.'

The narrower point of view, i.e. the tenet that nucleic acids or, more specifically DNA, are primary has been prevalent. This perspective was early fostered by Muller (1955), and it continues into recent times as can be seen in Crick's (1981) book *Life Itself*. Leaders in theoretical construction: Monod (1971), Eigen (Eigen and Schuster, 1979), Luria (1973) have all treated the subject so as to help strengthen the DNA-first concept. This thesis is aided also by companion assumptions of randomness, DNA self-replication and contexts that treat evolution as only selection, and life as only inheritance (cf. Fox, 1981c), none of which is justified by supporting treatments.

Controversy over these opposing views has erupted from time to time and Commoner's interpretations were 'shouted down', according to Hubbard (1982). One of the few authors to analyze each of the points of view with considerable impartiality is Lehninger (1975).

The proteinoid results have led directly to a view of proteins-first from the outset of evolution (Fox, 1960). But that experimentally derived mechanism is not a simple direct-line relationship. Rather, it is at least a two-stage development in which first arose *thermal* proteins (proteinoids) that gave rise to *protocells* that in turn had the power to make true cellular proteins, and also had the power to make polynucleotides. It is this *stepwise* sequence of events, modelled in large part by experiments, that made the general idea of proteins-first a viable one.

More recently, the work of Kornberg and others on DNA polymerase has highlighted the primacy of the enzyme proteins in modern systems. Kornberg (1980) has stated the position clearly with the words,

It was suggested in 1953 that A-, T-, C-, and G-containing precursors might orient themselves as base pairs with a DNA template and then be 'zippered' together without any enzyme action. However, to the biochemist it is implicit that all biosynthetic and degradative events are catalyzed by enzymes, making possible refinements of control and specificity, and rapid rates of reaction.

The key point is that the protein → nucleic acid sequence established by proteinoid investigations led directly into the protein → nucleic acid sequence now recognized for modern cells (cf. also Dillon, 1978). The protein → nucleic acid sequence is thus part of the protobiology → biology sequence.

A coding mechanism from proteins-first

The widely held view of non-enzymic 'zippering' of new bases stacked on already formed polynucleotides in modern cells has been dominant. This has undoubtedly been at least partly responsible for the transposed idea (e.g. Ponnamperuma, 1968) that a nucleic acid coding mechanism existed in the first cells. The physical models of protobiology, however, could not fit such a pattern; they led to the most critical questioning of the DNA-first assumption, as we have seen. What has become clear more recently (Commoner, 1968; Kornberg, 1980) is that the idea of the primacy of naked DNA in the *modern* biochemical economy is also incorrect, in line with Kornberg's critique of the problem. To recognize that DNA alone is not capable of carrying the burden of inheritance is certainly not to say, however, that it does not play a key role. It represents the blueprint through which the specifications of the living cell are bequeathed to the next generation.

How did evolution first incorporate DNA into its scheme of reproduction? A considerable part of this question is answered by experiment. First, however, it is necessary to examine the view which is forceful merely by virtue of the fact that it has been taught to numerous generations of students. This is the sophistry of 'DNA self-replication'.

No evidence exists that DNA replicates itself. All that has been shown is that DNA is replicated. DNA is thus passive, the result of transitive actions. No shred of evidence exists to demonstrate that DNA can 'make' anything, let alone more of itself. Approximately a century of biochemical investigation has taught us that the agents of manufacture in the cell are protein enzymes. Moreover, they do not operate alone. They require energy-rich substances, which are predominantly the ubiquitous ATP, occasionally GTP, and some other special compounds.

The phrase 'DNA self-replication', even though sometimes qualified in part, has grown in use (Watson, 1976, p.213; Luria, 1973; Eigen *et al.*, 1981). Most interesting is the fact that self-replication as a term is typically used in publications that explicitly refer also to the enzymes that carry out the processes (e.g. Eigen *et al.*, 1981). The difference in outlook is not an expression of intellectual territoriality. It seems, however, that only those students who were biochemically educated and have been exposed to the vast wonderful world of enzymes and all that they do, e.g. Kornberg (1980), acknowledge an as yet incompletely defined role of enzymes in inheritance.

If we recognize that DNA is made by enzyme proteins (Commoner, 1968;

Dillon, 1978; Kornberg, 1980) the question that properly concerns us here is whether it can be shown that the concept of proteins-first will lead into the mechanism of modern genetic coding.

This possibility has so far been recognized in mechanistic detail in only one way. The new perspective is the result of experiments in protobiochemistry. The first recognition of the possibility stemmed from the finding that a lysine-rich proteinoid catalyzes each of the two principal syntheses of biomacromolecules: the synthesis of peptide bonds and the synthesis of internucleotide bonds (Fox *et al.*, 1974). With the assumption that all lysine-rich peptides can synthesize peptides from all kinds of amino acid (Nakashima and Fox, 1980; Fox and Nakashima, 1980), a mechanism for the origin of the genetic code within cells resulted (Fox, 1981c,d).

The synthetic activity is found in thermal peptides rich (ca. 20–25 per cent) in any basic amino acid (Syren and Fox, 1982). Such activity is observed in aqueous solution, in phase-separated particles composed of basic and acidic proteinoids, and in phase-separated particles composed of basic proteinoid and polynucleotide. Laboratory protocells composed of basic and acidic proteinoids are stable at pHs above 7·0 (Fox and Yuyama, 1963b; Snyder and Fox, 1975) and are the kind that have been artificially fossilized (Francis *et al.*, 1978). They are also catalytically active (Fox and Nakashima, 1980). Their catalytic activity is due to basic proteinoid contained in the particles. This locale of the evolving cell itself was an ideal setting for the origins and evolution of the genetic code. The experimental findings thus bridge the component protobiological phenomena to what looks to be the first biological locale for the origin of the genetic mechanism, in which nucleic acid and protein syntheses were necessarily intertwined. Such direct interactions would then have evolved to mechanisms involving synthetases, adaptors, etc.

Self-ordering and self-organizing

Several statements have been made about the critical nature of self-organization in evolution (Fox, 1960; Matsuno, 1982; Nicolis and Prigogine, 1977; Eigen, 1971; Fox and Matsuno, 1983). The interdigitation of self-organization with other processes and within processes merits some consideration.

Self-sequencing is self-organizing
The significance of self-organization has been recognized by numerous evolutionists (Pasteur in Vallery-Radot, 1922; Fox, 1960; Eigen, 1971a,b; Nicolis and Prigogine, 1977). The reality of self-organization has been challenged by Yockey (1981) in a paper peppered with allusions to scripture

and suffering from failure to recognize that modern living systems are not primordial systems. Among those who recognize the significance of self-organization either to early steps in evolution or to biochemical steps in modern organisms, some (Eigen, 1971a,b; Crick, 1981) have failed to allow that self-sequencing of amino acids was an obligatory early step in self-organization. This is perhaps due to a basic commitment to the pervasive concept of the random matrix. The recognition of the larger process of self-organization was belated, requiring many decades since a definition of the problem by Pasteur in 1864 (Vallery-Radot, 1922), so it will not be surprising if general awareness of the principle of self-sequencing will need more time.

The evolutionary descendant of self-ordering is enzyme specificity

Writers such as Calvin (1969) and Dillon (1978) have discussed the phenomenon of self-sequencing (self-ordering) of amino acids in the primordial context. They have stated in addition that modern relicts of this phenomenon must be available.

Such a relict has been proposed (Fox, 1981c) to be the specificity of all evolved enzymes. The original self-ordering of amino acids is explained as due to the steric relationships of the various amino acids (Fox and Dose, 1977). This would largely be an expression of amino acid sidechains (Fox, 1980a). When these same amino acids became fixed in proteinoids, and later in proteins, the sidechains again supplied steric information for reactions with substrates in metabolism. According to this view, the relict of self-ordering is a very pervasive one, i.e. the specificities of all enzyme-substrate interactions.

Self-ordering is self-limiting

The significance of a self-limited array is that it is non-random, and can be very highly so. The difficulties of accomodating evolutionary fact to a random matrix has been commented on. The recognition of self-ordering as a key early step in evolution has allowed a quantitative valuation to be placed on dynamic non-randomness (Nakashima et al., 1977).

The couching of proteinoid evolution within cosmic evolution

The concept protobiology → biology suggests in turn that this transition fits into a larger transitional sequence. Especially has there been the need to understand how materials of the first cells could have become more ordered than those protocells. The experimental evidence for much non-randomness in both protocells and proteinoids has been accumulating for over twenty years. Much more recently, data from the Big Bang temperature studies have appeared (Davies, 1978). The evidence on events within the first 24 hours of the Big Bang suggests that the initial Universe (Eriksson *et al.*, 1982) was highly non-random. An earlier similar suggestion was that of Sakharov (1966), who inferred internal constraints in an early expanding universe.

When that picture is placed in tandem with the inferences on order in cellular precursors, and then in cells, the sequence appears to be a cosmic reaffirmation of the second law of thermodynamics (Fox, 1980b). In this perspective, a smooth transition of protobiology → biology, while still lacking much interstitial detail, is more easily defended. The relationship between order between galaxies and between molecules still needs to be spelled out, however.

The significance of the proteinoid evolutionary sequence is that it has revealed (a) the crucial nature of self-sequencing of amino acids to give non-random informed polymers, to yield (b) a temporally primary and dynamic proteinoid protocell and (c) the remarkable (Rutten, 1971; Price, 1974; Lehninger, 1975; Florkin, 1975) manifold (proto)biological functions of that unit of protolife (Fox, 1978). This sequence is rooted in the non-randomness of matter in the prebiotic era, both astronomical and geological.

Time-based instructive evolution of macromolecules

In two papers, Wassermann (1982a,b) has discussed principal consequences of recognizing internal, or molecular, selection relative to Darwin's natural selection. Wassermann uses the term molecular selection, which we had employed earlier (Fox and Dose, 1977, p.261) as a synonym for self-ordering or self-sequencing (in the case of amino acids). Wassermann points out that TIMA (Time-based Instructive Evolution of Macromolecules) 'would have an overwhelming advantage over the neo-Darwinian idea of exclusive evolution by environmental selection combined with random genesis of molecules. "Molecular selection" is an active selection which can dispense with the assumed highly improbable random neo-Darwinian origin of macromolecules'.

Wassermann relates TIMA and molecular selection to other topics such as DNA, RNA, germ-line cell genomes, and punctuated equilibria.

THE POPULAR AND PROTEINOID PARADIGMS

Matrix: random or non-random?

Most of the principal aspects in which the proteinoid theory differs from the popular paradigm are listed in Table 2·5. The tenets formulated from data

TABLE 2·5 Popular and proteinoid paradigms

Popular paradigm inferred from modern cells	Proteinoid paradigm inferred from experiments
Random matrix	Non-random matrix
Random polymerization of amino acids	Non-random polymerization of amino acids
Order out of chaos	Chaos out of order
Chance variation (outside)	Orthogenetic variation (endogenous)
Long prebiotic evolution	Brief prebiotic evolution (< 12h)
Intricate processes of origin	Simple process of origin
Replication at the molecular level first	Replication at the cellular level first
Enzyme-free replication of DNA	Enzymic replication of DNA
Emergence of cells after the inheritance mechanism (DNA-first)	Emergence of inheritance mechanism in cells (Protein-first)

from protobiological simulation are obtained only from proteinoid since, evidently, that is the only approach to have retraced the steps in molecular evolution. In the popular paradigm the neo-Darwinian premise is prominent and fundamental.

The random variation of neo-Darwinism begins from, and would indeed require, a random matrix. Almost without exception, leaders in the development of newer *gedanken* on the origin of life (Monod, 1971; Eigen, 1971a,b; Luria, 1973; Eigen and Schuster, 1979; Crick, 1981) build their concepts on the 'armchair' assumption of a random matrix.

In contrast, the proteinoid paradigm is based on a non-random matrix that functioned as the staging area for organic evolution. In this experimentally derived paradigm, the principal type of solid matter in the cell, structurally and functionally, was protein. As has been explained, the first protein was proteinoid, or 'thermal protein' as it is listed in Chemical Abstracts indexes.

The popular paradigm, as explained, identifies the first informational biomacromolecules as nucleic acid. However, the proponents of this view have not explained how the nucleic acid came into existence, nor how it could have come into existence without the agency or direction of prior specific protein. Nor have any of the proponents explained how nucleic acids could have been replicated without prior specific proteins.

The usual assumption for the origin of nucleic acid is that it arose from a random matrix by chance (*see* Introduction). According to the experiments, however, the proteins arose in a determinate and endogenous fashion, not by chance and not by the action of outside agents. *The matrix was non-random.* It was non-random because the proteins obtained instructions from the reactant amino acids.

Polymerization of amino acids

The possibility that life could have begun from a random matrix has been denied by a few (Eden, 1967; Salisbury, 1969; Wigner, 1961; Schutzenberger, 1967) on theoretical grounds. In a paper titled 'The Probability of the Existence of a Self-Reproducing Unit', Wigner (1961) pointed out that the usual 'assumption is that the Hamiltonian which governs the behavior of a complicated system is a random symmetric matrix, with no particular properties except for its symmetric nature'. In the same paper, Wigner states 'It is more likely that the present laws and concepts of quantum mechanics will have to undergo modifications before they can be applied to the problems of life'.

The fact that the assumption of randomness is central in both biology and physics helps to explain the hold that neo-Darwinism has in contemporary thought. In other words, some of *the assumptions of quantum and statistical mechanics in physics, not to mention the randomness intrinsic to information theory* (Fox and Matsuno, 1983), *reinforce the basic assumption of neo-Darwinism.*

Eden's (1967) computations gave a figure of 10^{325} for the random number of protein molecular types. He estimates the total number of protein molecules that have ever existed on Earth to have been 10^{52} while the number of species of protein molecules has been 'much smaller than this, say 10^{40}.' From this it follows that the protein matrix cannot have been random.

The answer proposed at the same conference of Eden's presentation is the experimental fact that proteins produced under geothermal conditions are sharply limited in number (Fox, 1967), which is of course non-random in fact, and antirandom in concept. Such self-constraining experimental results are well known to be highly reproducible.

Disorder ⟶ Order?

The concept of 'order out of chaos' has had a long and firm hold (Fox, 1973b) in popular and scientific thinking (Eigen *et al.*, 1981; Dyson, 1982). Dyson subscribes to the protein-first view, such as has emerged from experiments and to the need for early enzymes. In Dyson's proposal, however, this view is coupled with his assumption of evolutionary disorder → order which is incompatible with the experiments (Fox, 1981c,d).

A key concept presented by Dyson (1982) is that of 'islands' in evolution. Proteinoid microspheres fulfil that role, representing its experimental embodiment (Fox *et al.*, 1959, 1969; Przybylski *et al.*, 1982).

Exogenous or endogenous evolution (orthogenesis)

The studies of molecular evolution stand out for the support that they give to the perspective of orthogenesis (Haldane, 1966; Mayr, 1980). They dramatically illustrate the endogenous stepwise nature of evolution proceeding from amino acid molecules → copolyamino acid macromolecules → assembled macromolecules = cellular systems → templated cells. What arises at each step is quite precisely the product of the matrix from which it arose. This is the kind of evolution that has long been regarded as extreme in concept: orthogenesis. Orthogenesis is determinism-in action. When the successive changes studied are molecular they are most easily seen. They conform with the evolutionary reality of endogenous, or internal factors (Whyte, 1965). The burden of proof increasingly falls to anyone who argues that molecular orthogenesis does not lead to biological orthogenesis in its further hierarchical evolution. It is desirable to state that the connotations of orthogenesis in the discussion here do *not* include an absence of natural selection. Rather, natural selection is not a component of the endogenous variation but is operative subsequent to variation.

Period of prebiotic evolution

The popular paradigm visualizes the transition from prebiotic molecules to biotic systems as extremely prolonged (Calvin, 1969; Wald, 1954). The proteinoid experiments indicate a brief and rapid assembly, less than 12 hours in a warm place once a set of amino acids has arisen.

One consequence of the rapidity indicated is that a *de novo* spontaneous generation could occur within the dark of a diurnal cycle, not too different in length from that of a contemporary diurnal cycle. Under these conditions, concern about the instability of prebiotic protocell exposure to intense radiation is misplaced. Corliss *et al.* (1981) visualize the formation of *de novo* organisms at hydrothermal vents in part because the laboratory experiments suggest that the necessary steps could have occurred rapidly, overnight.

Simplicity of processes

Perhaps no attitude has delayed the modern era of experimental protobiogenesis as much as the belief that the essential processes were hopelessly complex. This view is easily inspired by the indisputable complexity of cellular mechanisms and the products of those mechanisms. That theoretical, or Aristotelian, reasoning leads readily to the inference of imponderability is testified to by the lengthy passages in tomes of Eigen and Schuster (1979) and Crick (1981).

The fallacy is in extrapolating from the complexity of mechanisms and structure to the operational level, which is extremely simple. Interactions quickly yield a geometrically large number of components (Fox, 1973c).

The level at which replication began

Replication in modern organisms occurs at two levels: the cellular level and the molecular level, DNA replication. It is seemingly logical that replication occurred first in evolution at a molecular level and thence at the more complex hierarchical level of reproduction in the first organism. The relationship of complexity to evolution is, however, an incompletely defined question (Saunders and Ho, 1976, 1981). What is most likely for any stage in evolution seems to be what can *most easily come into existence*, irrespective of its relative degree of complexity (cf. Saunders and Ho, 1981).

Experiments have revealed that cells could come into existence with the greatest of ease by the contact of water with proteinoids (Fox, 1976). Virtually any of the numerous kinds of thermal proteins quickly and efficiently yield cell-like structures. Collateral concepts that place this observation in perspective include the view that any such (proto)cell could and would evolve to a modern cell. But the experiments show that the protocell had much of the quality of a generalized modern cell. Of the various processes that constitute the phenomena of protobiology, this origin is the simplest operationally, albeit extremely complex mechanistically. The replication of DNA is an extremely complex process that requires scores of enzymes, each with its own specific role, plus repair enzymes, etc. (Kornberg, 1980). Modern knowledge leads to the recognition that cellular replication could arise first, as a far far simpler, if not universal process. Such recognition gives rise to a second collateral concept — that the originating cell would belong to a highly limited society. This is clearly demonstrated in experiments which showed that the processes prior to protocell genesis are self-ordering.

The outlines of the origin of DNA synthesis in reproductive protocells have been inferred from experiments (Fox, 1981a,b). That the replication at the cellular level came before DNA replication is a key distinction of the proteinoid from the popular paradigm.

Enzymic replication of DNA

The popular paradigm fosters a view of DNA-first. Proponents of this premise are open to the accusation that they believe that DNA arose as the result of instant creation. We have more than hints that protein enzymes are involved in the synthesis and replication of nucleic acids (Sumper and Luce, 1975; Kornberg, 1980; Kozu *et al.*, 1982).

Emergence of inheritance

Many neo-Darwinists fail to distinguish between life and inheritance. While the mechanism of inheritance is an essential part of the mechanism of life, life itself is a concentric set of mechanisms for carrying out its biochemical business as well as preparing the machinery for the next generation. Only when this was clearly realized, was it possible to focus on the problem of how the inheritance mechanism could have arisen from its noninheritive precursors.

SOME PERSPECTIVES FROM PROTEINOID EXPERIMENTS ON EVOLUTIONARY THEORY

Beyond neo-Darwinism

Although it is not easily documented, many evolutionists have long felt that a consistent theory of evolution might be developed through (a) clarifying the first steps, (b) disciplining component concepts by experiment. In the experimental approach *via* constructionism (Fox, 1977), both (a) and (b) simultaneously have been initiated.

These results have led, *via* repeatable experimental demonstrations, into a truly new overview of the Darwinian analysis. An evaluation denying the randomness of the evolutionary matrix is not an original contribution of the proteinoid experiments, as objections to randomism were made independently by others, as pointed out earlier. What is new is the experimental demonstration of non-randomness as a starting point for organic evolution for which the evidence for this phenomenon is extensive (Fox, 1980b). The high *degree* of non-randomness was especially contrary to general expectations.

Although much data that had been accumulated indirectly signalled non-randomness in thermal peptides (Fox, 1981b), the most significant results were obtained from sequence studies, which are a developed art (Rosmus and Deyl, 1972), by Nakashima *et al.*, (1977), and Hartmann *et al.*, (1981) and from terminal amino acid analysis (Melius and Sheng, 1975). The significance of the proteinoid paradigm is further enhanced in the light of the proposal that the Universe was highly non-random at the time of the Big

Bang (Fox, 1980b). The living world is a consequence of self-limiting processes in the evolution of the inanimate precursors.

The total number of protein molecules that could theoretically arise initially would depend upon the initial number of amino acid types. A reasonable number of the latter is twelve (Jukes, 1966; Fox and Windsor, 1970; Hayatsu et al., 1971). From twelve types, a plausible number of protein isomers is 10^{300} (Lehninger, 1975).

Despite that supraastronomical number, and even with three billion years of evolution, one estimate of the total number of protein molecules in today's organisms is (Lehninger, 1975) approximately 10^{10}–10^{12} (cf. also Eden, 1967). The self-ordering, therefore, of amino acids at the outset must have been extremely self-limiting. This kind of limitation was a design factor and would also have the quality of a creative force (Fox, 1980b). The experiments which define organismic evolution as natural selection acting on non-random variants explain the constraints.

Macroevolution

The importance of macroevolution, as against gradual evolution, is a problem that has received a renewal of interest (e.g. Stanley, 1979) in recent years. Studies of experimental protobiogenesis contribute to the dialogue of controversy by providing an example of the size of the first step in organic evolution.

The conversion of an amorphous powdery substance to uniformly organized, metabolically active, and structured cellular individuals is obviously a large leap—a macroevolutionary step. If the first step was a leap, it is reasonable to expect that some later steps in evolution would also have been large. The focus of attention in this kind of analysis is on creative variations in steps of *assembly*. This was evident in a symposium on assembly mechanisms, the participants of which included Oparin shortly before his death (Fox, 1980c). The evolutionary thread connecting the first spontaneous events in cellular assembly and later ones in macroevolution was commented on by several of the participants.

Heterotrophism

The significance of heterotrophs to initial cellular metabolism has received considerable theoretical attention (Oparin, 1957; Welch, 1972). Protobiological investigation also provides disciplinary perspectives on this question.

The formation of a protocell from its non-cellular components appears to be an act of pure heterotrophism, as that is usually defined. Similarly, early cellular reproduction, as defined by experiments, involves also a kind of heterotrophic growth. Reproduction occurs in at least four modes that appear to be evolutionary precursors of modern reproductive mechanisms (Fox, 1973a). In each of these cyclic mechanisms, the growth phase occurs through heterotrophic action, i.e. by accretion of preformed proteinoid. Both the origin and reproduction of protocells are thus macro events, and are mediated by heterotrophism.

The laboratory protocell, however, suggests that the first cell was not limited to heterotrophism. It was autotrophic in that it had the capacity for both anabolism and catabolism (Fox and Nakashima, 1980).

EPILOGUE

An evolvable laboratory protocell

The extensive investigations on proteinoid microspheres have led to the recognition that they represent spontaneously produced protocells on Earth (Follmann, 1982). The flow of chemical events, to recapitulate, is seen to proceed from amino acid mixtures containing aspartic and glutamic acids to proteinoids that readily formed cells capable of synthesis of both nucleic acids (Follmann, 1982; Fox, 1981c) and proteins, and therefore capable of further biochemical evolution. These protocells possessed, in contrast to long-held assumptions about inertness (Oparin, 1957), numerous catabolic activities as well as the primary anabolic activities mentioned (Fox, 1980a). (The synthesis of lysine-rich peptides by lysine-rich peptides *in seriatim* appears to be likely, but has yet to be examined through more than one cycle.)

The proteinoid microspheres are able also to undergo heterotrophic proliferation (Fox, 1973a). In these same proliferative cells, a coded genetic mechanism could have developed because the same agents synthesize both kinds of backbone bonds essential to proteins and nucleic acids from the monomers and ATP (Fox, 1981c).

A comprehensive theory of organismic evolution

The influence of protobiogenesis on subsequent evolution carries the sense of 'as the twig is bent, so grows the tree'. The initial twig in this metaphor is that of the first events in molecular evolution, and the tree is the familiar evolutionary tree.

Although there exist biologists who would deny an evolutionary link from macromolecular origins to the origin of species, this is a narrow practice suggestive of *closet creationism*. Salthe (1982) has expressed the evolutionary point of view:

> The evolutionary process is represented at the molecular level on the one hand and encompasses cosmogony on the other.

The theory for the first steps of molecular evolution is more easily discerned than for the principles operant in anything as complex as an evolved cell. The single most outstanding lesson from molecular evolution studies is that amino acid monomers order themselves during polymerization. The products are non-random, despite assumptions to the contrary.

Some evolutionists whose focus is typically organismal, assert that the connotation for 'random' means that the changes are undirected rather than that all mutations are equally likely. The new perspective that emerges from a molecular foundation is that selection occurs on variants that are neither random by a strict definition, nor random by the definition of undirectedness. The reality is rather that adaptation is directed by non-randomness, which is orthogenetic and deterministic (Morgan, 1932; Fox, 1980b).

This line of reasoning leads to the view that adaptation is directed by initial non-randomness, and provides a link between internal control and adaptation in terms of both direction and causality. The natural selection of non-random variants is beyond neo-Darwinism.

The newer synthesis

A new intellectual synthesis of evolutionary theory is now available due to physical synthesis in the laboratory. Indeed, proper evaluation of the significance of the experimental advance has been thwarted by the pervasiveness of the randomistic concept. The resolution of this intellectual difficulty lies ultimately in the experimentally observed non-random self-sequencing of amino acids during thermal polymerization.

A general failure to apprehend fully the distinction between modern cell and protocell is reminiscent of the failure to recognize that the biblical picture of instant creation has been replaced by an evolutionary picture that is necessarily stepwise and self-organizing.

The Aristotelian approach to explaining life's origin has a quality similar to that of the biblical one in that life arose from a random matrix which, while not a King James 'void', is periconceptual thereto. The concept of randomness attains its zenith in the need to explain the catalysis of DNA replication by a rare mineral on an unspecified alien planet, the resultant organisms having been sent to this planet by spaceship (Crick, 1981).

The non-random alternative does not require going beyond terrestrial events. One of the inferences derived from non-randomness is the easy formation of multiple copies of informational molecules, the thermal proteins, some of which are 'fitted' for catalytic activity involved in the formation of the internucleotide bonds of nucleic acids. Also included in their 'fitness' (cf. Henderson, 1913) is catalytic activity for the formation of peptides of the kind that can catalyze the polymerization of nucleotides in an aqueous milieu with the aid of ATP. This statement is made on the basis of experimental demonstration under conditions that still exist on the present Earth and therefore under conditions that, with virtual certainty, existed on the primitive Earth. No resort to mysterious catalysts on obscure planets is necessary, although something as far-flung would be needed if one were again to invoke randomness.

The non-random matrix does not lead to a sudden, journalistic 'synthesis of life in the laboratory'. Instead the emerging picture is one of aliveness by stages. For example, some degree of aliveness such as the ability to make protopeptides and protopolynucleotides would have preceded the first modern cells, whereas other kinds of aliveness such as the possession of cellular membranes would not have arisen until a protocell was formed by aggregation. How any of these kinds of aliveness would have arisen from randomness has not been defensibly indicated, whereas the sequence is clear in the proteinoid experiments.

Conceptual difficulties in reducing the proteinoid theory to a physico-mechanical theory have been identified (Matsuno, 1983). The physicist, who generally speaking has not thought in terms of evolutionary processes, must shift his mental gears in two ways. One is in the temporal mode, which requires that he distinguish between modern and primordial: to recognize that what occurred three billion years ago is discernibly related to, but not identical with, what occurs now. This is a common barrier for those who come fresh to problems of origins (Fox, 1981a).

The other kind of mental shift required is hierarchical. In evolutionary considerations, principles and phenomena at one level are discernibly related to those at another, but they cannot be identical. This staircase development, in fact, is known as 'emergent properties'. Recognition of emergent properties tends to be second-nature to true evolutionists, but it poses difficulties for others.

A third kind of difficulty concerns how one thinks of the quality of order. (A generally satisfactory definition of 'order' is elusive.) However, the analysis that must be applied concerns 'order within' versus 'order between', i.e. intraorder versus interorder.

The need to distinguish between intraorder and interorder is evident in the analytical discussion of Monod (1971). Monod sees the function of nucleic acids as providing the information for stamping out considerable numbers of the same protein molecule, i.e. interorder. He presents this process as obligatory. The process exists in fact, but it does not signify that the nucleic acid determination of sequence has been or is obligatory in the evolutionary sense. Monod also correctly infers modern protein intradisorder, e.g. he can find no order to the sequence within the insulin molecule. From these facts and inferences Monod back-extrapolates to originally random protein. Forward extrapolation (Fox, 1975) by experiment under relevant simulated geological situation emphasizes non-randomness. When we look beyond neo-Darwinism, we look beyond randomness.

The sequential steps of molecular evolution can best be understood as self-sequencing much as amino acids are self-sequencing. Amino acids lead to proteinoids which lead, with water, to protocells which lead to cellular macromolecule synthesis, etc. But this is orthogenesis, a concept that has had a checkered history (Haldane, 1966). It is a concept that fits best with the newer facts; evolution is thus endogenous. 'Outside agents' like DNA are either superfluous, or are internal products of the orthogenetic sequence (Fox, 1959; 1969), and thus not 'outside'.

In the absence of the assumption that variations in evolution may be considered as fluctuations derived from a random array, the question of selection by the environment can be examined anew. Variation is rationally explainable as due to the existence of determinate limits in the synthetic processes of evolution. No active contribution from the environment is essential in this view. The environment fulfils the role originally assigned for it—that of a passive sieve of determinate variations (Morgan, 1932). Not only is selection a limiting influence, it acts on an array much more narrowly limited than has been inferred from randomism (Fox, 1974). The fact that active contribution from the environment is indicated as superfluous to this thesis does not deny its participation (cf. Ho and Saunders, 1979).

The chemical answers to our questions on the mechanism of self-sequencing of molecules must reside in their shapes (which includes electronic configurations). Selection and direction have their first roots in the shapes of molecules (Fox, 1983b). Natural selection, which in this synthesis occurs in a fully passive environment operates on a narrow, determinate range of original variants (Fox, 1974). It is in this light that we infer new support for determinism.

Indeed a theme for determinism can be traced through the hierarchical steps of evolution. It begins in the physical particles as observed by an early proponent of determinism, Schrödinger (1946), and proceeds through the concept of genetic determinism introduced and developed by Morgan (1932). This was an antecedent of the genetic determinism of Wilson (1978), but Morgan's determinism, which is more dynamic than Wilson's, comports better with biochemical knowledge (Kenyon and Steinman, 1969).

This holistic fabric of determinism is that of an interdigitating sequence of interactions from physical particles (Schrödinger, 1946) → molecules → bio-polymers (Fox, 1980b) → behavior of lower organisms (Koshland, 1984) → behavior of humans (Bouchard, 1984).

Randomness *within* the protein molecule is by arbitrary definition a single state. Any departure from a total absence of interamino acid influence in the sequence is an act the information theorist calls a perturbation of intersymbol influence, is non-random. The non-randomness of our primary concern covers the theoretically widest possible range of states from almost-randomness to total determination of sequence by interactions of the symbols, the amino acids. A totally non-random order would thus yield a large number of a single *type* of protein molecule. Non-randomness in its variability is thus capable of the widest possible range. Exact characterization of the degree of non-randomness in the first, and later, populations of proteins, is a subject for future investigation. It is clear, however, that the degree of non-randomness must have been far, far closer to totally determined than to random (Fox, 1974). It is, indeed, of some significance that the evolutionary diversity of proteins would have developed from initial internal determinism as a boundary condition, as the experiments suggest, rather than from randomness as a boundary condition.

Holistic determinism is rooted in non-randomness at each stage, according to our interpretation of the evidence. In a comprehensive context, then, evolution, as we see it, is beyond the valley of neo-Darwinism, and consists of natural selection and preservation of non-random variants.

ACKNOWLEDGMENT

Much of the experimental study of proteinoids has been supported by Grant No. NGR 10-007-008 of the National Aeronautics and Space Administration.

Notes

(1) By extension of the perceptions of Allen (1978), the contribution of Morgan to the theory of evolution resembles his pervasive but quiet contribution to biological

education in the United States and elsewhere. At a time when leading fellow biologists did not share his vision (Allen, 1978), he initiated a profound melding of physical science with biology by moving from Columbia University to the highly technological, physical scientific, and mathematical atmosphere of what had been the Throop Institute of Technology, later to become the California Institute of Technology (Allen, 1978). Morgan was not a practicing physical scientist, nor was his mathematical prowess sufficient that he could ever even be considered for a professorship of population genetics. Morgan however had what was present in much smaller degree in the intellects of most of his biological peers; he had perspective and insight for the physical and mathematical realms. He could visualize the nature of matter and its interactions. While he might not generate any equations to describe events in genic recombination, he had a unique appreciation of the fact that genetics was a lonely example of an exact discipline within biology. This awareness was proudly stated in his conversations.

In the same vein, Morgan recognized and had an insight for the interactions of molecules in organisms. That such behavior is in accord with an ordered universe, one having a deterministic bias, he no longer felt he needed to question in his later years (Morgan, 1932). His view of evolution was thus of the broadest kind. The long-exercised conflict between determination and chance in science was not for Morgan a debatable choice (Morgan, 1932).

REFERENCES

Allen, G. (1968). Thomas Hunt Morgan and the problem of natural selection. *J. Hist. Biol.* **1**, 113–139.

Baltscheffsky, H. (1981). Stepwise molecular evolution of bacterial photosynthetic energy conversion. *BioSystems* **14**, 49–56.

Barghoorn, E. S. and Tyler, S. A. (1963). Fossil organisms from Precambrian sediments. *Ann. N.Y. Acad. Sci.* **108**, 451–452.

Barrett, P. H., Weinshank, D. J. and Gottleber, T. T. (1982). 'A Concordance to Darwin's Origin of Species'. Cornell University Press, Ithaca.

Bouchard, T. A., Jr. (1984). Twins reared together and apart: what they tell us about human diversity. *In* 'Individuality' (S. W. Fox, *Ed.*) Plenum Press, New York.

Brooke, S. and Fox, S. W. (1977). Compartmentalization in proteinoid microspheres. *BioSystems* **9**, 1–22.

Bryson, V. and Vogel, H. J., eds. (1965). 'Evolving Genes and Proteins'. Academic Press, New York.

Calvin, M. (1969). 'Chemical Evolution'. Oxford University Press, London.

Cloud, P. E. and Morrison, K. (1979). On microbial contaminants, micropseudo-fossils, and the oldest records of life. *Precambrian Res.* **9**, 81–91.

Commoner, B. (1968). Failure of the Watson–Crick theory as a chemical explanation of inheritance. *Nature* **220**, 334–340.

Copeland, J. J. (1936). Yellowstone thermal Myxophyceae. *Ann. N.Y. Acad. Sci.* **36**, 1–224.

Corliss, J. B., Baross, J. A., and Hoffman, S. E. (1981). An hypothesis concerning the relationship between submarine hot springs and the origin of life on Earth. *Oceanologica Acta,* 1981, NOSP, 59–69.

Crick, F. H. C. (1968). The origin of the genetic code. *J. Mol. Biol.* **38**, 367–379.

Crick, F. H. C. (1970). Central Dogma of molecular biology. *Nature* **227**, 561–563.

Crick, F. H. C. (1981). 'Life Itself'. Simon and Schuster, New York.

Crick, F. H. C., Brenner, S., Klug, A. and Pieczenik, G. (1976). A speculation on the origin of protein synthesis. *Origins Life* **7**, 389–397.

Darwin, C. (Undated). 'The Origin of Species & the Descent of Man'. The Modern Library, New York.

Davies, P. (1978). Smoothing primaeval chaos. *Nature* **271**, 506.

Dayhoff, M. O. 'Atlas of Protein Sequence and Structure 1972'. Vol. 5. National Biomedical Research Foundation, Washington.

Dillon, L. S. (1978). 'The Genetic Mechanism and the Origin of Life'. Plenum Press, New York.

Dose, K. (1983). The evolution of individuality at the molecular and protocellular level. *In* 'Individuality' (S. W. Fox, *Ed.*). Plenum Press, New York.

Dyson, F. (1982). A model for the origin of life. *J. Mol. Evol.* **18**, 344–350.

Eden, M. (1967). *In* 'Mathematical Challenges to the neo-Darwinian Interpretation of Evolution' (P. S. Moorhead and M. M. Kaplan, *Eds*), pp.5–12. Wistar Institute Press, Philadelphia.

Eigen, M. (1971a). Selforganization of matter and the evolution of biological macromolecules. *Naturwissenschaften* **58**, 465–523.

Eigen, M. (1971b). Molecular self-organization and the early stages of evolution. *Q. Rev. Biophysics* **4**, 149–212.

Eigen, M. and Schuster, P. (1979). 'The Hypercycle'. Springer-Verlag, Berlin.

Eigen, M., Gardiner, W., Schuster, P., and Winkler-Oswatitsch, R. (1981). The origin of genetic information. *Sci. Am.* **244(4)**, 88–118.

Erikkson, K.-E., Islam, S., and Skagerstam, B.-S. (1982). A model for the cosmic creation of nuclear energy. *Nature* **296**, 540–542.

Florkin, M. (1975). *In* 'Comprehensive Biochemistry', Vol. 29B. Elsevier, Amsterdam.

Florkin, M. and Mason, H. S. (1960). An introduction to comparative biochemistry. *In* 'Comparative Biochemistry'. Vol. I (M. Florkin and H. S. Mason, *Eds*) pp.1–14. Academic Press, New York.

Follmann, H. (1982). Deoxyribonucleotide synthesis and the emergence of DNA in molecular evolution. *Naturwissenschaften* **69**, 75–81.

Fox, S. W. (1956). Evolution of protein molecules and thermal synthesis of biochemical substances. *Am. Scientist* **44**, 347–359.

Fox, S. W. (1957). The chemical problem of spontaneous generation. *J. Chem. Education* **34**, 472–479.

Fox, S. W. (1959). The biological replication of macromolecules. *J. Chem. Education* **36**, 706A.

Fox, S. W. (1960). Self-organizing phenomena and the first life. *Yearbk. Soc. Genl. Systems Res.* **5**, 57–60.

Fox, S. W. (1965). Experiments suggesting evolution to protein. *In* 'Evolving Genes and Proteins' (V. Bryson and H. J. Vogel *Eds*), pp.359–369. Academic Press, New York.

Fox, S. W. (1967). Remarks *In* 'Mathematical Challenges to the Neo-Darwinian Interpretation of Evolution' (P. S. Moorhead and M. M. Kaplan, *Eds*), p.17. Wistar Institute Press, Philadelphia.

Fox, S. W. (1969). In the beginning . . . life assembled itself. *New Scientist* **41**, 450–452.

Fox, S. W. (1973a). Molecular evolution to the first cells. *Pure Appl. Chem.* **34**, 641–669.

Fox, S. W. (1973b). Origin of the cell: experiments and premises. *Naturwissenschaften* **60**, 359–368.

Fox, S. W. (1973c). The rapid evolution of complex systems from simple beginnings. *In* 'Proc. 3rd Int. Conf. From Theoretical Physics to Biology'. (M. Marois, *Ed.*), pp.133–144. S. Karger, Basel.

Fox, S. W. (1974). The proteinoid theory of the origin of life and competing ideas. *Am. Biol. Teacher* **36**, 161–172, 181.

Fox, S. W. (1975). Looking forward to the present. *BioSystems* **6**, 165–175.

Fox, S. W. (1976). The evolutionary significance of phase-separated microsystems. *Origins Life* **7**, 49–68.

Fox, S. W. (1977). Bioorganic chemistry and the emergence of the first cell. *In* 'Bioorganic Chemistry III. Macro-and Multimolecular Systems' (E. E. van Tamelen, *Ed.*), p.21. Academic Press, New York.

Fox, S. W. (1978). The origin and nature of protolife. *In* 'The Nature of Life' (W. H. Heidcamp, *Ed.*), pp.23–92. University Park Press, Baltimore.

Fox, S. W. (1980a). Metabolic microspheres. *Naturwissenschaften* **67**, 378–383.

Fox, S. W. (1980b). Life from an orderly Cosmos. *Naturwissenschaften* **67**, 576–581.

Fox, S. W. (1980c). Introductory remarks to the special issue on 'assembly mechanisms'. *BioSystems* **12**, 131.

Fox, S. W. (1981a). From inanimate matter to living systems. *Am. Biol. Teacher* **43(3)**, 127–135, 140.

Fox, S. W. (1981b). Copolyamino acid fractionation and protobiochemistry. *J. Chromatog.* **215**, 115–120.

Fox, S. W. (1981c). Origins of the protein synthesis cycle. *Int. J. Quantum Chem.* **QBS8**, 441–454.

Fox, S. W. (1981d). A model for protocellular coordination of nucleic acid and protein syntheses. *In* 'Science and Scientists' (M. Kageyama, K. Nakamura, T. Oshima, and T. Uchida, *Eds*), pp.39–45. Japan Sc. Soc. Press, Tokyo.

Fox, S. W. (1983a). The beginnings of evolution and behavior. *In* 'Evolution, Behavior and Levels' (G. Greenberg and E. Tobach, *Eds*). Erlbaum Publishers, Hillsdale, New Jersey.

Fox, S. W. (1983b). Creationism and experimental protobiogenesis. *In* 'Science and Creation' (A. Montagu, *Ed.*). Oxford University Press.

Fox, S. W. and Dose, K. (1977). 'Molecular Evolution and the Origin of Life', rev. ed. Marcel Dekker, New York.

Fox, S. W. and Harada, K. (1960). The thermal copolymerization of amino acids common to protein. *J. Am. Chem. Soc.* **82**, 3745–3751.

Fox, S. W. and Matsuno, K. (1983). Self-organization of the protocell was a forward process. *J. Theor. Biol.* **101**, 321–323.

Fox, S. W. and Nakashima, T. (1980). The assembly and properties of protobiological structures: the beginnings of cellular peptide synthesis. *BioSystems* **12**, 155–166.

Fox, S. W. and Windsor, C. R. (1970). Synthesis of amino acids by the heating of formaldehyde and ammonia. *Science* **170**, 984–986.

Fox, S. W. and Yuyama, S. (1963a). Abiotic production of primitive protein and formed microparticles. *Ann. N. Y. Acad. Sci.* **108**, 487–494.

Fox, S. W. and Yuyama, S. (1963b). Effects of the Gram stain on microspheres from thermal polyamino acids. *J. Bact.* **85**, 279–283.

Fox, S. W. and Yuyama, S. (1964). Dynamic phenomena in microspheres from thermal proteinoid. *Comp. Biochem. Physiol.* **11**, 317–321.

Fox, S. W., Winitz, M., and Pettinga, C. W. (1953). Enzymic synthesis of peptide bonds. VI. The influence of residue type on papain-catalyzed reactions of some benzoylamino acids with some amino acid anilides. *J. Am. Chem. Soc.* **75**, 5539–5542.

Fox, S. W., Harada, K., and Kendrick, J. (1959). Production of spherules from synthetic proteinoids and hot water. *Science* **129**, 1221–1223.

Fox, S. W., McCauley, R. J., and Wood, A. (1967). A model of primitive heterotrophic proliferation. *Comp. Biochem. Physiol.* **20**, 773–778.

Fox, S. W., McCauley, R. J., Montgomery, P. O'B., Fukushima, T., Harada, K., and Windsor, C. R. (1969). Membrane-like properties in microsystems assembled from synthetic protein-like polymer. *In* 'Physical Principles of Biological Membranes' (F. Snell, J. Wolken, G. J. Iverson, and J. Lam, *Eds*), pp.417–430. Gordon and Breach, New York.

Fox, S. W., Jungck, J. R., and Nakashima, T. (1974). From proteinoid microsphere to contemporary cell: formation of internucleotide and peptide bonds by proteinoid particles. *Origins Life* **5**, 227–237.

Fox, S. W., Adachi, T., Stillwell, W., Ishima, Y., and Baumann, G. (1978). Photochemical synthesis of ATP: protomembranes and protometabolism. *In* 'Light Transducing Membranes' (D. W. Deamer, *Ed*.), pp.61-75. Academic Press, New York.

Fox, S. W., Harada, K., and Hare, P. E. (1981). Amino acids from the Moon: notes on meteorites. *Subcell. Biochem.* **8**, 357–373.

Fox, S. W., Nakashima, T., Przybylski, A., and Syren, R. M. (1982). The updated experimental proteinoid model. *Int'l. J. Quantum Chem.* **QBS9**, 195–204.

Francis, S., Margulis, L., and Barghoorn, E. S. (1978). On the experimental silicification of microorganisms II. On the time of appearance of eukaryotic organisms in the fossil record. *Precambrian Res.* **6**, 65–100.

Fruton, J. S. (1982). Proteinase-catalyzed synthesis of peptide bonds. *Adv. Enzymol.* **53**, 239–306.

Gamow, G., Rich, A., and Ycas, M. (1956). Problem of information transfer from nucleic acids to proteins. *Adv. Biol. Med. Phys.* **4**, 23–68.

Gould, S. J. (1982). Darwinism and the expansion of evolutionary theory. *Science* **216**, 389–387.

Griffith, J. S. (1967). Self-replication and scrapie. *Nature* **215**, 1043–1044.

Haldane, J. B. S. (1966). 'The Causes of Evolution', p.12. Cornell University Press, Ithaca.

Hartmann, J., Brand, M. C., and Dose, K. (1981). Formation of specific amino acid sequences during thermal polymerization of amino acids. *BioSystems* **13**, 141–147.

Hayatsu, R., Studier, M. H., and Anders, E. (1971). Origin of organic matter in early solar system-IV. Amino acids: confirmation of catalytic synthesis by mass spectrometry. *Geochim. Cosmochim. Acta* **35**, 939–951.

Heinz, B. and Reid, W. (1981). The formation of chromophores through amino acid thermolysis and their possible role as prebiotic photoreceptors. *BioSystems* **14**, 33–40.

Henderson, L. J. (1913). 'The Fitness of the Environment'. Macmillan, New York.

Ho, M.-W. and Saunders, P. T. (1979). Beyond neo-Darwinism—an epigenetic approach to evolution. *J. Theor. Biol.* **78**, 573–591.

Hsu, L. L. and Fox, S. W. (1976). Interactions between diverse proteinoids and microspheres in simulation of primordial evolution. *BioSystems* **8**, 89–101.

Hsu, L. L., Brooke, S., and Fox, S. W. (1971). Conjugation of proteinoid microspheres: a model of primordial communication. *Curr. Mod. Biol.* (*BioSystems*) **4**, 12–25.

Hubbard, R. (1982). The theory and practice of genetic reductionism—from Mendel's laws to genetic engineering. *In* 'Towards a Liberatory Biology' (S. Rose, *Ed.*), pp.62–78. Allison and Busby, London.

Ishima, Y., Przybylski, A. T. and Fox, S. W. (1981). Electrical membrane phenomena in spherules from proteinoid and lecithin. *BioSystems* **13**, 243–251.

Jukes, T. H. (1966). 'Molecules and Evolution'. Columbia University Press, New York.

Kenyon, D. H. and Steinman, G. (1969). 'Biochemical Predestination'. McGraw Hill, New York.

Keosian, J. (1974). Life's beginnings—origin or evolution. *In* 'The Origin of Life and Evolutionary Biochemistry' (K. Dose, S. W. Fox, G. A. Deborin, and T. E. Pavlovskaya, *Eds*), p.225. Plenum Press, New York.

Kimberlin, R. H. (1982). Scrapie agent: prions or virinos? *Nature* **297**, 107–108.

Knoll, A. H. and Barghoorn, E. S. (1977). Archean microfossils showing cell division from the Swaziland system of South Africa. *Science* **198**, 396–398.

Korn, R. W. and Korn, E. J. (1971). 'Biology'. John Wiley, New York.

Kornberg, A. (1980). 'DNA Replication', pp.87–88. W. H. Freeman, San Francisco.

Koshland, D. E., Jr. (1983). Individuality in bacteria and its relationship to higher species. *In* 'Individuality' (S. W. Fox, *Ed.*). Plenum Press, New York.

Kozu, T., Yagura, T., and Seno, T. (1982). *De novo* DNA synthesis by a novel mouse DNA polymerase associated with primase activity. *Nature* **298**, 180–182.

Lehninger, A. L. (1975). 'Biochemistry', 2nd ed. Worth, New York.

Leo, R. F. and Barghoorn, E. S. (1976). Silicification of wood. *Bot. Mus. Leaflets, Harvard Univ.* **25**, 1–47.

Lima-de-Faria, A. (1962). Selection at the molecular level. *J. Theor. Biol.* **2**, 7–15.

Luria, S. (1973). 'Life, the Unfinished Experiment'. Charles Scribner's Sons, New York.

Matsuno, K. (1982). Natural self-organization of polynucleotides and polypeptides in protobiogenesis: appearance of a protohypercycle. *BioSystems* **15**, 1–11.

Matsuno, K. (1984). Open systems and origin of protoreproductive units. *In* 'Beyond Neo-Darwinism' (M. W. Ho and P. T. Saunders, *Eds*). Academic Press, London.

Mayr, E. (1980). Prologue: some thoughts on the history of the evolutionary synthesis. *In* 'The Evolutionary Synthesis' (E. Mayr and W. B. Provine, *Eds*), pp.1–48. Harvard University Press, Cambridge, Massachusetts, USA.

Mayr, E. and Provine, W. B. (1980). 'The Evolutionary Synthesis'. Harvard University Press, Cambridge, Massachusetts, USA.

Melius, P. (1977). Composition and structure of thermal condensation polymers of amino acids. *In* 'Bioorganic Chemistry' (E. E. van Tamelen, *Ed.*) Vol. III, pp.123–136. Academic Press, New York.

Melius, P. (1982). Structure of thermal polymers of amino acids. *BioSystems* **15**, 275–280.

Melius, P. and Sheng, J. Y-P. (1975). Thermal condensation of a mixture of six amino acids. *Bioorg. Chem.* **4**, 385–391.

Miller, S. and Orgel, L. E. (1974). 'The Origins of Life on the Earth'. Prentice-Hall, Englewood Cliffs, New Jersey, USA.

Monod, J. (1971). 'Chance and Necessity' (transl. by A. Wainhouse). Alfred A. Knopf, New York.

Morgan, T. H. (1919). 'The Physical Basis of Heredity'. Lippincott, New York.

Morgan, T. H. (1932). 'The Scientific Basis of Evolution'. W. W. Norton and Co., New York.

Mueller, G. (1972). Organic microspheres from the Precambrian of South-west Africa. *Nature* **235**, 90–95.

Muller, H. J. (1955). Life. *Science* **121**, 1–9.

Nakashima, T. and Fox, S. W. (1980). Synthesis of peptides from amino acids and ATP with lysine-rich proteinoid. *J. Mol. Evol.* **15**, 161–168.

Nakashima, T., Jungck, J. R., Fox, S. W., Lederer, E., and Das, B. C. (1977). A test for randomness in peptides isolated from a thermal polyamino acid. *Int. J. Quantum Chem.* **QBS4**, 65–72.

Needham, J. (1935). 'Order and Life'. Cambridge University Press, Cambridge.

Needham, A. A. (1965). 'The Uniqueness of Biological Materials'. Pergamon Press, Oxford.

Nicolis, G. and Prigogine, I. (1977). 'Self-Organization in Nonequilibrium Systems'. John Wiley and Sons, New York.

Oparin, A. I. (1957). 'The Origin of Life on the Earth', Third Edition. Academic Press, New York.

Oparin, A. I. (1971). Routes for the origin of the first forms of life. *Sub-Cell. Biochem.* **1**, 75–81.

Oshima, T. (1971). Catalytic activities of synthetic polyamino acids: models of primitive enzymes. *Viva Origino* **1**, 35–43.

Overton, W. R. (1982). Creationism in schools: the decision in McLean versus the Arkansas Board of Education. *Science* **215**, 934–943.

Pflug, H. D. and Jaeschke-Boyer, H. (1979). Combined structural and chemical analysis of 3,800-Myr-old microfossils. *Nature* **280**, 483–486.

Ponnamperuma, C. (1968). Origin of Life. *In* 'Encyclopedia of Polymer Science and Technology' (H. M. Mark, N. G. Gaylord, and N. M. Bikales, *Eds*), p.656. Interscience, New York.

Price, C. C. (1974). 'The Synthesis of Life'. Dowden, Hutchinson and Ross, Stroudsburg, Pennsylvania, USA.

Prosser, C. L. (1965). Levels of biological organization and their physiological significance. *In* 'Ideas in Evolution and Behavior' (J. A. Moore, *Ed.*), pp.358–390. Natural History Press, Garden City, New York.

Prusiner, S. B. (1982). Novel proteinaceous infectious particles cause scrapie. *Science* **216**, 136–144.

Przybylski, A. T., Stratten, W. P., Syren, R. M., and Fox, S. W. (1982). Membrane action, and oscillatory potentials in simulated protocells. *Naturwissenschaften* **69**, 561–563.

Rohlfing, D. L. and Fox, S. W. (1969). Catalytic activities of thermal poly-anhydro-α-amino acids. *Adv. Catal.* **20**, 373–418.

Root-Bernstein, R. S. (1983). Protein replication by amino acid pairing. *J. Theor. Biol.* **100**, 99–106.

Rosmus, J. and Deyl, Z. (1972). Chromatographic methods in the analysis of protein structure. The methods for identification of N-terminal amino acids in peptides and proteins. Part B. *J. Chromatogr.* **70**, 221–339.

Rutten, M. G. (1971). 'The Origin of Life by Natural Causes'. Elsevier, Amsterdam.

Sakharov, A. D. (1966). The initial stage of an expanding universe and the appearance of a nonuniform distribution of matter. *Soviet Physics JETP* **22**, 241–249.

Salisbury, F. (1969). Natural selection and the complexity of the gene. *Nature* **224**, 342–343.

Salthe, S. N. (1982). Original life. *Nature* **295**, 452.

Saunders, P. T. and Ho, M. W. (1976). On the increase in complexity in evolution. *J. Theor. Biol.* **63**, 375–384.

Saunders, P. T. and Ho, M. W. (1981). On the increase in complexity in evolution II. The relativity of complexity and the principle of minimum increase. *J. Theor. Biol.* **90**, 515–530.

Schrödinger, E. (1946). 'What is Life?' Macmillan, New York.

Schützenberger, M. P. (1967). Algorithms and the Neo-Darwinian theory of evolution. *In* 'Mathematical Challenges to the Neo-Darwinian Interpretation of Evolution (P. S. Moorhead and M. M. Caplan, *Eds*), pp.73–75. Wistar Institute Press, Philadelphia.

Scott, Foresman and Co. (1980). 'Biology'. Glenview, IL.

Snyder, W. D. and Fox, S. W. (1975). A model for the origin of stable protocells in a primitive alkaline ocean. *BioSystems* **7**, 22–36.

Stanley, S. M. (1979). 'Macroevolution'. W. H. Freeman, San Francisco.

Stebbins, G. L. (1982). 'Darwin to DNA, Molecules to Humanity', p.176. W.H. Freeman, San Francisco.

Sumper, M. and Luce, R. (1975). Evidence for *de novo* production of self-replicating and environmentally adapted RNA structures by bacteriophage QB replicase. *Proc. Nat. Acad. Sci. USA* **72**, 162–166.

Syren, R. M. and Fox, S. W. (1982). Unpublished experiments.

Tobach, E. and Schneirla, T. C. (1968). The biopsychology of social behavior of animals. *In* 'Biologic Basis of Pediatric Practice' (R. E. Cooke and S. Leven, *Eds*), pp.68–82. McGraw Hill, New York.

Vallery-Radot, P. (1922). 'Pasteur', p.328. Masson, Paris.

Vegotsky, A. and Fox, S. W. (1962). Protein molecules: intraspecific and inter-specific variations. *In* 'Comparative Biochemistry', Vol. IV (M. Florkin and H. S. Mason, *Eds*), pp.185–244.

Wald, G. (1954). The origin of life. *Sci. Am.* **191(2)**, 44–53.

Wassermann, G. D. (1982a). TIMA Part 1. TIMA as a paradigm for the evolution of molecular complementarities and macromolecules. *J. Theor. Biol.* **96**, 77–86.

Wassermann, G. D. (1982b). TIMA Part 2. TIMA-based instructive evolution of macromolecules and organs and structures. *J. Theor. Biol.* **96**, 609–628.

Watson, J. D. (1976). 'The Molecular Biology of the Gene'. W. A. Benjamin, Menlo Park, California, USA.

Weissbach, H. and Pestka, S. (1977). 'Molecular Mechanisms of Protein Biosynthesis'. Academic Press, New York.

Welch, C. (1972). The heterotroph hypothesis and high school biology. *In* 'Molecular Evolution' (D. L. Rohlfing and A. I. Oparin, *Eds*), pp.443–447. Plenum Press, New York.

Whyte, L. L. (1965). 'Internal Factors in Evolution'. George Braziller, New York.

Wigner, E. (1961). The probability of the existence of a self-reproducing unit. *In* 'The Logic of Personal Knowledge' (E. Shils, *Ed*.), pp.231–238. The Free Press, Glencoe, Illinois, USA.

Wilson, E. O. (1978). 'On Human Nature'. Bantam Books and Harvard University Press, Cambridge, Massachusetts, USA.

Wright, S. (1967). Comments on the preliminary working papers of Eden and
   Waddington. *In* 'Mathematical Challenges to the neo-Darwinian Interpretation
   of Evolution'. (P. S. Moorhead and M. M. Kaplan, *Eds*), p.117. Wistar Institute
   Press, Philadelphia.
Yockey, H. P. (1981). Self organization origin of life scenarios and information
   theory. *J. Theor. Biol.* **91**, 13–31.

3

Open Systems and the Origin of Protoreproductive Units

KOICHIRO MATSUNO

Abstract

Evolution of matter is a mode of matter constraining itself by itself, not an outcome selected by something else.

The molecules in the primordial soup spontaneously form aggregates open to material flow if the energy of molecular association does not much exceed the thermal agitation energy and if the dissociation energy is not much less than the thermal energy. Once these aggregates are formed, they maintain the continuity of material flow by the adjustment of interaction in the medium. Material flow equilibration as a process for the maintenance of the continuity always induces successive equilibration due to the fact that the interaction change propagates in the medium with a finite velocity. Material flow equilibration is an organizing principle underlying the evolution of self-generated open systems and all biosystems. Open systems in the process of material flow equilibration generate only those variations that fulfil the continuity of material flow *a posteriori* and in turn serve as the generator of further non-random variations. The genesis of variations always goes along with endogenous constraints. Material flow equilibration makes biosystems the producers of evolutionary changes as well as the resultant of them. Variation and selection are not separable in the evolutionary process, in sharp contrast to the neo-Darwinian scheme. Material self-organization is consonant with the generation of non-random variations whereas self-organization due to random variations is conceptually inadmissible.

Material flow equilibration is certainly a process underlying a realization of the second law of thermodynamics.

BEYOND NEO-DARWINISM
ISBN: 0-12-350080-X

INTRODUCTION: THE ORIGIN OF LIFE AND OPEN SYSTEMS

The origin of life, the construction of biological systems from atoms and abiotic molecules, is a long-standing major scientific problem. It involves material self-assembly or self-organization (Fox, 1980) in which the elements are constrained and regulated internally. The extent of self-organization is such that one cannot but suppose that a regulative principle is at work. It is crucial however to recognize that the organizing principle, if it exists, rests upon matter itself and nowhere else (Webster and Goodwin, 1982; Smith and Morowitz, 1982). If atoms and abiotic molecules do not contain within themselves the capacity to regulate, then as Haeckel observed more than a hundred years ago (Dose, 1981), material self-organization in its literal sense would not follow.

Evolution also exhibits an increase in complexity with time (Saunders and Ho, 1976, 1981). Although the term complexity is difficult to define, we assign to it the connotation of symmetry-breaking among constituent elements both in space and in time. A system that is heterogeneous in space is less symmetrical than the one that is homogeneous. The increase in complexity suggests that the regulative principle constraining an aggregate of material elements always creates changes and leaves freedom for new ones. Moreover, the new freedom is further constrained by the regulative principle itself. The regulative principle inherent in matter is thus self-generative in the sense that the constraint acting upon material organization always generates a set of new relationships which is further subject to the constraint due to the regulative principle itself.

The first step towards uncovering the nature of the regulative principle is to spell out what fundamentally important character is common to the emergence of biosystems and what distinguishes biosystems from abiotic ones. A strict definition of biosystem is not needed for this purpose. Only observed facts are necessary.

A biosystem has a boundary formed through its interaction with the exterior. A biosystem is also open to material flow crossing the boundary in either direction.

Openness to material flow, which is common to all biosystems, is a consequence of a regulative principle acting upon molecules which are themselves not open to material flow. Even if it were possible to enumerate the chemical character of all the molecules found in biosystems, their construction could not be fully understood unless the process leading to an emergence of open systems is made explicit.

The molecular biology of present-day biosystems, although it certainly provides a rich catalogue of contemporary biochemical processes

(Monod, 1971; Jacob, 1974), is not the basis of the open systems which emerged and evolved. The intricate chemical process of DNA transcription presumes an underlying biosystem which is open to material flow (Commoner, 1968). Nevertheless, it does not necessarily follow that such a system possesses the DNA transcriptive machine from the start. The real question is whether a chemical precursor of the DNA transcriptive mechanism appeared simultaneously with or was preceded by the origin of an open system or a protocell (Fox, 1981a).

An open system equipped with a DNA transcriptive chemical automaton is much more constrained than open systems in general. Constraining the extent of an open system is a process. If one supposes that the emergence of a DNA transcriptive mechanism is due to chance events involving rare material aggregation, one would reject the regulative capability of atoms and molecules at the onset. This would in turn imply an invalidation of the regulative principle altogether because once it is set in motion, it never ceases to function. The regulative principle provides a nexus to maintain material aggregation over arbitrary intervals along evolutionary time. Its failure would inevitably lead to material dissociation. The regulative principle underlies both the origin and the evolution of the DNA apparatus.

The omnipresence of the regulative principle can explain why material self-organization can maintain its own material configuration, although its full content remains to be determined. This perspective leads to the observation that the origin of open systems could have preceded the emergence of the DNA apparatus, since the latter is a much more constrained special case of the former.

Natural selection of random variations within the neo-Darwinian scheme is inadequate as a candidate for the regulative principle inherent in matter (Eden, 1967; Schützenberger, 1967; Grene, 1974). In that scheme an evolving object cannot occupy the role of regulative agent; only the environment assumes the role of both active agent for natural selection and the generation of random variation. The evolving object is forced to remain passive or at most adaptive. However, the separation of the biosphere into an object and the remaining environment is arbitrary. This arbitrariness originates in an anthropomorphic perspective which necessarily leads to the antinomy that a material system, which was once perceived only as a passive object, can transform itself into an active agent by immersing itself in the environment. It is logically unacceptable that an evolving object claims to be at once an active and a non-active agent. One more difficulty associated with the neo-Darwinian scheme is its silence about how variations are generated in the environment.

What must be critically examined about the neo-Darwinian scheme is whether natural selection could clearly be distinguished from the generation of variation as has been claimed. Random variation within the context of 'the natural selection of random variation' is a very difficult concept to define. The irony about random variation is that it needs a fixed frame of reference with respect to which one can define what is random. The frame of reference is already a particular one selected out of many alternatives. Random variation requires a predetermined selection. For instance, the random substitution of amino acid in a polypeptide presumes an already selected form of protein synthesis process (Kimura, 1968; King and Jukes, 1969). The randomness of amino acid substitution is permissible only under the predetermined context, which remains unchanged. Such a context-dependent random variation (Lewis, 1980) cannot be responsible for constructing the context itself, namely, for giving rise to the process of protein synthesis *ab initio* from an evolutionary perspective.

At one possible extreme, one could imagine an emergence of the protein synthesis process through the mere random coming together of monomers in the primordial soup. The fixed frame of reference upon which randomness is defined is the primordial soup of monomers. As long as one tries to look for an evolutionary pathway in such terms, the primordial soup always remains as a fixed reference for the later evolution. One cannot penetrate into how it came into existence from the evolutionary perspective, although it should definitely have been a result of the earlier evolutionary process. This is the price one must pay for employing random variation as the only creative agent.

The framework of generating random variation, once fixed, remains unchanged thereafter. This rigidity again forces us to pay another price. If one tries to explain the emergence of contemporary nucleic acids from the most primitive ones through the random coming together and random substitution of four different kinds of nucleotides over roughly 4×10^9 years of terrestrial evolution, the real evolutionary pathways turn out to be no more than $10^{11 \sim 12}$ fractions out of 10^{40} cases tried. The latter were supposedly randomly chosen from $5^{10^{11}} \sim 10^{6 \times 10^{10}}$ equally likely possibilities (Noda, 1982). Contemporary biosystems could not be explained except as a sequential accumulation of extremely improbable events. Random variation is thus conceptually inadequate as the starting point for evolution. We must look for an alternative.

One obvious alternative is that of non-random variation. Once this receives the attention which it deserves, it will become clear that selection and variation are no longer separable. Non-randomness already implies an intrinsic selective capability. The generation of non-random variations is always under the influence of the ones generated previously because of the internal selective capability of the generation. An earlier non-random variation becomes a cause for a later one, and consequently non-random variation acts as an evolutionary agent. In fact, the internal generation of non-random variations underlies the regulative principle for material self-organization. Evolutionary mechanism comes into existence through evolution itself.

ORIGIN OF OPEN SYSTEMS AND THE REGULATIVE PRINCIPLE

Atoms and abiotic molecules in the primordial soup (Florkin, 1975) can aggregate and dissociate. The combined process of aggregation and dissociation can give rise to material units which are open to material flow provided that the material aggregates once formed, can participate in further association with other elements during their inevitable partial degradation.

The maintenance of material aggregates which are open to material flow depends very much upon the three characteristic energies of association, dissociation and thermal agitation. Material aggregation is unlikely to occur if the energy needed for material association is much greater than the thermal energy of the primordial soup. Also, if the energy needed for material dissociation is much greater than the thermal energy, a material aggregate, once formed, could avoid degradation. On the other hand, if the dissociation energy is much smaller than the thermal energy, the aggregates, whenever they are formed, will soon disintegrate due to thermal agitation. Only if the dissociation energy is of the order of thermal energy and if the association energy does not much exceed thermal energy, could one expect the maintenance of material aggregates which are open to material flow (Matsuno, 1980a).

When the three energies of material association, dissociation and thermal agitation are found within the appropriate range, open systems appear as the result of interactions internal to the constituent material elements and not as the result of external action. The boundaries separating open systems for their surroundings are self-generative so long as the energy window for the appearance of open material aggregates is realized by some external means. It should be emphasized that the external means to provide the energy window for the genesis of open systems does not specify how each open system is phase-separated from the primordial soup. It provides no more than a weak macroscopic constraint and leaves enough room for internal determination. It is a necessary though not sufficient condition for the emergence of open systems in the primordial soup. Nevertheless numerous experiments have already shown that a simulated primordial soup including various amino acids and thermal polyamino acids can give rise to microsystems or proteinoid microspheres open to material flow (Fox and Dose, 1977; Fox, 1978). This demonstrates that the emergence of open systems is not a mere theoretical surmise, but a repeatedly observable fact.

Open systems with self-generative boundaries are unique in the sense that on the one hand their macroscopic contours must be shaped in a consistent manner with the underlying microscopic configurations of constituent atoms and molecules (Matsuno, 1981a), and on the other the microscopic configurations are subject to influences from the macroscopic entity which they constitute. This interplay between the microscopic configurations of constituent elements and their macroscopic contours is the key to the genesis of open systems. The microscopic interactions among atoms and molecules is always constrained by what they have accomplished. This constraint is suggestive of the regulative principle we are seeking.

In order to visualize the dynamic process underlying the interplay between the microscopic and the macroscopic, let us take a material aggregate open to material flow that is phase-separated in the primordial

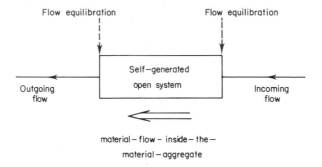

FIG. 3·1 Material flow equilibration of a self-generated open system. It is noted that the self-generated open system between the in-coming and out-going material flows is in a state of flow, that is, material-flow-inside-the-material-aggregate.

soup. It is in fact in a state of material-flow-inside-the-material-aggregate since any material element within the aggregate can be interchanged with others from the exterior (Fig.3·1).

The material flow entering the aggregate equilibrates with the flow-inside-the-material-aggregate, which in turn equilibrates with the flow out of the aggregate. The continuity of material flow must be satisfied at both interfaces. The interaction between the material aggregate and the exterior is responsible for the maintenance of the continuity of material flow. However, it is important to notice that the material aggregate maintains by itself the continuity by adjusting its interaction with the exterior.

Regulation of the interactions in an open system is self-generative. Suppose that the continuity of material flow is first satisfied at both interfaces, and then the flow entering the aggregate happens to vary due to some change in the interaction with its exterior. Unless a proper adjustment is made continuity between the flow into the aggregate and the flow-inside-the-material-aggregate would be destroyed. Material flow equilibration at the interface forces the flow-inside-the-material-aggregate to change so as to preserve continuity with the in-coming flow. The change in the flow-inside-the-material-aggregate propagates toward the interface with the out-going flow. Material flow equilibration at the interface now induces a change in the out-going flow so as to recover the continuity by adjusting the interaction with the surroundings. The change in the interaction between the material aggregate and its surroundings accompanying equilibration at the interface with the out-going flow again induces a variation in the in-coming flow as the interaction change propagates in the medium.

Most fundamental to the maintenance of the flow continuity through a material aggregate is that equilibration at the interface between the in-coming flow and the flow-inside-the-material-aggregate does not occur concurrently with the similar equilibration at the interface between the flow-inside-the-material-aggregate and the out-going flow. This is because the change in the medium is propagated with a finite velocity. An immediate consequence is that material flow equilibration at one interface always becomes a cause for the similar equilibration at other interfaces in a successive manner. Material flow equilibration in open systems continues indefinitely once it has begun. The cause of material flow equilibration in an open system is always the preceding equilibrations.

The *a posteriori* result of material flow equilibrium is the continuity of material flow. However, material flow equilibration implies more than simple flow continuity does. It includes the action for the continuity of material flow *simply as a consequence of the fact that the change in interaction propagates in the medium with a finite velocity*. If the velocity were infinite, the difference between material flow equilibration and simple flow continuity would vanish.

We shall summarize the process of material flow equilibration in a mathematical form. We denote the in-coming material flow by $f^{(in)}$ and the flow-inside-the-material-aggregate at the interface with $f^{(in)}$ by $f_1^{(int)}$. The flow disequilibrium at this interface is:

$$\Delta f_1 \equiv f^{(in)} - f_1^{(int)}$$

Similarly, the flow disequilibrium between the out-going material flow $f^{(out)}$ and the flow-inside-the-material-aggregate $f_2^{(int)}$ at the interface with $f^{(out)}$ is:

$$\Delta f_2 \equiv f_2^{(\text{int})} - f^{(\text{out})}$$

Flow continuity simply asserts the equalities

$$\Delta f_1 = \Delta f_2 = 0$$

However, the internal processes realizing the flow continuity differ according as to whether the propagation velocity of interaction change in the medium is finite.

If the propagation velocity of interaction change remains finite as it should, the removal of flow disequilibrium at the two interfaces cannot occur simultaneously. The removal of the flow disequilibrium $\Delta f_1 \neq 0$ successively causes a new flow disequilibrium $\Delta f_2 \neq 0$ to be removed subsequently. Again, the removal of the flow disequilibrium $\Delta f_2 \neq 0$ turns out to be a new cause for the flow disequilibrium $\Delta f_1 \neq 0$. A schematic representation of the process underlying the removal of the flow disequilibrium gives:

$$\Delta f_1 \neq 0 \leftarrow \Delta f_2 = 0$$
$$\downarrow \qquad\qquad \uparrow \qquad \text{Adjustment}$$
$$\Delta f_1 = 0 \rightarrow \Delta f_2 \neq 0$$
Sequential propagation of effect

This demonstrates the internal action for the maintenance of continuity of material flow.

On the other hand, if the propagation velocity of interaction change were considered infinite, the removal of flow disequilibria $\Delta f_1 \neq 0$ and $\Delta f_2 \neq 0$ would take place simultaneously. The internal cause for generating the flow disequilibrium would vanish, and material flow equilibration then becomes indistinguishable from the simple continuity of material flow.

Material aggregates under the influence of material flow equilibration induce adjustments of interaction with the surroundings. The preceding adjustments incessantly supply endogenous causes for the later ones in a successive manner, thus manifesting as a form of variation imposed upon the configuration of interaction among material elements constituting the open system. The variations are not arbitrary or random as only those fulfilling the conditions for material flow equilibration are realizable.

Open material aggregates once formed in the primordial soup, keep constraining their own construction so as to fulfil material flow equilibration. The regulative capability for generating constrained variations endogenously is inherent in such systems. Material flow equilibration is a regulative principle inherent in matter and is certainly a physical expression of what Waddington (1959) has called 'a producer of evolutionary change as well as a resultant of it' with its inevitable non-randomness (Ho and Saunders, 1979).

The genesis of variations and the constraints acting upon them are not separable in open material aggregates. This is in direct contrast to the neo-Darwinian scheme in which the genesis of random variations is supposed to be separable from the constraint known as natural selection. If the constraint acting upon variation is independent of the genesis of variation, a difficult problem would arise as to how the constraint or mechanism of selection could come into existence. If one looks for the origin of such constraint in variations that have not been constrained by any means, one would be forced into trying to explain how the least likely event could really happen. This problem, however, dissolves once it is noticed that endogenously generated variations in open material aggregates are already constrained. Material flow equilibration in an open system underlies both the genesis of and the constraint acting upon variation.

Material flow equilibration brings about a decrease in the number of degrees of internal freedom in the underlying material aggregate because of the successive constraints to be imposed internally. The decrease in the number of degrees of internal freedom is associated with a symmetry-breaking among the constituent material elements and also with an increase in structural complexity. The structuring of matter in the sense of symmetry-breaking thus results from material flow equilibration. It should be emphasized that the decrease in the number of degrees of internal freedom does not lead to the decrease in the number of internal microscopic states. On the contrary, it can increase the number of distinguishable microscopic states. Symmetry-breaking increases the number of distinguishable states because of the dissolution of the degeneracy observed at the higher symmetry stage. The increase of constraint is thus synonymous with the structuring of matter.

THE REGULATIVE PRINCIPLE AND NON-EQUILIBRIUM THERMODYNAMICS

A simplified theory about the genesis of constrained non-random variation in open systems is available from non-equilibrium thermodynamics. There, the genesis of variations and the constraint acting upon them are hypothesized to be separate (Haken, 1977). This separation is paradigmatic in its origin, being specific only to the thermodynamic theorization.

The constraints which non-equilibrium thermodynamics proposes typically include boundary conditions fixed externally. In the modelling of the primordial soup, for instance, the thermodynamic variables are the concentrations of the reacting chemical species. Once the reacting chemical species and their reaction scheme are identified, one can write down a kinetic equation with fixed reaction rate coefficients as a macroscopic constraint. The thermodynamic variables and the fixed kinetic equation result from the theoretical constraints imposed arbitrarily upon the system. Within the scheme of non-equilibrium thermodynamics, the evolutionary process cannot proceed without variations, the genesis of which is free from thermodynamic constraints. The source of primary variations or fluctuations in this scheme is sought in the environment external to the system under consideration. Although it is legitimate to say that the wandering motion among the attractors seems non-random and internal to the system, the primary cause of wandering is still sought in external fluctuations. Unless external fluctuations are present, the motion will be trapped by one of the attractors even if it is a 'strange attractor' allowing a chaotic trajectory.

Thermodynamic variations or fluctuations are applied to the thermodynamic constraints identified in terms of the kinetic equation of thermodynamic variables. Since the kinetic equation, except for pathological cases, is non-linear with regard to the thermodynamic variables, fluctuations can sometimes destabilize the current thermodynamic constraint (Nicolis and Prigogine, 1977). When the equation destabilizes, a modified kinetic equation with a new set of the thermodynamic variables can be written down. The emergence of new thermodynamic variables is due to a macroscopic enhancement of microscopic fluctuations (Nicolis and Prigogine, 1977). A new chemical species can then establish itself in the modified kinetic equations. The structure of the kinetic equations can change with time only through fluctuation-induced instability. The fluctuations themselves originate outside the thermodynamic constraints which govern the kinetic equation. This tells us that the thermodynamic constraints aided by external fluctuations only partially recognize the regulative principle working in open systems. Incessant endogenously produced changes in the interaction configuration of an open system are impossible within the thermodynamic scheme.

The interaction configuration fixed by the thermodynamic constraint remains unchanged until an instability is induced. The fixed reaction rate coefficients which constitute mass action approximations in chemical reaction kinetics would be acceptable only if we could identify a physical agent such as a flow reactor which artificially fixes the interaction configuration.

However, open systems which are predominant during material evolution are self-generative in both their structure and function. Unless interference from the surroundings is so overwhelming as to rule out internal determination, the regulative principle provides the internal impetus for the evolution of interaction configurations in open systems.

Experiments intended to demonstrate how the regulative principle could work in open systems must not be so restrictive in their boundary conditions as to nullify the autonomy of the system but must be specific enough to be repeatable. In fact, we already have examples of experiments which verify the working of the regulative principle in open systems: the self-generation of proteinoid microspheres — from thermal proteins on contact with water — where the formation of boundaries is internally regulated. These experiments must be distinguished from those in which changes in the interaction configuration are produced in response to varying control parameters. In the latter, the thermodynamic constraint influencing the interaction configuration of the system is controlled externally (Nicolis and Prigogine, 1977; Haken, 1977). Variations therefore occur only through either fluctuation-induced instability or changes in controllable so-called 'bifurcation' parameters.

In self-generated open systems such as proteinoid microspheres thermodynamic constraints cannot be imposed arbitrarily. Any self-generated open system produces non-random variations constrained by itself due to material flow equilibration. What is unique about material flow equilibration is that although it provides a constraint acting upon variations, the constraint itself never ceases to change its own content with time. Constraint and variation are not separable. Constraint generates variation, and variation generates constraint. Consequently, the imposition of externalized thermodynamic constraints contradicts the working of the regulative principle.

The thermodynamic variables within the thermodynamic constraints are aggregated macroscopic variables lacking any role as active agents. The microscopic elements involved in an aggregated macroscopic variable are forced to behave in a uniform fashion. This necessarily leads to the assumption that the propagation of interaction change in the aggregate of microscopic elements constituting a macroscopic variable proceeds at an infinite velocity, thus depriving the thermodynamic variables of their capability for generating variations internally.

The regulative principle in self-generated open systems, on the other hand, accepts that the propagation of interaction change proceeds at a finite velocity. Consequently, an aggregated macroscopic variable such as the molecular concentration of self-generated open system in the primordial soup is an active agent changing its own value with time. The regulative principle provides us with macroscopic variables whose values can change autonomously in distinct contrast to the thermodynamic variables whose motion always assumes impetus from something else.

Whenever macroscopic variables are defined under the premise that the propagation of interaction change proceeds in the medium at a finite velocity—as it actually does—the internal changes in the interaction configuration are maintained and in turn vary the values of macroscopic variables indefinitely. Only when some external means of breaking the dynamic sequence of changes between the interaction configuration and the associated macroscopic variable is available, can the active agents perceived in the macroscopic variables be phased out.

A common device for preventing macroscopic variables from becoming active agents is the so-called adiabatic approximation. Any material system involves a wide spectrum of kinetic processes ranging from the very slow to the very fast. One can separate the entire spectrum into two subgroups, the slower processes and the faster ones. The adiabatic approximation claims that the characteristic velocity of the faster processes could be put at infinity if only the slower processes are of interest. This way of thinking, in one form or another, is found in all branches of physics. The adiabatic approximation thus eliminates the regulative principle.

Physics equipped with the adiabatic approximation conspires to make matter inanimate. If we free ourselves from adiabatic approximation, matter can be seen as an active agent simply by virtue of the fact that interaction changes propagate at a finite velocity. If it is to be applied to material evolution, non-equilibrium thermodynamics must be extended so that it can rid itself of the fetters imposed by the adiabatic approximation. Matter aggregated in self-generated open systems is not inanimate. Physics does not necessarily render matter inanimate—physicists might.

It may also be emphasized that material flow equilibration is a process underlying a realization of the second law of thermodynamics which simply states the empirical fact that any system cannot return by itself to its original state. Material flow equilibration is an endogenous constraint on an open system through which the system loses its capability to return to the original state. The number of degrees of internal freedom decreases with time during the process. The self-limiting constraint due to material flow equilibration is thus an indication of irreversibility.

CONSTRAINT AND VARIATION

The self-generated open systems in the primordial soup continue both to generate variation and to constrain it. Material flow equilibration is an agent for the endogenous generation of non-random variation. The preceding variation always generates and constrains the succeeding variation. Constraint and variation are inseparable. This implies that the specificity of open systems naturally increases with time, on account of the multiplicative accumulation of constraints, thus conferring a time's arrow on evolution. The multiplicative constraint due to material flow equilibration starts to operate from the onset of self-generated open systems. However, the generation of non-random variation is not restricted only to material flow equilibration.

Energized monomers in the primordial soup can constrain their sequence during polymerization. In fact, thermal polymerization of amino acids reveals that favored amino acid sequences are generated in a higher yield than other sequences (Fox and Harada, 1958; Nakashima, et al., 1977; Hartmann, et al., 1981). The self-sequencing of amino acids during thermal polymerization demonstrates a sort of multiplicative constraint that differs from material flow equilibration. What is at work during polymerization is specific chemical interactions among monomers and between monomers and synthesized polymers. The stereochemical interaction for example is multiplicative in constraining the polymerization of monomers. During its elongation an earlier polymer becomes a carrier which constrains the monomers to be polymerized subsequently through specific chemical interactions. The resultant constrained polymers now provide raw materials for the assembly of aggregates open to material flow. Multiplicative constraints upon the formation of polymers are necessary for the origin of open systems.

The multiplicative constraints acting upon polymerization as distinct from the constraints on self-generated open systems discussed previously do not contribute to the successive generation of variation autonomously. This does not mean that matter in the state of a simple polymer lacks the capability for becoming an agent for generating variation. It is just that the size of the material aggregate is sufficiently small so that the adjustment of interaction configuration internal to the aggregate can be accomplished nearly instantaneously without affecting its interaction with the exterior; the propagation of interaction change could thus be deemed as proceeding at an infinite velocity. As the size of the material aggregate grows, it reaches a stage at which the internal adjustment of interaction configuration may require a long enough time so that the interaction with its exterior takes place before the completion of its own adjustment. Beyond this stage, the multiplicative constraint acting upon the material aggregate can become internal so that the degree of constraining will sharply increase. There are thus two different constraints at work. One may be termed the external multiplicative constraint acting on the formation of specific polymers. The other is the internal multiplicative constraint inherent in a self-generated open system.

Evolutionary constraining of material coordination sharpens as internal multiplicative constraint sets in. In other words, molecular evolution as an expression of what has been constrained cannot further increase its specificity without cellular process. Hence cellular process is a prerequisite of molecular evolution. Internal multiplicative constraint working in open systems makes cellular and molecular evolution inseparable.

External multiplicative constraint working in polymerization gives rise to so-called chemical selection. The term selection, however, does not imply the separate existence of selector and selectee. It denotes only the microscopic constraints on the specific sequence of monomers during their polymerization. The microscopic constraints originate in the specificity of interacting monomers and their expression depends upon which monomers are available. Material aggregates subject to external multiplicative constraints are passive towards the exterior in the sense that the microscopic constraints do not come into effect before molecular collision takes place; it works only during the collision. So long as only the external multiplicative constraint is allowed to operate, or if the self-generated open system has not yet come into existence, the microscopic constraint expressed as a specific sequence of monomers to be polymerized does not constrain the molecular flow coming into the interaction region. It is the preceding molecular collisions which influence the incoming molecular flow. The kinetic energy or motion of participating elements is a principal impetus for the microscopic constraint.

Internal multiplicative constraints which operate only after the appearance of self-generated open systems, on the other hand, get their impetus from the immediately preceding constraint. The constraint that keeps constraining its own content is macroscopic since it applies to an open system as a whole. The macroscopic constraint, being internally multiplicative, now constantly changes the environment under which the microscopic constraint works. Monomers and polymers inside are always placed under the influence of the macroscopic constraint as the result of material flow equilibration. Consequently, the microscopic constraint whose partial expression is seen in the sequence of monomers to be polymerized is incessantly forced to change so as to keep itself consonant with the macroscopic constraint. One thus notes that the macroscopic constraint becomes dominant over the microscopic. As the macroscopic constraint increases in specificity during evolution so does the microscopic constraint, in the form of molecular specificity.

The emergence of specific polymers inside self-generated open systems can be seen as molecular evolution proceeding in protocells. The macroscopic constraint acting upon protocells is internally multiplicative and keeps inducing changes in the conditions which influence the process of constraining the sequence of polymerization.

The macroscopic constraint inherent in protocells incorporates both its action and its expression. If one concentrates only on its expression, the difference between macroscopic constraints observed at two adjacent time points can be seen to be the variation generated in the protocell during the same time interval. The variation can be viewed as a result of the macroscopic constraint. On the other hand, if one singles out the action of constraining, the variation will be identified with the action itself. The macroscopic constraint can be seen to be a result of the variation. In protocells constraint and variation are indistinguishable. The functional aspect of the macroscopic constraint is seen in the generation of non-random variation; the structural aspect is seen in the specificity of the molecular configuration in protocells. The specificity sharpens with time. The number of degrees of internal freedom in molecular configurations decreases with time. What is being constrained is protocell, not molecule. Molecular evolution in molecular terms is no more than an expression of what has been constrained in protocells or cells. Cellular processes, on the contrary, maintain in themselves the capability for constraining themselves indefinitely.

As variation associated with protocells is the process of providing the entity which could further be constrained, the dichotomy between selector and selectee as presumed in the neo-Darwinian natural selection of random variation is no longer tenable. If variation were an object to be selected among many alternatives, it would have to make its own internal content explicit and to leave no room for indefiniteness. Otherwise, what has to be selected would remain indefinite and selection itself could not proceed. Random variation within the neo-Darwinian scheme must be explicit in its content in the face of natural selection. This hypothetical explicitness of random variation, however, is not a feature of the variation in self-generated open system. The real variation never exhausts its concrete content during development. To assume random variation is to imply that all variations can, in principle, be exhausted and specified explicitly. The real variation, on the other hand, is always in the process of defining itself, and hence can never be fully described, otherwise it would cease to constrain itself beyond what would have already been described. Despite this it should be emphasized that material flow equilibration which underlies both constraint and variation is a real physical process.

THE ORIGIN OF PROTOREPRODUCTIVE UNITS

Protocells in the primordial soup maintain themselves through material flow equilibration. Constraints due to material flow equilibration give rise to the successive decrease in the number of degrees of internal freedom found in protocells in the primordial soup. Hence, internal multiplicative constraints working in protocells both enhance local material accumulation and increase the specificity of molecular configuration there. The increase in material accumulation cannot go beyond a certain limit. When the limit is exceeded, so that a new interaction change with its exterior could arise before the preceding change in interaction has been settled in the interior, the protocell would fragment into the smaller units each of which could interact coherently with the exterior (Fox *et al.*, 1967; Kaplan, 1981; Whatley, 1981; Matsuno, 1981b). Budding of smaller protocells and the generation of endoparticles inside a parental protocell are in fact ubiquitous in experiments with proteinoid microspheres as a model of protocells (Fox *et al.*, 1967).

Internal multiplicative constraints acting upon protocells always keep suppressing the number of degrees of their internal freedom beyond what has already been constrained. When the constraints are applied to the generation of smaller individual protocells, those of the later generation are more constrained in their molecular configuration than the earlier counterparts. The difference in molecular configuration between the successive generations of protocells could decrease. The succession of protocellular generation is protoreproductive in the sense that there is hereditary transmission from one generation to the next, but the fidelity is not sufficient to qualify as true reproduction. The fidelity of protoreproduction could increase with the succession of generation, because of the successive decrease in the number of degrees of internal freedom, that is to say, the successive increase in structuring.

A major problem in evolution is how the first reproductive unit could come into existence and how reproductive variants could emerge. By definition, reproduction presumes almost perfect fidelity between successive generations. However, the spontaneous emergence of a reproductive unit with such generational fidelity is not to be expected within the scheme of internal multiplicative constraints acting upon protocells. Although random variation could in theory give rise to a reproductive unit spontaneously, we have seen that the real variation is never random and becomes more constrained with time. Reproductive units with perfect generational fidelity do not arise miraculously. What is possible in reality is the never-ending enhancement of fidelity of protocell generation after generation. Even if RNA or DNA replication happened to emerge by chance it could not maintain itself unless accompanied by cellular processes.

Proteinoid microspheres make their own peptides if energized free amino acids are available in the presence of basic thermal polyamino acids or proteinoids (Fox and Nakashima, 1980). Catalytic polymerization of amino acids can proceed on or inside proteinoid microspheres. Internal multiplicative constraints working in proteinoid microspheres induce both an enhancement of material accumulation and an increase in the specificity of molecular configuration there. A possible way of achieving this at an early stage of evolution is the transition from open-looped to close-looped catalytic polymerization of amino acids. If the catalytic polymerization is open-looped, no polyamino acid serving as a catalyst for the formation of a different polyamino acid can be made either by its own products or by their derivatives. However, if a certain catalyst is made either by its own products or by their derivatives, the polymerization constitutes a close-looped cycle thus making the overall polymerization autocatalytic.

Once autocatalytic process sets in, material accumulation inside proteinoid microspheres increases exponentially with time, at least initially. Autocatalytic polymerization provides a more constrained protocellular process because of the decrease in the number of degrees of internal freedom due to the increase in specific material accumulation. Material flow equilibration underlies the transition from acyclic to cyclic polymerization. The appearance of cyclic polymerization is by no means spontaneous nor is its fixation a result of free competition between acyclic and cyclic polymerizations. Rather, protocells are intrinsically capable of producing molecular species that could further constrain the internal molecular configuration. After the transition to cyclic polymerization, both the growth rate and the structuring due to the decrease in the number of degrees of internal freedom increase. Those cyclic polymerizations with the smaller growth rates are phased out thus increasing the specificity of molecular configuration in the remaining cyclic polymerizations with the greater growth rate.

Cyclic polymerization is a primitive form of molecular multiplication which is in turn dependent on the protoreproductive process. If molecular multiplication proceeded without the accompanying cellular process, there would be no enhancement of structuring.

A more constrained protocellular process would result if polynucleotides participate in the cyclic polymerization of amino acids (Matsuno, 1980b, 1982a), as explained below. Nucleotides can be polymerized on or in the protocells made of polyamino acids. This process is suggested by the work of Jungck and Fox (1973). The resultant polynucleotides can in turn influence amino acid sequence in peptides (Nakashima and Fox, 1981).

Polymerization of nucleotides in modern organisms requires polyamino acids as catalysts (Follmann, 1982) and at the same time a polyamino acid requires for its own production both a catalyst in the form of a polyamino acid, and a polynucleotide as a directive factor. Hence, the number of polyamino acids needed to make one polyamino acid will turn out to be more than one in the case of cyclic polymerizations involving both polyamino acids and polynucleotides, since polynucleotide also needs polyamino acid as a catalyst in its own making. If we call this number the *order of autocatalysis* in peptide synthesis, the simple cyclic polymerization of amino acids is first order while that of a catalytic cycle containing both polyamino acids and polynucleotides is greater than one. The protocellular process involving autocatalysis of higher order enhances its own structuring by further increasing local material accumulation (Matsuno, 1982a). This is because the rate of production of polyamino acids is proportional to the concentration of catalysts raised to the power of the order of autocatalysis.

There are two subclasses of cyclic polymerization involving amino acids and nucleotides. In one the translation of polymeric information proceeds from polyamino acids to polynucleotides (Fox, 1978). In the other, the translation is from polynucleotides to amino acids (Eigen, 1971). In contemporary peptide synthesis translation is from polynucleotides to polypeptides. Nevertheless, this does not imply that there might not have been reversed translation during the course of evolution (Crick, 1970; Fox, 1981a; Hartmann, *et al.*, 1981; Follmann, 1982; Root-Bernstein, 1982). The simple catalytic cycle of first order consisting only of polyamino acids can increase, though only slightly, the order of autocatalysis by incorporating a polynucleotide in the cycle. The immediate evolutionary successor to simple catalytic polymerization thus permits the translation of information from polyamino acids to polynucleotides. As the number of different polynucleotides participating in the cycle increases, inversion of translation could occur. Translation from polynucleotides to amino acids could materialize if the average number of polyamino acids needed for instructing the synthesis of one polynucleotide becomes greater than that of polyamino acids needed for the synthesis of one polyamino acid (Matsuno, 1982a,b). This is because as the average number of polyamino acids needed for the synthesis of a particular polymer increases, the information content involved in the synthesis will also increase and because the resulting product will become much more inclusive informationally.

The inversion of the translation between polyamino acids and polynucleotides enhances the constraint upon protocells in the form of the increase in local material accumulation there. The contemporary mechanism of polypeptide synthesis could be the result of the multiplicative constraint acting upon the original polyamino acid-polynucleotide complex.

Molecular multiplication requires cellular processes. Protoreproductive cellular units serve as the first carriers of molecular multiplication. Protoreproduction increases reproductive fidelity between successive generations. Therefore, what has been observed as reproductive selection among protocells is not external to protocells. Those protocells with the greater replication rate coefficients will be more likely to survive if the environment remains unchanged. This argument, however, is incomplete because the values of the replication rate coefficients are self-determined. If one ascribes the cause of variation in the replication rate coefficient to an external agent one would invoke the neo-Darwinian argument of random variation originating in the environment.

In the neo-Darwinian scheme, reproductive selection is the result of action on the part of the environment, although how the environment acts is not spelled out except that the effect is invariably a change in the replication rate coefficient (Wimsatt, 1970).

Reproductive selection in the real evolutionary process results from the never-failing enhancement of reproductive fidelity. Reproductive units are active agents in enhancing their reproductive fidelity. Their replication rate coefficients are merely theoretical derivatives from the underlying active process of enhancing their reproductive fidelity. There is no active agent that could directly manoeuver the replication rate coefficient. In essence, the replication rate coefficient of protoreproductive unit is what must be explained, not what can explain something else.

THE EVOLUTION AND DEVELOPMENT OF PROTOREPRODUCTIVE UNITS

Protocells in the primordial soup undergo reproductive multiplication if their local material accumulation exceeds a certain limit. This dynamic sequence of growth and multiplication may be regarded as a developmental process, as well as an evolutionary one. Protocellular evolution refers to those changes observed over a long period ranging over many protocellular generations, and protocellular development refers to those observed within a single generation. The question thus arises as to how the processes of evolution and development relate to each other.

If reproduction is assumed to occur with perfect fidelity in the absence of external disturbances, development could be viewed as already programmed in spite of its superficial change with time. The source of evolutionary change must therefore be external disturbance, i.e. random variations originating in the environment. For if these were not random, the environment would be affecting the developmental process in a predetermined way and the sharp separation between reproductive units and the environment would dissolve. In the neo-Darwinian scheme of the natural selection of random variations, the direction of influence is always from evolution to development. Spontaneous change in the reproductive units with a change in chemical imprints comes first and development varies accordingly.

However, once due attention is paid to the gradual enhancement of reproductive fidelity in protocells, the one-way influence from evolution to development turns out to be no longer tenable. It is the developmental process within each protocellular generation that enhances the trans-generational fidelity of reproduction. The sharp distinction between evolution and development dissolves. Sequential accumulation of developmental structuring is the essence of the evolutionary process.

We have noted that protoreproductive units continue to constrain themselves during their development. There are two types of constraints: repetitive and non-repetitive. A repetitive constraint is one that remains the same between successive protocellular generations, the non-repetitive one differs between generations. The trans-generational enhancement of reproductive fidelity refers to the non-repetitive constraints acquired during development. What is unique to self-generated protoreproductive units is that the non-repetitive constraint never vanishes because of its internal multiplicative nature due to material flow equilibration.

Non-repetitive constraints acquired during development of the protocell bring about a more constrained phenotypic expression. Moreover, the constraint once acquired can be transmitted to the succeeding generations in the form of organized chemical complex in the body of the parental protocell. Even if the frequency of non-repetitive constraints with discernible phenotypic changes is quite rare, this does not invalidate the working of internal multiplicative constraint. Non-repetitive developmental constraining can be seen as the canalization of development (Waddington, 1961, 1974) or a self-limiting of the evolutionary highway (Fox, 1981b).

A non-repetitive developmental change in protocells represents a change in the macroscopic constraint which regulates the phenotype of the protocells. This in turn induces a change in the environment which conditions the microscopic constraint upon the sequence of monomers during polymerization. As a result, a non-repetitive developmental change in a protocell can induce an irrevocable change in chemical imprints transmitted to the succeeding generations. A new polymeric species thus emerges inside the protocell due not to chance but to a structuring process dependent upon the preexisting protocells and their molecular configuration. In this sense, the evolutionary pathway is built into the protocells through their developmental process much as the epigenetic landscape (Waddington, 1957) provides a built-in evolutionary pathway through development. Development therefore underlies evolution, in sharp contrast to the neo-Darwinian scheme in which evolution underlies development.

The non-repetitive constraining of reproductive protocells can give rise to one or more classes of protoreproductive units. The similarity between the members of each class is enhanced with successive generations, although this does not exclude the possibility that one class of protocells can fragment into more than one subclass. Each member of a particular class can be seen as the result of evolutionary selection. However, the unit of evolutionary selection is not the individual protocell, but a class of protocells exhibiting similar constraints. This is a natural consequence of the increase in reproductive fidelity in successive generations. In contrast to the neo-Darwinian scheme in which the evolutionary unit is each reproductive individual, the key to the present argument is that a class of individuals exhibiting similar properties is the evolutionary unit.

The emergence of new classes of reproductive individuals results from the acquisition of further constraints upon the preceding evolutionary units. This always proceeds with successive interaction changes with the exterior in which all the other evolutionary units are involved, so that a further emergence of new classes would result from the response of some predecessor to the exterior. If we regard the process of incorporating changes in the exterior into the interior as adaptation, then non-repetitive constraints can be seen as agents of adaptation. Evolutionary adaptation is by no means a consequence of something adapting itself to something else. It is only a name assigned to the process which would almost look like that (Hahlweg, 1981). In fact, internal multiplicative constraint due to material flow equilibration, especially its non-repetitive part observed during development, underlies the adaptive process.

Protocells are by no means the only self-generated open systems. Any living system satisfies the definition. Therefore, internal multiplicative constraints due to material flow equilibration should apply to any living system. One consequence is that development underlies evolutionary change (Waddington, 1974) as suggested by epigenetic processes. Another is that the evolutionary unit is not each reproductive individual but a class or species of individuals exhibiting similar constraints (Gould, 1980), as suggested by punctuated equilibria (Eldredge and Gould, 1972; Gould, 1982). Epigenetic processes and punctuated equilibria have a common denominator in material flow equilibration. The emphasis is on the origin of variation by the process of material flow equilibration. In this connection we recall Darwin's remark (quoted in Fox, 1981a):

A grand and almost untrodden field of inquiry will be opened, on the causes
and laws of variation . . .

The neo-Darwinian viewpoint forces the environment to be the sole agent
for generating variations and requires in addition that those generated are
random. This is a consequence of employing anthropomorphic dichotomies
between the environment and the evolving object, between selector and
selectee, and between adaptation and variation. Once material flow
equilibration in self-generated open system is recognized as an organizing
principle, it is not necessary to invoke the neo-Darwinian dichotomy to
account for the origin of variations and for the evolutionary process in
general.

CONCLUSION: RANDOMNESS AND NON-RANDOMNESS

The concept of randomness enters into evolutionary thought as two
fundamental assumptions about the origin of variation. One is that a frame
of reference is available which makes it possible to generate variation. The
other is that each variant is equally likely to appear. In the genesis of
random variation a complete separation is assumed between the generation
of variations and the variations generated, with the generator of random
variations remaining unchanged. If it could change in a random way, this
would be to presume a higher frame of reference for the random change.
On the other hand, if the generator can change in a predetermined way, this
would deny random variation itself. The assumption that every variation is
equally probable is nothing more than a mere reflection of the ignorance of
the observer who cannot examine all of the possibilities in an exhaustive
manner.

Random variation could be acceptable only if the generator of variation
remains unchanged. A good example of this kind is statistical mechanics at
or near thermal equilibrium. What equilibrium statistical mechanics teaches
us is that the most likely generator of variation is the one that can maximize
the number of equiprobable variants. This is a static argument, and cannot
account for how the generator of variation could develop with time. Hence,
statistical mechanical theorization is applicable only to those generators
which can remain stable (Lochak, 1981). Although it might be possible to
extend statistical mechanics to the non-equilibrium case including open
systems, the basic paradigmatic statement remains unaltered. Even in the
non-equilibrium scheme, how the generator itself develops is left
untouched.

One could imagine that both the neo-Darwinian natural selection of random variation and the non-Darwinian random neutrality hypothesis (Kimura, 1968; King and Jukes, 1969) have their own supportive facts so long as the generator of variation is supposed to remain unchanged. However, a fixed generator of variation is nothing more than a theoretical artifact. In particular, whether the generator remains unchanged is not a matter of hypothesis, but a real question that must be answered. What we have recognized is that so long as any living system and its precursors are identified as self-generated open systems, the generator of variation keeps restricting and constraining its products. What is more, a self-generated open system is the generator of variation itself.

A self-generated open system acting as a generator of variation always keeps transforming itself into a more constrained form. Once a self-generated open system appears, an enhancement of constraining in the generation of variation makes evident the irreversibility of evolutionary time. Random and non-random variations are sharply distinguishable if one refers to time. Random variation means simply that every possible variant is equally likely to occur, independent of the time when it really occurs. In contrast, non-random variation due to material flow equilibration keeps constraining the members constituting the ensemble of possible variants. The likelihood of each variant depends on *when* it occurs. Non-randomness always refers to time. For instance, suppose hypothetically an identical variant happens to occur at two different time points. What this non-randomness suggests is that the one that occurred at the later time point has the greater likelihood. This observation indicates that the carrier of non-randomness is not each variant, but the self-generated open system itself as the generator of variation. It is time which can bring about the uneven likelihood of occurrence among possible variants. Such a non-randomness inevitably implies something predetermined since evolution does in retrospect appear to be unidirectional. What is predetermined is the process of self-constraining in the generation of variation.

If non-randomness were static in the sense that the distribution of uneven likelihood among possible variants is fixed and conceivable in advance, there would be no evolution since the most likely variant can appear at any arbitrary time with the same likelihood. This denies the historical character of evolution. Evolution is always under the influence of what it has already accomplished. For this reason, we must refer to the generator of variation, not to variations generated, in order to explicate non-random variations in evolution. What is evolutionary is the generator of variation. The static generator of variation, conceived within the neo-Darwinian scheme, is not evolutionary in this regard. The dynamic non-randomness of variation points to the generator of variation as the active agent. The origin and evolution of such an active agent are in accord with a physically permissible process that reduces to material flow equilibration in self-generated open system.

Variation always materializes within a certain constraint in self-generated open system which keeps restricting its own extent with time. It should be stressed that even if we have not yet had much experience in dealing with self-limiting constraints or self-varying boundary conditions in physics, it would be premature to say that physics precludes the possibility of self-limiting constraints. What we have observed in evolution in fact is just the opposite.

One notes that any constraint provides a reference with regard to which the information content of each variant within the same constraint can be measured. The quantitative estimation of information is thus guaranteed if a fixed frame of reference is available. Recalling the definition of information as a measure of the degree of ambiguity on the part of the receiver or observer, one first needs a fixed reference with respect to which one can measure if and by how much the ambiguity decreases. Therefore, if the constraint acting upon variation changes with time as it does, a straightforward definition of information will be impossible (c.f. Saunders and Ho, 1976; 1981). What one can say about the moving constraint from an information-theoretic viewpoint is that the generator of variation decreases the number of degrees of internal freedom with time because of its on-going self-limitation, i.e. the information content of what has been expressed increases with time.

The self-sequencing of amino acids in the primordial soup is a generator of variation. It has the internal capability to increase its information content with time. If one views the self-sequencing of amino acids as a variation generated within a fixed constraint, it might be said (Yockey, 1981) that the self-sequencing of amino acids could not be an evolutionary precursor of contemporary protein because of its lesser information content per each amino acid residue (Vegotsky and Fox, 1962). This is, however, a wrong answer to a wrong question (Matsuno, 1982b; Fox and Matsuno, 1983). The self-sequencing of amino acids is a most primitive form of the self-constraining generator of variation which evolves by material flow equilibration. It is thus of utmost importance to pay due attention to the generator of variation, and not to the variations generated, in order to theorize about how living systems came into existence and evolve.

Natural selection of non-random variation is but the evolution of the generator of variation. The incessant self-limiting capability is intrinsic to the generator of variation. This necessarily implies that one cannot complete the description of the generator of variation at any moment. What is left undescribed is the capability for self-constraining to appear later than at the time when the description is undertaken. The present incompleteness is due totally to the inadequacy of the language one has to employ (Pattee, 1978). The neo-Darwinian scheme does not suffer from such an inadequacy thanks to the fixed constraint acting upon the generation of variation. But, precisely for the same reason, it fails to identify the real mechanism of evolution.

We thus conclude that a timeless description of evolution is unacceptable or, more precisely, incorrect. The inclusion of terms with indefinite implication is inevitable, as regards the generator of variation with its self-constraining capability. Evolving matter always behaves as an active agent. This, however, by no means implies that evolving matter is a free agent like the one envisaged from the vitalistic viewpoint. It is an active agent simply by virtue of its possession of the internal freedom to be fixed by itself in accord with permissible physical processes. Any material aggregate adjusts its own state by itself while the interaction change propagates in the medium. Since the propagation velocity of interaction change remains always finite, the material aggregate changes its own implication during the propagation of interaction change even though it is still called by the same name. Material flow equilibration in any self-generated open system is an agent for realizing evolutionary novelties. What characterizes living systems is the organizing principle founded upon the physical fact that the propagation of interaction change proceeds in the medium always at a finite velocity. Evolution of matter is a mode of matter constraining itself by itself, not an outcome selected by something else.

ACKNOWLEDGEMENT

This work was supported in part by the National Foundation for Cancer Research.

REFERENCES

Commoner, B. (1968). Failure of the Watson–Crick theory as a chemical explanation on inheritance. *Nature* **220**, 334–340.

Crick, F. (1970). Central dogma of molecular biology. *Nature* **227**, 561–563.

Dose, K. (1981). Ernst Haeckel's concept of an evolutionary origin of life, *BioSystems* **13**, 253–258.

Eden, M. (1967). Inadequacies of neo-Darwinian evolution as a scientific theory. *In* 'Mathematical Challenges of Neo-Darwinian Interpretation of Evolution'

(P. S. Moorhead and M. M. Kaplan, *Eds*), pp.5–12. Wistar Institute Press, Philadelphia, USA.

Eigen, M. (1971). Selforganization of matter and the evolution of biological macromolecules. *Naturwissenschaften* **58**, 465–523.

Eldredge, N. and Gould, S. J. (1972). Punctuated equilibria: an alternative to phyletic gradualism. *In* 'Models in Paleobiology' (T. J. M. Schopf, *Ed.*), pp.82–115. Freeman and Cooper, San Francisco, USA.

Florkin, M. (1975). Ideas and experiments in the field of prebiological chemical. *In* 'Comprehensive Biochemistry', (M. Florkin and E. H. Stotz, *Eds*), Vol. 29B, pp.231–260. Elsevier, Amsterdam, Holland.

Follmann, H. (1982). Deoxyribonucleotide synthesis and the emergence of DNA in molecular evolution. *Naturwissenschaften* **69**, 75–81.

Fox, S. W. (1978). The origin and nature of protolife. *In* 'The Nature of Life' (W. H. Heidcamp, *Ed.*), pp.23–92. University Park Press, Baltimore, USA.

Fox, S. W. (1980). Introductory remarks to the special issue on 'Assembly Mechanisms'. *BioSystems* **12**, 131.

Fox, S. W. (1981a). Copolyamino acid fractionation and protobiochemistry. *J. Chromatography* **215**, 115–120.

Fox, S. W. (1981b). Origin of the protein synthesis cycle. *Int. J. Quant. Chem.* **QBS8**, 441–454.

Fox, S. W. and Harada, K. (1958). Thermal copolymerization of amino acids to a product resembling protein. *Science* **128**, 1214.

Fox, S. W. and Dose, K. (1977). 'Molecular Evolution and the Origin of Life,' rev. ed. Marcel Dekker, New York, USA.

Fox, S. W. and Nakashima, T. (1980). The assembly and properties of proto-biological structures: the beginnings of cellular peptide synthesis. *BioSystems* **12**, 155–166.

Fox, S. W. and Matsuno, K. (1983). Self-organization of the protocell was a forward process. *J. Theor. Biol.* **101**, 321–323.

Fox, S. W., McCauley, R. J. and Wood, A. (1967). A model of primitive heterotrophic proliferation. *Comp. Biochem. Physiol.* **20**, 773–778.

Gould, S. J. (1980) The evolutionary biology of constraint. *Daedalus* **109** no 2, 39–52.

Gould, S. J. (1982). Darwinism and the expansion of evolutionary theory. *Science* **126**, 380–387.

Grene, M. (1974). 'The Understanding of Nature'. Reidel, Boston, USA.

Hahlweg, K. (1981). Progress through evolution? An inquiry into the thought of C. H. Waddington. *Acta Biotheoretica* **30**, 103–120.

Haken, H. (1977). 'Synergetics'. Springer, Berlin, West Germany.

Hartmann, J., Brand, M. C. and Dose, K. (1981). Formation of specific amino sequences during thermal polymerization of amino acids. *BioSystems* **13**, 141–147.

Ho, M. W. and Saunders, P. T. (1979). Beyond neo-Darwinism — an epigenetic approach to evolution. *J. Theor. Biol.* **78**, 573–591.

Jacob, F. (1974). 'The Logic of Living Systems'. Allen Lane, London, UK.

Jungck, J. R. and Fox, S. W. (1973). Synthesis of oligonucleotides by proteinoid microspheres acting on ATP. *Naturwissenschaften* **60**, 425–427.

Kaplan, R. W. (1981). Origin of life via hypercyclic and simpler protobionts. *Biol. Zbl.* **100**, 23–32.

Kimura, M. (1968). Evolutionary rate at the molecular level. *Nature* **217**, 624–626.

King, J. L. and Jukes, T. H. (1969). Non-Darwinian evolution. *Science* **164**, 788–798.

Lewis, R. (1980). Evolution: a system of theories. *Perspectives Biol. Med.* **23**, 551–572.

Lochak, G. (1981). Irreversibility in physics: reflections on the evolution of ideas in mechanics and the actual crisis in physics. *Found. Phys.* **11**, 593–621.

Matsuno, K. (1980a). Operational description of microsystems formation in prebiological molecular evolution. *Origins Life* **10**, 39–45.

Matsuno, K. (1980b). Compartmentalization of self-reproducing machineries: multiplication of microsystems with self-instructing polymerization of amino acids. *Origins Life* **10**, 361–370.

Matsuno, K. (1981a). Material self-assembly as a physicochemical process. *BioSystems* **13**, 237–241.

Matsuno, K. (1981b). Self-sustaining multiplication and reproduction of microsystems in protobiogenesis. *BioSystems* **14**, 163–170.

Matsuno, K. (1982a). Natural self-organization of polynucleotides and polypeptides in protobiogenesis: appearance of a protohypercycle. *BioSystems* **15**, 1–11.

Matsuno, K. (1982b). A theoretical construction of protobiological synthesis: from amino acids to functional protocells. *Int. J. Quant. Chem.* **QBS9**, 181–193.

Monod, J. (1971). 'Chance and Necessity'. Alfred A. Knopf, New York, USA.

Nakashima, T. and Fox, S. W. (1981). Formation of peptides from amino acids by single or multiple additions of ATP to suspensions of nucleoproteinoid microparticles. *BioSystems* **14**, 151–161.

Nakashima, T., Jungck, J. R., Fox, S. W., Lederer, E. and Das, B. C. (1977). A test for randomness in peptides isolated from a thermal polyamino acid. *Int. J. Quant. Chem.* **QBS4**, 65–72.

Nicolis, G. and Prigogine, I. (1977). 'Self-Organization in Nonequilibrium System'. Wiley, New York, USA.

Noda, H. (1982). Probability of life. Rareness of realization in evolution. *J. Theor. Biol.* **95**, 145–150.

Pattee, H. (1977). Dynamic and linguistic complementarity in complex system. *Int. J. Gen. Syst.* **3**, 259–266.

Root-Bernstein, R. S. (1982). On the origin of the genetic code. *J. Theor. Biol.* **94**, 895–904.

Saunders, P. T. and Ho, M. W. (1976). On the increase in complexity in evolution. *J. Theor. Biol.* **63**, 375–384.

Saunders, P. T. and Ho, M. W. (1981). On the increase in complexity in evolution II. The reality of complexity and the principle of minimum increase. *J. Theor. Biol.* **90**, 515–530.

Schützenberger, M. P. (1967). Algorithms and the neo-Darwinian theory of evolution. *In* 'Mathematical Challenges to the Neo-Darwinian Interpretation of Evolution' (P. S. Moorhead and M. M. Kaplan, *Eds*), pp.73–75. Wistar Institute Press, Philadelphia, USA.

Smith, T. F. and Morowitz, H. J. (1982). Between history and physics. *J. Mol. Evol.* **18**, 265–282.

Vegotsky, A. and Fox, S. W. (1962). Protein molecules: intraspecific and interspecific variations. *In* 'Comparative Biochemistry' (M. Florkin and H. S. Mason, *Eds*), Vol. 4, pp.185–244. Academic Press, New York, USA.

Waddington, C. H. (1957). 'The Strategy of Genes'. Allen and Unwin, London, UK.

Waddington, C. H. (1959). Behaviour as a product of evolution. *Science* **129**, 203.

Waddington, C. H. (1961). Genetic assimilation. *Adv. Genet.* **10**, 257–290.

Waddington, C. H. (1974). A catastrophe theory of evolution. *Ann. N. Y. Acad. Sci.* **231**, 32–242.

Webster, G. and Goodwin, B. C. (1982). The origin of species: a structuralist approach. *J. Soc. Biol. Struct.* **5**, 15–47.

Whatley, F. R. (1981). Problems in the rate of evolution in biological systems. *Phil. Trans. Soc. (London)* **A303**, 611–623.

Wimsatt, W. C. (1970). A book review on 'Adaptation and Natural Selection: A Critique of Some Current Evolutionary Thought' by G. C. Williams. *Phil. Sci.* **37**, 620–623.

Yockey, H. P. (1981). Selforganization origin of life scenario and information theory. *J. Theor. Biol.* **91**, 13–31.

4

On the Increase in Complexity in Evolution

JEFFREY S. WICKEN

Abstract

This chapter examines the evolutionary process from a thermodynamic perspective, and attempts thereby to find unifying principles that account for its overall complexifying, anamorphic character. The second law can be conceived as contributing to evolutionary direction in two general ways, that correspond roughly with the Darwinian principles of variation and selection. In its variational or generative role, the second law promotes chemical complexification through the randomization of potential energy and of material configuration. In its evaluative or natural selective role, the second law, together with the physicochemical conditions under which it operates in the biosphere, prescribes the general character of the thermodynamic flow patterns, including organisms, that are likely to be successful in the competition for energy resources.

The generative power of the second law in evolution, and for that matter the very possibility of a self-organizing universe, depends on the character of nature's fundamental forces. Since these are for the most part associative ones, the dissipation of potential energy tends to be achieved through integrative processes. The second law is in this respect a principle of potency, that provides for the generation of structure through the dissipation of potential. This principle of dissipation-through-structuring operates in the origin and evolution of the biosphere by providing for the utilization of the Earth's impressed energy gradient in ordered pathways of entropy production. Chemical potential is generated by photon absorption, converted to heat through chemical bonding and higher-order associations, which is then released to the sink of space. The randomization of atomic constituents also contribute to complexification, since configurational entropy is created by the linkage of atoms in new sequences. These ideas are developed here with the aid of information theory, which provides a common

BEYOND NEO-DARWINISM
ISBN: 0-12-350080-X

language for treating thermodynamic processes and the concept of 'complexity'.

The thermodynamic basis for natural selection, discussed briefly toward the end of this chapter, lies in the character of organic systems as self-organizing patterns of thermodynamic flow integrated with the overall flows of their ecosystems. The preservation of the organizational type through survival and reproduction thus involves exploitative and homeostatic strategies for promoting and stabilizing thermodynamic flows.

INTRODUCTION

As the web of bureaucracy ramifies into more aspects of our everyday life, the notion that organizational complexities should increase in time seems utterly commensurable with common experience. Complex organizations, after all, can *do* more than simple ones and organizations of all kinds grow and establish themselves by expanding their competencies and scope of activities. Since natural selection favors the competent, it does not seem unreasonable that the complexity of organic nature, like that of the sociosphere, should tend to increase over the course of evolutionary time. If the issue of complexity were this simple, it would have been settled long ago. Instead, laws of complexification have been notoriously undemonstrable. Herbert Spencer's grand schemes have faded into obscurity since the work of Willard Gibbs. This chapter aims to reconcile Spencer with Gibbs.

Far from having the status of 'law', the complexifying movement of evolution is not even acknowledged as real by many within the neo-Darwinian community (*see* Saunders and Ho's (1976) article and a critique of it by Castrodeza (1978)). Much of the reluctance to accept an increase in complexity in evolution has to do with the theoretical (and metaphysical) structure of Darwinism, which has trouble accommodating any systematically directional parameters in evolution, including that of complexity.

Some of this we can attribute to the historical conditions in which Darwinism established itself as a reputable science within the materialist programme. Like the development of mechanics two centuries earlier, Darwinism was established on a commitment to certain notions of cause and effect that explicitly excluded teleology. Lamarck failed, after all, not because of his assumption that acquired characters can be genetically significant — a notion that Darwin also held — but because of his 'perfecting principle', which enables organisms to advance ever upward through higher levels of organization. This simply did not square with hard-won materialist tenets. Darwinism succeeded by being epistemologically anti-Lamarkian, and hence any directive principle in evolution is vigorously denied.

Yet there is a tendency towards complexity in organic evolution, not as an inviolable directive principle, but as an overall trend. Before we identify the sources of this complexifying tendency, let us assess the degree to which complexification is commensurable with at least the spirit of Darwinism as a 'research programme' (Popper, 1976) rather than as a theory.

As research programme, Darwinism involves above all a metaphysical commitment to the idea that evolutionary change occurs according to the principles of random variation and natural selection, as opposed to internal orientation. These are hardly 'mechanisms' of evolution, as is sometimes maintained, but merely certain ground rules to which evolutionary explanation must be faithful, i.e. that variation—whatever that might mean in terms of physical theory—must be undirected with respect to its eventual survival value, and that all evolutionary changes must in some way be adaptively sanctioned.

There is of course no reason to assume that because evolution is materialistic it must be causally structured in this way. Neo-Darwinism is often accused, with some justification, of operating within a materialism of classical physics rather than of modern science. To the extent that the understandings of modern science provide for a vastly more pluralistic ground for physical causation than made available by the 'blind motions' of classical physics, one might also expect this pluralism to express itself in evolution.

The spirit of Darwinism is, in fact, causally pluralistic with respect to the sources of evolutionary change. Variation was viewed by Darwin as undirected with respect to survival ends, but at the same time, he allowed for environmental influences and for the hereditary assimilation of acquired traits. Much of this openness simply expressed Darwin's acknowledged ignorance of life's hereditary and developmental basis. The ushering in of neo-Darwinism by Weismann's demonstration of the germ plasm's isolated continuity, and the subsequent assimilation of Mendelian genetics into the theoretical framework closed some of these options. In neo-Darwinism, the genotype–phenotype dichotomy is sharply drawn in a one-way flow of causation (The Central Dogma). Variation became a random error generator unaffected by environmental (or somatic) conditions, and differential reproductivity became the only source of evolutionary direction. The question is, can these 'meta-principles' of variation and selection be sufficiently fleshed out by understandings drawn from the physical, biological, informational and cybernetic sciences to make the neo-Darwinian programme theoretically successful in explaining evolutionary trends? Or is a more pluralistic causal mix required, one that goes beyond the variation-selection scheme?

There are indications that thermodynamic influences may operate beyond the random-generator, natural–selective level in giving the environment itself a formative role in self-organization (Glansdorff and Prigogine, 1971; Ho and Saunders, 1979)—a notion quite incommensurable with even an expanded neo-Darwinism.

My approach in this chapter will be largely thermodynamic, and even here a qualifiedly pluralistic perspective seems warranted. When treated thermodynamically, variation and natural selection are able to explain a great deal of the complexifying movement of evolution. I will content myself with considering the extent to which evolutionary complexification can be understood within a thermodynamic framework that expands on essentially neo-Darwinism categories.

ORDER, COMPLEXITY AND ORGANIZATION

We live in a world of form and potency, both essential conditions for evolutionary self-organization. The very possibility of form depends on the operation in nature of certain associative forces, acting to arrange matter into structured spatial relationships. A statically-structured cosmos is opposed by irreversible production of thermal energy. If this energy could be somehow systematically drained from the cosmos, it would crystallize into a vast, multielemental snowflake. Energy, and the modes in which it is carried, make the cosmic order a dynamic one requiring entropic dissipation for its realization.

The expansion of the cosmos maintains it in a condition of perpetual disequilibrium, in which potential energy converts irreversibly to the kinetic energy of radiation and microscopic motions. Since the forces of nature are for the most part associative ones, this conversion correlates with the buildup of structure. There is a certain materialist teleology to this, a means–ends relationship that permeates the whole of evolution: atomic nuclei aggregate from protons and neutrons as a means of dissipating the potential energy of the separated nucleons that results from the strong nuclear force; nuclei and electrons form atoms to dissipate electrostatic potential energy; and so on—through molecules, through supra-molecular structures, through life itself. Dissipation is the driving force of the Universe's building-up or integrative tendency; nature's forces give it form in space and time (Wicken, 1981).

The relationship between thermodynamic driving forces and evolutionary direction has been obscured over the years by a certain semantic looseness regarding what, in fact, was dissipated during dissipative processes—a common perception being that the second law was 'disordering' or 'disorganizing' (often the two were conflated) in its effects—with the implication that the self-organizing processes of evolution were, if not exactly anti-entropic, then at least subject to other integrative principles. In fact, dissipative processes can be integrative as well—a truth that ties evolution to the total flow of nature. So before proceeding further, it will be useful to consider the meanings of concepts, such as structure, order and organization, that are invoked to describe the integrative side of nature, and the sense in which these can be causally connected to its dissipative side.

'Structure' can be simply defined as constraint on form, as the imposition of certain law-like relationships in a medium or set of elements over space and time. Dissipation through structuring is an evolutionary first principle, made available by nature's associative forces and the cosmic asymmetry between potential and kinetic forms of energy.

'Complexity' and 'order' each involve structure; they differ with respect to 'pattern'. The order and the complexity of a physical system can be quantitatively assessed by mapping it into a mathematical sequence and applying to it the concepts of information theory. The complexity of a sequence is the minimum algorithm required for its umambiguous specification (Chaitin, 1966), whereas its order is its degree of internal patternedness as measured by its algorithmic compressibility (Chaitin, 1975). Complex structures tend to have low orders of patternedness and are hence relatively incompressible. Viewed from their own hierarchical level, complex structures appear to be *random* in their assembly. But from a higher, functional level they may in fact be highly organized (e.g. enzymes appear random at the level of chemical description, but organized at the level of metabolic function).

'Organization' is a functional rather than a structural concept. Organizations are always *for* something, even if this be only the perpetuation of its organizational relationships (i.e. a self-organization). At the same time, organization is built on complexity, on the possibility of interconnecting elements in ways that are not prescribed simply by their physical properties. So although one does not ordinarily speak of a 'quantity' of organization, one can at least measure its *complexity*, and this, as Saunders and Ho (1976) have argued, is therefore the parameter of interest in evolutionary direction. Contributing to the algorithmic length or complexity of a physical system are the number and variety of its elements, and the number and aperiodicity of their interconnections. Each of these ingredients tends to increase through the randomizing directiveness of the second law.

The entropy principle promotes evolutionary complexification in both generative and evaluative ways. In its generative role, the entropy principle underlies the phenomenon of random variation and imposes on it a complexifying direction in time. In its evaluative role, this principle provides a basis for a thermodynamics of natural selection which accommodates comfortably the 'progressiveness' of evolution. This topic, which relates profoundly to evolutionary self-organization, will be taken up in the final sections of this chapter. The following sections will consider the intrinsic, pre-selective drive toward complexification imparted to the principle of random variation by its macro-statistical interpretation.

MODES OF DISSIPATION

Evolutionary structuring and complexification depend upon the overall character and direction of the biosphere's thermodynamic flows. Information theory provides a convenient formalism for representing these relationships (Wicken, 1978; 1979), as summarized in the following development.

The macroscopic information content I_M of a thermodynamic state is related to its statistical probability P_i through the equation

$$I_M = -k \ln P_i \tag{1}$$

The problem is to represent P_i in terms of measurable or theoretically-calculable parameters. As open systems, permeable to and dependent upon exchanges of matter and energy with their surroundings, organisms themselves are difficult to deal with in terms of probability distributions. So the evolutionary 'laws' formulated here will be concerned with those approximately closed regions of the biosphere (ecospheres) that cycle an essentially fixed pool of matter through one-way fluxes of energy. For such systems the statistical probability of a given thermodynamic state occurring is given by Gibbs' canonical ensemble equation

$$P_i = W_i \exp(\bar{A} - E_i)/kT \tag{2}$$

Here E_i represents the internal energy of that state, W_i the number of microstates contributing to that state, $\bar{A}$ the ensemble average value of the Helmholtz free energy at equilibrium and T the absolute temperature of the system. Since the various microstates contributing to this state all have the same energy, each occurs with equal probability.

Substituting this equation into equation 1, one obtains

$$I_M = -k \ln W_i + (E_i - \bar{A})/T \tag{3}$$

The information content of a thermodynamic state is thus separable into two components. One depends only on the number of microstates in which a macrostate can be expressed and will be referred to as that state's *negentropic information*, I_W. The second is a function of internal energy and temperature and will be referred to as *energetic information*, I_e. These informational parameters are not independent properties of a thermodynamic state, but *modes* in which its relative statistical improbability is expressed.

Negentropic information can be resolved still further, into configurational and thermal contributions. Each microstate of a thermodynamic system consists of a particular spatial configuration of chemical entities and a particular allocation of kinetic energy among their various thermal quantum states. For ideal systems each microscopic configuration is energetically equivalent, having the same density of thermal quantum states. In such cases the number of microstates associated with a given macrostate can be expressed as a product of configurational and thermal factors ($W = W_c \cdot W_{th}$), and one may separately define configurational and thermal entropies and informational components accordingly to the equations

$$I_c = -S_c = -k \ln W_c \qquad (4)$$
$$I_{th} = -S_{th} = -k \ln W_{th}$$

so that I_W separates into configurational and thermal terms. Equation (3) now becomes

$$I_M = I_c + I_{th} + I_e \qquad (5)$$

It should be noted that the assumption of ideality on which the explicit definition of I_c and I_{th} depends is quite reasonable for the gas-phase and dilute liquid-phase processes that were involved in the early stages of probiotic evolution, but somewhat less so for more advanced stages of evolution involving concentrated reactants and structured biological systems. But computational utility is not at issue; the intent in resolving I_w into configurational and thermal contributions is simply to distinguish general modes in which negentropic potential occurs. Even when I_c and I_{th} are not explicitly definable, they still contribute implicitly to I_w. The question is, how do these parameters change in irreversible processes and what do these changes mean in terms of the biosphere's evolutionary structuring?

INFORMATIONAL CHANGES IN IRREVERSIBLE PROCESSES

All irreversible processes occurring in the biosphere act to increase the entropy of the universe. This is the ultimate isolated system to which the second law applies. On the other hand, these irreversible processes occur in limited regions of the universe, so it is customary to relate entropy changes to a limited system in which the processes of interest actually occur and a reservoir of essentially infinite capacity with which this system can exchange materials and energy. For such cases the second law may be written

$$\Delta S_u = \Delta S_s + \Delta S_r > 0 \tag{6}$$

where ΔS_u, ΔS_s and ΔS_r respectively represent changes in the entropies of the universe, the limited system under consideration, and the reservoir. Here, the limited systems of interest are those, such as the biosphere and its component ecospheres, that can be treated effectively as closed.

For an irreversible process occurring in any such system, the second law takes the form

$$\Delta S_u = \Delta S_s + Q/T > 0 \tag{7}$$

where Q is the heat delivered to the surroundings as a consequence of the process and T is the absolute temperature at which it occurs. If no work is performed on or by the system in the course of this process, Q results entirely from a reduction in the system's internal energy, so that $Q/T = -\Delta E/T = -\Delta I_e$. Replacing ΔS_s by $-\Delta I_w$ and dividing the latter into configurational and thermal components, one may express the second law in informational terms as follows

$$\Delta I_e + \Delta I_c + \Delta I_{th} < 0 \tag{8}$$

Evidently, any of these parameters is permitted to increase providing that there is a compensating reduction in another. For reasons to be discussed presently, dissipation cannot occur from each mode independently, but only through patterned flows of I_M from one mode to another.

The three components of I_M each represent a mode of statistical improbability in the distribution of the basic stuff of the universe: matter, motions, and potential—and hence a particular source of potency for irreversible change. Changes in the value of I_{th} reflect changes in the overall level of a system's chemical structuring. Specifically, I_{th} increases (toward less negative values) during processes that increase chemical interconnections. This becomes evident when one considers that structuring reactions are accompanied by the movement of thermal energy from translational modes to less densely spaced vibrational modes. Increases in structuring therefore involve reductions in kinetic freedom, which are reflected thermodynamically in W_{th} reductions and hence I_{th} increases. Using I_{th} as a gross index of structuring, it is convenient to express equation (8) in the form

$$\Delta I_{th} < -\Delta I_e - \Delta I_c \qquad (9)$$

At this point, the randomizing directive of the second law begins to make contact with the integrative movement of evolution. I_e and I_c provide two modes of thermodynamic potency whose dissipation (designated for heuristic convenience as 'energy-randomization' and 'matter-randomization') is coupled to the complexifying, self-organizing evolution of the biosphere. The question we must now consider is the nature of the thermodynamic constraints prevailing in the biosphere that require reductions in one informational mode to be coupled to increases in another.

There is nothing first of all in the character of the second law itself that mandates either increases or decreases in complexity with time. What it does mandate is increases in randomness, which can be achieved through reductions in any or all of the three modes of I_M. However, the constraints under which the second law is obliged to operate, imposed both by the laws of nature and also by the particular chemical conditions prevailing in the biosphere, strictly limit the possible routes along which I_M can be dissipated. Under these constraints, few reactions are completely dissipative in the sense of bringing about reductions in all three modes. In general, reductions in one mode will occur through increases in another. For example, the production of H_2O from H_2 and O_2 is powered by the reduction in I_e that accompanies the formation of the very stable $O-H$ covalent bonds. But as a consequence of the increase in molecular size and macroscopic homogeneity resulting from this reaction, I_{th} and I_c must both increase. The thermodynamic basis for evolutionary structuring is the fact that certain physicochemical conditions, and particularly those prevailing in the earth's prebiotic atmosphere and oceans, tend to promote increases in I_{th} and the expense of its other two modes, as the following sections will consider.

ENERGY RANDOMIZATION

The chemical evolution of the biosphere has been powered by the dissipation of energetic information deriving ultimately from solar radiation through *chemical bonding*. Covalent bonding promotes the development of molecular skeletons; weaker interactions (hydrophobic, etc.) provide for the stabilization of macromolecular conformations and for the formation of supramolecular aggregates such as membrane systems and organic microspheres. As Black (1973) has pointed out, self-assembly in the prebiotic milieu results largely from hydrophobic forces favoring phase separation. The strength of hydrophobic interactions derives primarily from the additional hydrogen-bonding between water molecules that occurs when organic solutes are squeezed together, and secondarily from the van der Waals forces between these solutes. These interactions serve as effective energy randomizers in the self-aggregation of both macromolecules and microspheres in an aqueous milieu. The general result of these associations has been to reduce the number of discrete chemical entities of the biosphere over the course of its evolution and to increase their average sizes and complexities.

Putting two smaller chemical species together to form a larger one always dissipates I_e, for the simple reason that chemical bonds form through the movement of electrons into potential energy wells; conversely, breaking apart a chemical entity into fragments always increases I_e. For this reason, there tends to be a reciprocal relationship between thermal and energetic information, with increases in one mode ordinarily occurring at the expense of reductions in the other. The second law itself has nothing to say about the direction of this information flow: The fact that it occurs from I_e to I_{th} in the evolution of the biosphere is due to the thermodynamic conditions that prevail there, which happen to be quite general of this universe.

The biosphere has no exogenous source of I_{th}, but I_e is created continually from influxes of radiant energy which generate high-energy chemical species through electronic excitations, ionization, or free-radical formation. Because of this generation of I_e from exogenous sources, irreversible processes tend, on average, to bring about the dissipative flow of I_e to I_{th}, and hence, structuring. Given the availability of associative pathways, the continual input of radiant energy into the biosphere requires its structuring. I_{th} therefore tends to increase with evolutionary time. This principle has its most straightforward application in prebiotic evolution, for the responses here are to impressed gradients rather than potential energy stores. But with the emergence of autocatalytic systems, thermodynamic flows are competitively *pulled* into self-organization, and any build-up of thermodynamic potential becomes fair game for

I_{th}—reducing exploitation—by, for example, invasion of a foreign micro-organism. So, the foregoing are trends and propensities rather than inviolable laws.

MATTER-RANDOMIZATION AND COMPLEXITY-GENERATION

Organization requires complexity, and the complexity of a chemical structure depends not only on its number of elements and interconnections, but on their variety and aperiodicity. The generation of I_{th} at the expense of I_e is therefore a necessary but insufficient condition for complexification. Increasing elemental variety and interconnection aperiodicity requires the dissipation of configurational information. This mode of dissipation promotes two kinds of reactions, referred to here as 'dispersive' and 'permutative', that are essential to molecular evolution. Dispersive reactions involve the formation of large varieties of biomonomers such as amino acids and nucleotides from relatively small sets of simpler reactants; permutative reactions involve the generation of ensembles of alternative molecular sequences from a given basis–set of monomeric elements (e.g. the polymerization of nucleotides). In both cases the thermodynamic driving force derives from the opening up of new configurational microstates whenever a new chemical species is formed.

The number of configurational possibilities available to a thermodynamic system is given by the permutational equation

$$W_c = N!/N_a! \, N_b! \ldots N_i! \tag{10}$$

where N is the total number of molecules in the system and the various N_i represent the number of molecules of each distinct chemical species. This equation indicates that W_c is an increasing function of both molecular number and variety. But variety is always the dominant consideration: matter-randomization always favors the generation of new chemical species, regardless of whether N increases or decreases as a consequence. Provided atoms *can* be hooked together in new, more complex combinations, the attendant production of configurational entropy requires that this occur, at least to some extent. This can be demonstrated as follows.

Irreversible processes dissipate I_M along particular reaction routes, such that

$$I'_M = dI_M/d\xi < 0 \tag{11}$$

where ξ is the 'extent of reaction' parameter, representing the fractional extent to which a given process or reaction has proceeded to completion along the composition axis. The value of I_M' at any point along this axis represents the potential of the system for irreversible change toward equilibrium.

Under ideal conditions I_M' can be separated into configurational, energetic, and thermal components. Of these, I_{th}' and I_e' are both constant, or nearly so, over the entire reaction coordinate. The parameter of interest here, the one that establishes a point of equilibrium for all reactions, is I_c'. The magnitude of this parameter, which is highly sensitive to the value of ξ, represents the matter-randomizing drive of the system. The sign and magnitude of I_M' at any point along the ξ-axis represent the system's thermodynamic tendency to move in a particular direction along this axis. It is the strong composition-dependency of I_c' that assures that any kinetically-feasible reaction will occur to *some* extent, regardless of its energetic favorability.

For the general reaction shown below

$$aA + bB \rightarrow cC + dD$$

it can be readily shown (Wicken, 1979) that

$$\bar{I}_c' = -(R/a)\ln[1 + b_o/a_o + r\xi/a)^r(1-\xi)^a(b_o/a_o - b\xi/a)^b/(c\xi/a)^c(d\xi/a)^d] \quad (12)$$

In this equation a_o and b_o represent the initial concentrations of A and B; $\bar{I}_c'$ is the molar value of I_c', calculated on the assumption that A is the limiting reactant in the system ($\bar{I}_c' = I_c'/a_o$), and $r = (c + d - a - b)$.

It should be noted that the value of $\bar{I}_c'$ is negatively infinite when $\xi = 0$ and positively infinite when $\xi = 1$. Matter-randomizing considerations therefore promote *all* reactions through some compositional range, and reciprocally, assure that no reaction can proceed entirely to completion.

Equation 12 simplifies considerably when applied to specific reactions. The reaction $A + B \rightarrow C$ may be taken as a simple paradigm case of an associative, complexity-generating reaction. For this case equation 12 becomes

$$\bar{I}_c' = -R \ln((1-\xi)^2/\xi(2-\xi)) \quad (13)$$

provided that $a_o = b_o$. The value of $\bar{I}_c'$ is zero, and I_c is minimized for this system when $\xi = 0.29$. States of the system in which little or no product has been formed experience a very high, negative value of I_c' which promotes the formation of C molecules. This thermodynamic tendency toward states of higher configurational randomness is therefore a highly creative force in molecular evolution, favoring the generation of new, more complex molecular species even in energetically-unfavorable reactions.

PREBIOTIC EVOLUTION

These principles can be applied readily to account for the progressive, complexifying character of prebiotic evolution leading toward the emergence of life. The following stages can be distinguished in this emergence: (a) the formation of simple molecules; (b) the formation of biomonomers; (c) the formation of biopolymers; (d) the aggregation of biopolymers into 'protocells'. The first two stages are strongly divergent ones (Kuhn, 1972), in which matter assumed new, more highly differentiated forms. Matter-randomization was an essential generative force in these stages. Convergent evolution began to express itself in stage (c). But in its pre-selective, generative phase, polymerization was a permutative process, also promoted extensively by I_c dissipation.

If one has a large pool of monomers (A, B, C, D) which have no preferred sequence of polymerization so that the 16 possible dimers are formed in equal amounts, and if the number of molecules of each species of monomer is identical: $(N_a = N_b = N_c = N_d = M)$, the number of configurational possibilities available to the system for any given concentration of dimers is given by the equation

$$W_c = 2M(2 - \xi)! / [M/(1 - \xi)]!^4 [M\xi/8]!^{16} \tag{14}$$

from which the value of I_c' at any point along the composition axis can be calculated to be

$$I_c' = -R \ln [4(1 - \xi)^2/(2 - \xi) (\xi)] \tag{15}$$

This function is zero (I_c minimized) when $\xi = 0.553$. That is, even though total molecular number decreases during dimerization, matter-randomization through permutation favors a dimerization level of over fifty percent. To achieve this level of dimerization it is necessary that the increase in I_{th} be offset by the dissipation of thermal entropy to the system's surroundings, requiring a compensating decrease in I_e. In general, energetically-favorable reactions are those that meet this condition. But even if the reaction is not energetically favorable, *some* level of dimerization will be promoted by matter-randomizing considerations alone. Again, the value of I'_e is negatively infinite when no dimer has been formed, indicating an infinite thermodynamic reaction potential toward dimer formation at that point. So, at low values of ξ dimerization is strongly favored by matter-randomization.

This argument can be expanded to mandate the formation of larger oligomers as well. Each additional molecular species that can be formed provides new microstates for the further randomization of matter in the system. I_c dissipation thus assures that various degrees of polymerization will occur in nature, because with a stable basis-set of monomers permutative reactions provide a route for the generation of new configurational microstates.

A boundary condition on such permutative processes is provided by the fact that the biosphere is for all practical purposes a closed system; as such its atomic constituents remain virtually unchanged with the passage of time. They therefore constitute a stable basis-set of elements that can be combined and permuted within the limits of chemical possibility to generate a vast range of molecular and supramolecular structures, but which cannot be broken down by chemical processes. The early evolutionary bias toward increasing molecular complexity has its thermodynamic source here: by permuting this atomic basis-set into larger, more heterogeneous structures new microscopic configurations of matter can be generated. Each new structure formed provides new microstates for the biosphere, and will accordingly be promoted at least to some extent by reductions in I_c, the actual extent being determined by the I_e changes that accompany the process. The evolutionary route, then, to I_c dissipation in the biosphere is by putting its atoms together in new, more complex ways; and any quasi-stable set of structures so generated (e.g. amino acids, nucleotides, proteins) becomes in turn the raw material for the production of still other, higher-order structures.

MUTATIONAL ORIENTATION

The dissipation of I_c in the context of a reproducing organism generates genetic variations through mutation, whose phenotypic correlates provide raw material for natural selection. Mutations of all kinds are promoted by this principle—from point mutations involving base substitutions to insertions, deletions, and chromosome rearrangements. It is customary to regard mutations as 'mistakes' in the replicative process, but this misconstrues somewhat their law-like nature when regarded from the macro-statistical perspective. All these various mutational occurrences provide new configurational possibilities for the biosphere, and thereby reduce its I_c content. Although individual mutations are unpredictable, mutations as classes of events occur as means for dissipating I_c. While their frequency can be minimized by optimizing the specificity with which the various components of the hereditary system interact, their elimination is strictly forbidden by the second law.

The question here is, has the thermodynamic bias toward increasing complexity that prevailed in prebiotic evolution extended to its biotic phase as well? It is of course axiomatic in neo-Darwinism that mutations are *undirected*, in the sense of being statistically neutral with respect to eventual functional utility. The presumed correctness of this axiom does not mean, however, that mutations cannot be biased in afunctional ways toward increased genome size or aperiodicity. The possibility of a size-increasing bias is difficult to assess, but might conceivably have operated in the early phases of evolution; if so, it would go far to explain the coming-to-be of genomes whose sizes seem to far exceed the requirements of function. Such a mutational bias, if it existed at all, could not have lasted long into the metazoic explosion. In prebiotic evolution, polymerization occurred in part for the entropic payoff of arranging monomeric alphabets in long sequences. But with increasing molecular size this payoff diminishes quickly and that for fragmentation becomes progressively greater. In any case, there is no systematic correlation between DNA length and order of evolutionary appearance of the various metazoic taxa. In contrast, the entropic drive toward increased aperiodicity is at the heart of evolutionary change, both the phylogeny of proteins and in the hierarchical regulation of their patterned production.

The matter-randomizing drive constantly promotes the assortment of genes, and fragments of genes, in new structural connections. This drive to genome diversification has profound implications with respect to the emergence of hierarchically organized patterns of gene expression, one DNA sequence acquiring by its positioning and protein interactions regulatory controls over groups of other sequences (Britten and Davidson, 1969; Zuckerkandl, 1976). This scheme seems borne out by recent indications that the eukaryotic genome is a highly dynamic structure, quite unlike the static beads-on-a-string model, with mobile sections of DNA between genes and within genes able to migrate about the genome. Evolution has provided controls that make this better than a haphazard process, but its ultimate rationale and necessity would seem to reside in the principle of matter-randomization.

THE SELECTION OF ORGANIZATIONAL COMPLEXITY

Matter-randomization is an essential driving force of evolutionary complexification at the molecular level. But it must be noted that its operation is highly limited by the general conditions of matter flow in closed systems. The biosphere has neither an exogenous source of I_c nor a sink for configurational entropy. For these reasons, I_c dissipation makes no systematic contribution to the thermodynamic arrow and contributes far less to the overall direction of evolution than does the dissipation of I_e. The latter is fundamental to the thermodynamics of natural selection and bears strongly on the conditions for the functional complexification of self-organized systems.

The idea that natural selection had a predictive thermodynamic basis was first explicitly articulated by A. J. Lotka (1922), who argued that those organisms that were most successful in channeling energy flows through themselves, and at the same time increasing total flow through their ecosystems, would necessarily be selected for. This mutually-supportive connection between organism, population, and ecosystemic flow expresses the holistic character of natural selection. Ecosystems cycle matter and process energy under the impress of a free energy gradient in which high-grade radiant energy is photosynthetically trapped as I_e and dissipated as thermal entropy into the sink of space. Irreversible processes are those that act to relieve this gradient, and ecosystem self-organization, both of organisms and of flow networks among organisms, constitute the evolutionary strategy for accomplishing this. Indeed, each stage in the grand movement of evolution from its prebiotic beginnings through its socio-technological developments can be regarded as a new way of concentrating and processing I_e. This movement tends, in a meandering way, to be progressive: there is always 'room at the top' for levels of organization that can provide the biosphere with additional pathways of entropy production that don't competively interfere with the lower levels. Higher patterns tend rather to be built on the lower, with the result that the biosphere's overall I_M flow from solar radiation to thermal entropy becomes routed through a progressively more complex and hierarchically integrated network of energy-transforming systems with the passage of time. Such systems include not only organisms, but societies, technologies, and world systems as well.

The tendency of useful, ecomass-supporting energy flow to increase and ramify in complexifying ways over the course of evolution is implicit in the second law and the physicochemical circumstances of its operation in the biosphere. Dissipative pathways tend to elaborate over the course of evolution because high-grade forms of energy entropically 'seek' to be dissipated. Any ecosystemic novelty that taps theretofore unutilized sources of potential energy will have found a thermodynamic niche and be selectively preserved. Photosynthetic plants increase flux by diverting energy into their biosynthetic processes, thus competing with reflection and convection for a share of solar radiation utilization. This development in turn creates a corresponding niche for heterotrophs able to exploit these resultant I_e stores. In turn, by restoring CO_2 and H_2O to the ecosystem, the latter promote photosynthetic fixation, thus establishing a positive feedback loop that increases total energy flow through the system (*see* Odum, 1971).

The stability and power of the total ecosystemic flow is built on the stabilities and capacities of its sub-patterns (e.g. organisms, populations) and on their mutually-facilitating interdigitation. Organisms are themselves patterns of thermodynamic flow and interlace both horizontally and in hierarchically ascending, self-regulative network patterns. As systems of self-producing kinetic relationships, organisms are tied integrally to nature's dissipative dynamics. Since biological organization is identical with the flow pattern it mediates, the issue of organizational stability, or of the natural selection of certain organizational types over others, is very much a thermodynamic one. If there is a natural selection for organizational complexity at all, its basis must be in the power and stability of the organizational type as flow pattern.

This relationship between ecosystem self-organization and thermodynamic flow expresses a fundamental dual-referentiality to all evolutionary processes, that allows us to break into the survival-of-the-fittest tautology. The Darwinian referent, of course, is the stabilization and propagation of the organizational type itself; the other, ecosystemic referent is the pattern of thermodynamic flow it is involved in—not only of the individual organizational type, but of the entire patterned network in which it participates. This relationship between ecosystem self-organization and thermodynamic flow, whereby strategies for organizational stabilization and propagation are constrained in their possibilities by the tendency of thermodynamic flow patterns to increase in total power and interdigitation, gives evolutionary process a law-like character above and beyond the 'chance' dimension emphasized in neo-Darwinian treatments. Thermodynamic flow patterns are, along with the genes they serve to propagate, essentially selfish; but by their participation in higher-order flow patterns the selfish interests of individual organizational types are, on average, best served by adaptive strategies that contribute to the higher ecosystem good—i.e. the power and complexity of ecosystemic flow.

There are two general strategies by which organizational types stabilize and propagate themselves. One involves reproductive fecundity; the other involves homeostasis, the preservation of the self and the securing of its autonomy in the face of environmental fluctuations and stress. The latter strategy will interest us most, since it provides an important rationale for the selection of organizational complexity and for progressive evolution in general. Slobodkin (1964) has rightly identified homeostasis as an increasing parameter in evolution, as opposed to adaptation or fitness. The latter are conditions of life, not measurable attributes that can be compared among different species in different kinds of environments. Homeostasis, on the other hand, is an operational property of a system, one that can be measured or at least qualitatively assessed independently of its adaptive correlates. In this sense, it is like 'complexity'; indeed, increasing homeostasis is the functional correlate of increasing organizational complexity. Each new level of homeostasis contributes to the stability of the organizational type, and consequently to the stability of the flow pattern of which genotypic propagation is an integral feature. Homeostasis is achieved through organizational complexification—through the elaboration of anatomical systems, discriminating behavioral programmes, societies. The latter can in fact be regarded as higher-order 'selves', as centers of valuation and homeostatic regulation that act to preserve their own organizational relationships and the informational programmes that support them without primary regard for individual members.

ORGANIZATIONAL STABILITY

The selective premium on progress in evolution is, on the face of it, almost transparently easy to appreciate: complex, highly-integrated organizations can *do* more than simple, loosely-integrated ones. Thus Maynard Smith (1969) has suggested that since the first organisms were minimal systems containing the smallest number of components needed for a self-maintaining metabolism, the only way for evolution to possibly proceed was toward increasing complexity. Evolution, if it is to occur at all, must be biased in this direction.

Bronowski (1970) has expanded on this same theme, arguing that the stability of an organism is 'stratified', with stable units at one level of organization providing raw materials for the composition of the next, hierarchically-superior level. Matter is regarded as having a 'potential of stability' that can only be evoked in complexifying steps: from elementary particles to atoms, molecules, organisms and societies. This is a worthwhile argument, if used with judicious appreciation for its metaphorical tone and evident failure at universal exemplification. Regressive mutations do, after all, occur and are in fact quite advantageous under certain circumstances where full behavioral repertories are superfluous and metabolically burdensome—parasitism being the main case in point.

In Saunders and Ho's (1976) treatment of this problem, organizational complexification is seen as an evolutionary principle quite independent of natural selection. The functional coherence of any organism, as a parsimoniously organized whole, will almost always be drastically disrupted by any mutation that causes a random elimination of some element or interconnection. Mutations leading to random additions of parts will also tend to have negative fitness correlates, but these at least preserve the core of functional interconnections that make the organism work, and moreover, provide a structural substrate in which still more functional interconnections might be honed through natural selection.

The drive toward complexity is thus constrained in this view by fitness considerations, while being itself independent of them—fitness being simply the condition to which all evolutionary innovations must tautologically be faithful. This recognition of fitness as constraint rather than orienting principle squares well with what we know of evolution's creativity. From its inception, Darwinism has been wed to the metaphorical framework of Malthusian economics, with its emphasis on competition for scarce resources. Economic theory has itself long since transcended these 'dismal science' roots by recognizing that resources are not given in nature, but are invented by the economic process from nature's raw materials. Evolution has an analogous resource-inventing power that allows life to take the initiative in evolution through the behavioral exploration of dissipative opportunities. In each case, fitness applies as constraint on possibility. In microevolution, such constraint occurs within the intense competitions of given Malthusian economies to prevent genuine innovation. In macroevolution, it acts to assess the capacity of an innovation to open up a new economy. The thermodynamic view advanced here has certain affinities with that of Lamarck. Behavior is primitive to nature, and owes to the nature of living systems self-active, self-producing *processes* in exploitative interaction with thermodynamic resources. The context of discovery always precedes the context of fixation, in evolution as with human ideas. The plasticity for discovery lies in behavior, and the plasticity for adaptation lies in the morphological possibilities of the organizational type. Only then should one expect the sanction of genetic hard-wiring.

BEHAVIOR AND PROGRESS

Nature provides its 'niches' for exploitative behavior not in some preformed, static array, but as a multidimensional hierarchy, each level of which provides ingredients for other, higher-order levels. Lower-order niches are characterized above all by immediacy of energy access, and lower organisms are characterized above all by the very small behavioral component that goes into the exploitation of these niches. Living systems emerged to utilize the energy stores of the primeval oceans in stable patterns of I_M flow. The filling of this niche by metabolically self-sufficient microbes provided other, hierarchically superior ones for their own exploitation. Thus eukaryotes, and thus the metazoic explosion — each new development involving the establishment of hierarchical controls on previous levels of organization, and each new niche having a greater behavioral requirement for its negotiation.

And so it proceeded, the biosphere growing outward as a kind of organic *plenum*, analogous in ways to the 'Great Chain of Being' (Lovejoy, 1942) that figured so prominently in Lamarck's evolutionary vision. But the analogy cannot be pressed too far: the organic *plenum* is not a one-dimensional escalator that carries generations of organisms ever upward toward higher levels of perfection, but a multidimensional envelope that spreads outward in opportunistic ways from a core of simple organisms and tested kinetic pathways, a mobile and adaptively interactive system of thermodynamic flows. Evolutionary progress occurs because of the opportunities for homeostatic stabilization to be found at the outer surface of the *plenum*. There is always ecological room for energy-transforming systems that can behaviorally circumscribe their environments.

This is not to say that behavioral control of the environment in any way substitutes for chemical access. Evolutionary process unfolds cumulatively in time, so that the past is always at the core of the present — not just in the recapitulationist context of ontogenic development, but also in the broader sense of carrying with it the essential conditions of life as staked out by the first proto-organisms. Life is based on informed kinetic stabilizations requiring chemical flows between environment and self. The strategy of anagenesis is to secure the stabilities of these flows through complex anatomical organization and behavior, such that the environment is brought into the homeostatic field of the self. The physiology of an advanced organism has the function of preserving, in homeostatically stable internal environments, the essential conditions of life. This stabilization of internal environment can only be achieved by a freeing of life from dependency on biochemicals in nutrient broths in favor of *resources* that can be selectively exploited, i.e. by an evolutionary movement from direct chemical access to exploitative behavior. Anagenesis can be regarded as the creation of behavioral space between a resource and its ultimate chemical accessibility, or better, as the creation of resources themselves through behavioral repertories that extend the metabolic center of life.

So the organic manifold is an historical structure, inventing itself in time through the conditions of the past and the opportunities of the present. Each new species adds another feature to the ecological topography and introduces into it a new set of exploitative possibilities, including those for parasitism and symbiosis, which contribute to the density of the *plenum* but not to its volume. Parasitism is an evolutionary strategy for achieving organizational type-stability while avoiding the behavioral and organizational requirements for autonomous existence. But since each new opportunity for parasitism requires a previously established species that has solved the problem of organizational stabilization in autonomous, homeostatic ways, parasitic evolution can only occur in the wake of anagenesis.

Anagenesis and parasitism are thus two opposite solutions to the problem of organizational stability. One involves adaptation to narrow, homeostated environments that provide immediacy of metabolic access without the intervention of extensive behavioral programs; the other involves increasing control over complex and metabolically remote environments through correspondingly complex and discriminating forms of behavior. One leads in the direction of contingency, the other toward autonomy. Between these extremes falls a range of possibilities for parasitic, symbiotic and predatory relationships that sometimes shade imperceptibly into each other.

Whereas adaptive openings occur at all levels of the organic *plenum*, it is at its surface that the evolutionary process is its most vital. It is here that Huxley's (1954) characterization of a progressive adaptation as one that creates possibilities for future adaptations most holds true. When adaptations are to environmental specifics, evolutionary flows clog and rigidify with specialized anatomical baggage and stereotyped behavioral responses—the 'excesses of adaptation' (de Beer, 1958) that dispose most evolutionary lines to eventual extinction in the face of environmental change. Adaptation may be opportunistic and without foresight, but nature provides niches for just those adaptations that look, after the fact, as though they did possess foresight. Behavioral control of the environment is the best long-term guarantee of evolutionary success, and the vital force of evolution is maintained through plastic behavioral programs that are not indissolubly wed to particular environmental conditions. The progressive movements of evolution can be connected with just those innovations that increase freedom and discriminating behavior: the central nervous system, the lensed eye, language, social organization. In each case, adaptive potentialities are barely touched by immediate applications.

THERMODYNAMIC TRADE-OFFS

There are some specific thermodynamic payoffs for the evolution of behavioral complexity. With the establishment of prokaryotic populations in the biosphere, two general strategies were made available to the evolutionary process which have worked synergistically to maximize total dissipative flow. The plant strategy has been to support a basic plan of metabolic autonomy by extending structurally into the environment through roots, leaves and supports, such that proximity to physicochemical sources is always maintained. The animal strategy, in contrast, has been to deal with the environment through physical mobility and discriminating modes of behavior.

In animals, metabolic autonomy has been systematically compromised in the interest of behavioral range. Mammals, for example, are relatively insufficient metabolically, having lost the ability to synthesize a sizable fraction of the twenty essential amino acids as well as a variety of metabolic co-factors, which then assume the status of 'vitamins'. Metabolic self-sufficiency and behavioral range cannot be jointly optimized. One favors fixity and physicochemical proximity to the environment; the other requires mobility and behavioral control of the environment. The basic incommensurability between plant and animal strategies is that metabolic and behavioral autonomy are purchased at costs to each other, due to the unavoidable competition between biosynthesis and organized contractile motion for limited I_e stores. Each set of abilities requires a certain minimum energy investment that eliminates options for other investments; so each presents itself as an alternative strategy for evolutionary success, defining in general terms the difference between plants and animals.

The investigations of Zamenhof and Eichorn (Pauling and Zuckerkandl, 1972) into the effects of losses in biosynthetic function on microbe evolution seem especially germane to understanding the thermodynamics of this strategic trade-off. It turns out that such losses are accompanied by tremendous reproductive advantages for the genetically diminished mutant, provided that its growth medium is supplemented with the lost metabolite. This is not really surprising when one considers that the maintenance and utilization of superfluous biosynthetic equipment diverts I_M flows away from reproductive activity. Nor is it surprising that the extent of the reproductive advantage depends on the location of the mutation in the biosynthetic pathway, being proportional to the number of metabolic steps that it eliminated. Genetic deletions, as opposed to point mutations, proved the most advantageous of all.

Pauling and Zuckerkandl (1972) have deduced from this that evolution will occur whenever organisms in a given environment are provided with a source of nutrients that they ordinarily make for themselves. When this occurs, it is in the interest of metabolic economy for the organisms to divest themselves of the now-redundant machinery needed to synthesize these substances. Under conditions of environmental plenty, selection will favor reduced metabolic comprehensiveness: the richer the environment, the leaner the metabolic machinery can profitably become. This of course is the recipe for parasitic devolution, rather than for evolution in its progressive sense. Yet the same thermodynamic considerations favoring biosynthetic minimality in parasitism are at work in anagenesis as well. While successful organisms are necessarily those that effectively funnel thermodynamic flows into the propagation of their own organizational types, the constraints under which this occurs in nature make the evolutionary process diverse in its elaborations. Where the environment is rich and metabolically proximate, organisms can become metabolically spare and behaviorally simple. But where it is less salubriously supportive of genetic propagation, metabolism and behavior must intervene to secure it. The gains in reproductive efficiency that accrue to the jettisoning of redundant metabolic pathways are crucial to the development of the behavioral dimension of life. For the environment rarely provides its fruits *gratis*, in the manner of a laboratory scientist. Amino acids have not been found free in nature since the exhaustion of the primal soup. They occur in proteins, as parts of other organisms. For the jettisoning of metabolic machinery to be a productive move in these environmental circumstances, it must be accompanied by appropriate behavioral modifications that provide for exploiting the metabolic labors of other organisms.

The above experiments indicate the thermodynamic feasibility of this. While the bacterial growth medium is a very weak model for an actual environmental energy niche, it does very clearly show the advantages of avoiding biosynthetic work. The great gains in reproductivity that accompany metabolic simplification show that the energy required to run behavioral programs is in fact available to organisms provided they relinquish some measure of metabolic autonomy. In this way, the existence of populations of metabolically autonomous organisms opens up thermodynamic niches for others able to substitute behavioral exploitation for biosynthesis. The principle of 'plenitude' in nature requires that this behavioral component of adaptation find every possible exemplification commensurable with thermodynamic constraints, through every variety of parasitism, symbiosis, and predatory activity. Each represents a solution to the problem of organizational stability. The parasitic solution leads in the direction of increased contingency within the energy-rich interstices of the organic *plenum*. Anagenesis occurs at its outer surface, where environmental conditions provide new opportunities for behavioral control.

REFERENCES

de Beer, G. (1958). The Darwin–Wallace Centenary. *Endeavor* **XVII** 61–76.
Black, S. (1973). A theory on the origin of life. *In* 'Advances in enzymology' (A. Meister, *Ed.*), Vol. 38, pp.193–234. Academic Press, New York, USA.
Britten, R. and Davidson, E. (1969). Gene regulation for higher cells: a theory. *Science* **165**, 349–357.
Bronowski, J. (1970). New concepts in the evolution of complexity: stratified stability and unbounded plans. *Zygon* **5**, 18–35.
Castrodeza, C. (1978). Evolution, complexity and fitness. *J. Theor. Biol.* **71**, 469–471.
Chaitin, G. (1966). On the length of programs for computing finite binary sequences. *J. Assoc. Comp. Mach.* **13**, 547–569.
Chaitin, G. (1975). Randomness and mathematical proof. *Sci. Am.* **232**, 47–54.
Glansdorff, P. and Prigogine, I. (1971). 'Thermodynamic theory of structure, stability and fluctuations'. Wiley and Sons, New York, USA.
Ho, M.-W. and Saunders, P. T. (1979). Beyond neo-Darwinism — an epigenetic approach to evolution. *J. Theor. Biol.* **78**, 573–591.
Huxley, J. (1954). The evolutionary process. *In* 'Evolution as a process' (A. Hardy and E. Ford, *Eds*), pp.1–23. Allen and Unwin, London, UK.
Kuhn, H. (1972). Self-organization of molecular systems and the evolution of the genetic apparatus. *Angew. Chem. Inter. Ed.* **11**, 798–820.
Lotka, A. (1922). Contribution to the energetics of evolution. *Proc. Nat. Acad. Sci.* **8**, 147–155.
Lovejoy, A. (1942). 'The great chain of being'. Harvard University Press, Cambridge, Massachusetts, USA.
Maynard Smith, J. (1969). The status of neo-Darwinism. *In* 'Towards a theoretical biology' (C. H. Waddington, *Ed.*), Vol. 2, pp.82–89. Edinburgh University Press, Edinburgh, UK.
Odum, H. (1971). 'Environment, power and society'. Wiley and Sons, New York, USA.
Pauling, L. and Zuckerkandl, E. (1972). Chance in evolution — some philosophical remarks. *In* 'Molecular evolution: prebiological and biological'. (D. Rohlfing and A. Oparin, *Eds*), pp.113–126. Plenum Press, New York, USA.
Popper, K. (1976). 'Unended quest' pp.167–180. Open Court, La Salle, France.
Saunders, P. T. and Ho, M.-W. (1976). On the increase in complexity in evolution. *J. Theor. Biol.* **63**, 375–384.
Slobodkin, L. (1964). The strategy of evolution. *Am. Sci.* **52**, 342–357.
Wicken, J. (1978). Information transformations in molecular evolution. *J. Theor. Biol.* **72**, 191–204.
Wicken, J. (1979). The generation of complexity in evolution: a thermodynamic and information-theoretical discussion. *J. Theor. Biol.* **77**, 349–365.
Wicken, J. (1981). Evolutionary self-organization and the entropy principle: teleology and mechanism. *Nature and System.* **3**, 129–142.
Zuckerkandl, E. (1976). Programs of gene action and progressive evolution. *In* 'Molecular anthropology' (M. Goodman, R. Tashian and J. Tashian, *Eds*), pp.387–447. Plenum Press, New York, USA.

Section III

PATTERN AND PROCESS

5

Patterns in the Fossil Record and Evolutionary Processes

ELISABETH S. VRBA

Abstract

Patterns in the fossil record provide a severely limited view of phenotypic variation in space and time. Yet they can be used to test hypotheses of microevolutionary and macroevolutionary processes. In particular monophyletic groups of sexually reproducing organisms, hypotheses of which parts of fossil patterns may represent species, can be tested. Phylogenetic trees from cladistic analysis, which include all known extinct and extant species, plus estimates of past and present geographic distribution of species, are the best possible data for testing models of evolutionary patterns and rates and hypotheses of the causal processes. I discuss the predictions for patterns in space and time which flow from various concepts of process. The neo-Darwinian synthesis has traditionally regarded two kinds of processes occurring within populations as of pre-eminent importance in evolution. (1) The introduction of genic and phenotypic variation is held to come from random gene and chromosome mutations and recombination. (2) Such variation is seen as sorted mainly by the deterministic process of natural selection or by the random one of genetic drift. As a consequence the patterns of variation, from those among genes to trends among species, are generally interpreted in the adaptive mode under the synthesis. In contrast, in this paper I view the biota as a genealogical hierarchy of nested sets of individuals, genomic constituents to species in monophyletic groups. An hierarchical approach includes wider possibilities of causation within additional levels, as well as upward and downward causation between levels. It acknowledges that many patterns of variation may not be correctly described as adaptive. Several hypotheses of causes of macroevolution are reviewed: the hypothesis of intrinsically directed introduction of phenotypic variation, and the synthetic, directed speciation, effect macroevolution, random and species selection hypotheses. Some evidence from the Miocene-Recent evolution of African mammals is discussed to illustrate the various hypotheses, predictions and conclusions, especially as they relate to among-species evolution.

BEYOND NEO-DARWINISM
ISBN: 0-12-350080-X

INTRODUCTION

Palaeontologists study the fossilized hard parts of organisms, that is, only one subset of the set of morphological characters which in turn is only a part of what we understand by 'phenotype'. At best they gain a very rough estimate of how character variation is distributed in time and space. This paper is about such estimated patterns and about how one might use them to study evolutionary processes. However restricted and imprecise such estimates may be, it is obvious that certain hypotheses of macroevolution are testable only via the fossil patterns in monophyletic groups. (Monophyletic groups in the precise sense are implied. That is, the group includes all known species, and only those species, hypothesized to have descended from a common ancestral species, which latter is itself included in the group; *see* Nelson, 1971, after Hennig, 1966. I define microevolution broadly as changes in genomic and phenotypic variation within populations; macroevolution as large changes in variation and/or species composition in monophyletic groups.) A fruitful approach to hypotheses of macroevolution, within particular monophyletic groups which include extant survivors, involves simultaneous tests using studies from the living organisms together with those from palaeontological patterns. In my discussions I shall make special reference to the Miocene-Recent evolution of African mammals, which includes some of the most spectacular mammalian radiations. This evidence offers a bewildering vista of diversity, persistence, multiplication and turnover of morphologies. Some questions in relation to these data, as to fossil data in general, are these: What are the genealogical patterns? How do they distribute in phenotypic and geographic space and in time? How do they relate to long-term directional phenotypic divergence, to the distribution of rates of change along lineages, to species and species diversity? What can they tell us about evolutionary processes and their causes?

PATTERNS

Phylogeny

At the start of the kind of studies addressed in this paper we need an hypothesis of the pattern of ancestry and descent among organisms—a phylogeny. It is important to arrive at this basic hypothesis (a cladogram) without reliance on particular notions of evolutionary mechanism, such as those of the synthetic theory or particular models of speciation or the nature of species, and without reference to time and geography. Many recent publications have discussed this; for instance Gaffney (1979) who points out that only two hypotheses at the highest level of generality (that is, which have survived testing for some time) are required to develop and test a cladogram: (1) that evolution has occurred and (2) that new taxa may be characterized by new features.

In approaching evolutionary process one can achieve quite a lot with a cladogram (which does not require specific hypotheses of what is an ancestor of another taxon, merely that pairs of taxa share a closer common ancestry with each other than with any other taxon). But one can go even further with a phylogenetic tree (Eldredge and Cracraft, 1980; Eldredge, 1979a). If one plots a phylogenetic tree against some phenotypic axis (or axes) and against time, one gains an estimate of the 'shape' of the tree — the distribution of rates of change along lineages in relation to long-term phenotypic divergence, such as trends, and lineage splitting events (*see* Fig. 5·1). When biologists discuss trends they usually agree that they are talking about evolution in monophyletic groups. But they may differ in the

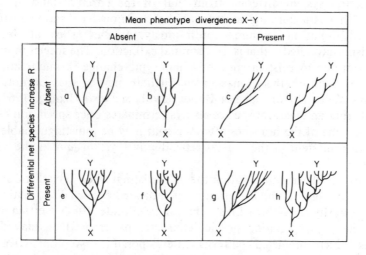

FIG. 5·1 Phylogenetic trees. One or more phenotypic characters vary along the horizontal axes; time along the vertical axes. Cases b, d, f, h show a pattern of punctuated equilibria; a, b are not trends; c, d are trends only in terms of phenotypic divergence; e, f only in terms of differential R; in trends g, h divergence occurs together with increase in R. In the examples used here for e–h, increasing 5 overwhelms increasing S to result in differential, positive R values. But cases e–h could also have been drawn to show increasing R as a result of decreasing E overwhelming S (after Vrba, 1983: Fig. 1).

emphasis on particular components of trends. Reference may be only to phenotypic divergence between early and late lineage endpoints. For example one can note a trend from the short, straight, smooth horns of some Miocene antelope species to the long, elaborately twisted and keeled ones of the extant kudu *Tragelaphus strepsiceros*. Another aspect of trends is the differential increase in numbers of separate lineages. Thus the kudu genus is much more diverse than its eland sister clade *Taurotragus*, and one can refer to a trend towards more branches with features that characterize *Tragelaphus*. Figure 5·1 shows how mean divergence XY in particular characters may theoretically occur separately from or together with differential lineage splitting and extinction, and therefore with diversity.

Species

As the word species will be used a lot in this paper it is best to define what I mean. I shall be referring to species as entities which contain sexually reproducing organisms. They are ontological individuals (Ghiselin, 1975; Hull, 1976) at a high level of the genealogical hierarchy. A species is an individual because it is spatio-temporally bounded (unlike a class): it has a birth when the lineage splits consequent upon a subpopulation evolving a fertilization system different from that of the parental (and all other) species. It has a duration as long as the common fertilization system of included organisms endures continuously. It dies when all included organisms have died — that is, by terminal extinction. The species individual may continue to exist in spite of phenotypic change through time, and irrespective of whether it has given birth to new species or not. Here I follow Wiley (1978), but not Bonde (1981; *see also* Hennig, 1966) who regards that an ancestral species always terminates once speciation occurs, although one of the branches after the split may be indistinguishable from the ancestral stem in the crucial reproductive system and in the general phenotype.

A further relevant point is that the units distinguished in cladistic analysis need not be species, and the branchings indicated by a cladistic analysis are not necessarily equatable with speciation (*see* Bonde, 1981). Certain models of species and speciation predict character patterns which may delimit species. Such predicted patterns, mere possible symptoms from the underlying nexus, are our only hope for hypothesizing and testing that bits of pattern in the fossil record belong to individual species. To state the obvious: the hypotheses of extinct species sprouting on phylogenetic trees, so useful in testing for macroevolutionary causes, are unfortunately heavily laden with assumptions.

Punctuated equilibria

The model of punctuated equilibria was proposed by Eldredge and Gould (1972) as an hypothesis of among-species *pattern*, resulting from systematically differential distribution of rates and modes of phenotype evolution along lineages. Punctuated equilibria postulates that evolution within species differs from that which establishes new species during lineage splitting. Relatively more phenotypic change is thought to be associated with the latter (the punctuation), and less net change to occur within species (the equilibrium). As an hypothesis of pattern it is silent on particular process, and merely insists on systematic combinations of processes with splitting and non-splitting evolution. The authors contrasted their model with the traditional one of 'phyletic gradualism': rate of phenotypic change is independent of lineage splitting and thus more evenly distributed along lineages (*see also* Gould and Eldredge, 1977; and review in Vrba, 1980).

The simple statement of punctuated equilibria is that something different must be happening during the establishment of new species than during the remaining species' lifetime. This bit of new information has far-reaching consequences for explaining macroevolutionary patterns such as trends. The interesting twist introduced into the range of macroevolutionary hypotheses is that punctuated equilibria links the divergence and diversity components (Fig. 5·1). To the extent that a punctuated pattern predominates in a phylogency, we cannot extrapolate notions of directional evolution within species to explain a divergence trend. Rather the latter must depend on variation in speciation and extinction rate and/or speciation direction. (In Fig. 5·1(d) the trend XY is accounted for by a bias in the direction of speciation events, in (h) by a net increase in species from left to right.) In contrast divergence trends evolved by 'phyletic gradualism' are not a function of speciation and extinction rates (Fig. 5.1(c), (g)). Directional selection and/or directed origin of variation at the organism level suffices to account for the trend. Thus a possibility of punctuated equilibria forces us to consider not only the potential causes of origin and sorting of variation at the level of organismal phenotypes, but also those at the among-species level. To put it simply, if punctuated equilibria predominates in the phylogenetic trees of life, then the synthetic explanation of long-term evolutionary directions is predominantly wrong.

Punctuated equilibria introduced the notion of a possible 'decoupling' of processes at the among-species level from those among organisms within populations. In so doing it broached the more general notion (it did for me and perhaps for others) of quasi-independent processes churning over at *each* level of the genealogical hierarchy — a view of life which considers a much richer range of potential causes of evolutionary pattern than does the mainstream of the synthetic tradition. This brings me to the next section on processes and hierarchy, and on the predictions of pattern which flow from particular notions of process.

PROCESSES

Hierarchy and process

Vrba and Eldredge (1984) explored some possibilities which arise if the biota is an hierarchically arranged system of individuals. At each level there is variation, and change through time in variation among individuals. The question is what causes that variation. In asking it, the focus is on one level at a time, let us call it the *focal level*, but the enquiry as to causes of that variation considers other levels as well. We introduced (*op. cit.*) a simple classification of processes responsible for variation at a focal level: (1) introduction of variation, in distinction to its sorting by (2) differential birth and (3) death rates of individuals. It is important to distinguish clearly between *de novo* introduction of variation at a particular level on the one hand, and birth process within that level on the other. For instance, a bias towards repeated independent expression of an epiphenotype is one way of affecting variation among organisms. The differential birth rates of such epiphenotypes (in a process of assimilation and thereafter) is another way of biasing variation. The first is *introduction*, the second is *sorting* of variation by differential birth rates. The repeated formation, in a species, of partly isolated groups of organisms characterized by a particular 'altruistic' allele which was fixed by random drift, *introduces* among-group variation. Any differential birth rate or extinction of such groups, once they have been introduced, is *sorting* of among-group variation, and so on.

Now one should at least be prepared to consider theoretically that the direct causes of both introduction and sorting of variation at one particular level, e.g. the level of organismal phenotypes, may be lodged *at that focal level*, or alternatively that such causes may derive *from either lower or higher levels*. The latter two principles have been christened respectively 'upward causation' and 'downward causation' by Campbell (1974).

An example of focal level causation, of deterministic birth and death processes at that level, is of course selection; whether it be among 'selfish' genomic constituents (e.g. Doolittle and Sapienza, 1980), among organismal phenotypes (the determinism of evolutionary direction regarded as paramount by most neo-Darwinists), among populations (the familiar group selection) or among species (e.g. Eldredge and Gould, 1972;

Stanley, 1979). An important and generally acknowledged point is that there is no need to speak of adaptation and selection at a particular higher level when lower level characters and processes suffice to explain sorting at the higher level. For instance there is widespread agreement (e.g. Williams, 1966; Lewontin, 1970; Maynard Smith, 1976) that if a new mutant increases fitness of organisms, then its spread in some or other groups and any resultant differential death and/or birth of such groups need not be attributed to anything other than selection of genotypes and phenotypes. That sorting of variation at each level, whether genomic, organismal, population or species, may be random has of course also been variously explored in the literature.

Upward causation in broad outline is intuitively obvious. Lower level events may affect any or all higher levels by cascading upward causation. For instance proliferation of non-coding DNA may affect the higher level of organismal phenotypes (and for that matter the levels of populations and species) merely by altering among-organism variation of the included genomic 'sum-of-the-parts' kind. But such a non-coding DNA sequence may also insert into a gene to result in altered gene function and in selection among organisms. Here the upward causation is of a more profound kind, to the introduction of characters and to sorting processes which enter *de novo*, or are peculiar to, the level of organismal phenotypes. The same distinction in possible upward causes can be exemplified at the organism–population interface. Take the case where selection among phenotypes promotes a character, such as shellbanding in snails, in some environments but not in others. Here upward causation affects only characters held by groups as sums of the characters of included organisms. Alternatively selection of organisms may promote a particular social behavior with the incidental but deterministic effect of characteristic small group size. Here upward causation from sorting among organisms may result not only in the establishment of a character that enters at the higher level of populations, but also in group selection (e.g. if group size facilitates the fixation by random drift of altruistic alleles) and perhaps in characteristic speciation rate at the still higher level of species.

Downward causation is similarly a common sense notion. One immediately apparent source of downward determination of lower level variation is the operation at higher levels of differential birth and death processes. For instance as natural selection sorts among organisms it willynilly 'downward causes' the sorting of all included lower entities, be they non-coding DNA sequences, genes, cells, chromosomes, etc. There is also a more subtle form of downward causation. The structural or organizational aspects of a particular higher individual may downward determine the introduction and sorting of variation among lower level entities included in its 'body'. For instance some developmental biologists have suggested that the phenotypic characters and processes in epigenesis are such that they bias the *introduction* of lower level variation (e.g. Oster and Alberch, 1982). In such a case the 'missing phenotypic variants' were not edited away by selection. Rather downward causation precluded their genotypic and cellular basis from arising at all. At least theoretically there may exist similar possibilities of downward causation from the levels of populations and species. To return to the previous hypothetical example of group selection: the structural group character of group size may downward cause the sorting of variation among included phenotypes (e.g. the spread of altruistic phenotypes by random drift, against the force of natural selection, which requires small population size; *see* Maynard Smith, 1976).

Such an expanded notion of genealogical hierarchy and multiple within-level and inter-level causation is simply missing from neo-Darwinian explanations. The mainstream of the synthesis addresses two kinds of basic processes occurring within populations as of pre-eminent importance in evolution. (1) The introduction of genic and phenotypic variation is held to come from random gene or chromosome mutations and recombination. (2) Such variation is seen as sorted mainly by the deterministic process of natural selection or by the random one of genetic drift. This system of input and sorting at the levels of genes and individual organisms is widely regarded by neo-Darwinists to account not only for the evolution of populations but also, by extrapolated small-step change, for the evolution of species and of long-term patterns among species. Thus the inter-level causations on which the synthesis has overwhelmingly concentrated is upward causation from genotypic mutation to sorting among phenotypes in populations, and the crude version of downward causation from selection among organisms as it affects the distribution of elements included in organisms.

Intriguing new possibilities of upward and downward causation involving the genome have been raised during the last decade (*see contributions in* Dover and Flavell, 1982). The neo-Darwinian solution on how genotypic and phenotypic novelty originates and evolves in populations, the very level exhaustively addressed by the synthesis, has long been questioned among developmental biologists. Recent publications from this quarter argue anew that intrinsic factors, arising from the organization of the epigenetic system, may be more important in determining directional evolution than admitted by the synthesis (e.g. Riedl, 1978; Rachootin and Thomson, 1981; Løvtrup, 1982; Oster and Alberch, 1982; Ho and Saunders, 1982; Zuckerkandl, 1983). The preoccupation of neo-Darwinians with population evolution, and its extrapolation across the 'evolutionary board', has resulted in an almost total neglect of a range of hypotheses on why there should be differential species diversification. Species richness in ecosystems has received much attention, and is often explained in terms of selection and adaptation of organisms. But differential net rates of increase in numbers of species *in monophyletic groups* has received almost no consideration. It is this aspect on which I shall especially focus in the ensuing discussions of alternative hypotheses of macroevolution and of their predictions for phenotypic patterns.

Some hypotheses of process and predictions for pattern

I am particularly addressing process and pattern in monophyletic groups of sexually reproducing organisms. The very fact of sexual reproduction immediately suggests some predictions for macroevolutionary patterns. Sex is a complex system of combinatorial information. A built-in characteristic of sexual systems is that particular divergent phenotypes will be unable to mate with the parental ones. Thus in a sense, the diversification of species is a 'built-in necessity' of sexual systems. By this I am not saying that there is *selection for speciation*, merely that speciation is an expected secondary consequence of the presence of sex and of evolution at the organism level. Apart from the prediction that biparental lineages should speciate there is a simple further prediction that can be made without much strain: that related lineages in a monophyletic group should intrinsically differ in characteristic probability of speciation. In other words macroevolutionary patterns of differential species diversity may result incidentally by upward causation from lower levels, intrinsically determined from the characters of organisms and their genomes (the effect hypothesis of macroevolution; *see below and* Vrba, 1980, 1983).

Particular concepts of species and speciation entail particular predictions for pattern. Let us take as an example the model of Paterson (1978, 1982a). He argues in detail that a biparental species should be defined as comprising organisms which *recognize* each other for the purpose of mating and fertilization (the Recognition Concept), rather than being defined in terms of reproductive *isolation* (the Isolation Concept widely accepted within the synthetic paradigm). He further argues that (1) divergence of the Specific-mate Recognition System (SMRS) in an isolated daughter population is crucial to speciation (namely that the evolution of post-mating isolation is unlikely to result in the establishment of new species); that (2) reproductive isolation and speciation result as mere incidental *effects* (*sensu* Williams, 1966) and can in no sense be seen as selected for or adaptive *per se*; and that (3) the SMRS even more so than other phenotypic characters should be closely adapted to the species' preferred habitat, and subject to strong stabilizing selection.

Paterson's model entails a number of predictions for process and patterns, some of which are testable even in the fossil record. Many biologists have noted that the biological species concept, i.e. the Isolation Concept (e.g. Mayr, 1969, p.26 'Species are groups of interbreeding natural populations reproductively isolated from other such groups.'), presents operational difficulties in evaluating whether allopatric populations belong to the same species or not. Under the Recognition Concept, provided one succeeds in identifying the crucial characters of the SMRS, one can test an hypothesis that allopatric populations belong to one species.

Furthermore Paterson's concept (1978, 1982a) predicts a pattern of punctuated equilibria (he predicted it independently, Paterson and James, 1973, and by personal communication of the articulation by Eldredge and Gould, 1972). Indeed, there is the expectation of particularly narrow variances in SMRS characters in space and through time, relative to other phenotypic characters in a species. Insofar as the skeletons of extant survivors are found to include characters of the SMRS, this hypothesis is testable by examination of the fossil record of that monophyletic group. Paterson's model explicitly suggests that equilibrium should result from stablizing selection and not as a function of gene flow—again testable in monophyletic groups with a good fossil record and extant survivors (discussed below). In such cases where mating communication is primarily visual a breakdown of communication between populations, i.e. speciation, should involve a shift in morphology. In such groups, rates of morphological change and speciation, under the punctuation model, should be positively correlated. Even relatively closely related species of visual 'communicators' would be expected to be morphologically distinct. This explanation may be appropriate for the morphological differentiation between chimpanzee and man (King and Wilson, 1975) and other higher primates, in speciose groups of *Haplochromis* (Fryer and Iles, 1969) and the homoallozymic *Drosophila heteroneura* and *silvestris* (Carson, 1976). Conversely, uniform species' flocks (for example minnows, Avise and Ayala, 1976, and some fruitflies) may have an SMRS dominated by olfactory, auditory and/or other non-visual signals. The observation that frogs, with their bias towards auditory communication, include many sibling species would support this argument. Among mammals particularly numerous sibling species have been recorded for rodents. Distinctive auditory and behavioral SMRSs have been reported for morphologically identical rodent species (Gorden and Dennet, 1979). The implication of this for the palaeontologist is obvious: in monophyletic groups including extant species whose visual communication involves skeletal characters, the probability of successfully estimating species and species diversity patterns in the fossil record may be quite good. In contrast it is hopeless in the case of groups of species which are sibling in terms of their hard parts. Finally the Recognition Concept predicts that speciation in total allopatry should be ubiquitous; and that any distribution of variation, such as might be expected to result from speciation by reinforcement or from sympatric speciation, should not be found in past and present life patterns (Paterson, 1978, 1982a).

This example illustrates how particular notions of microevolution can predict pattern and thus be tested, even via the fossil record. The problem is that often several rival models of microevolution make precisely the same predictions for aspects of pattern. For instance, it is amazing how many different notions, from different subdisciplines, predict a pattern of punctuated equilibria. Thus ontogenetic macromutations, random drift in small populations, or just directional selection of organisms in isolated populations have been discussed by various authors as important in promoting rapid change which results in speciation; while ontogenetic constraint, co-adapted genotypes, species-level homeostatic properties, high gene flow among populations, or stabilizing selection of phenotypes feature among suggestions of how species equilibrium may come about. Most recently a phenotypic pattern of punctuated equilibria has been predicted from studies of the dynamics of genome evolution (Dover et al., 1982). So we are faced with several possibilities even if we do manage to test for an hypothesis of punctuated equilibria and find it corroborated. Among patterns which would support the model of punctuated equilibria the following may be briefly mentioned (for more expanded arguments see Vrba, 1980; 1983(a))

(1) the demonstration of specific identity across barriers which can be documented to have separated the subspecific populations for millions of years (e.g. the Panama land bridge; see Vrba, 1980);
(2) sudden appearances and subsequent persistence of complex morphologies in a well-sampled (both temporally and geographically) fossil record;
(3) the association of high speciation rates with high phenotypic divergence between lineage endpoints, and vice versa (as in Fig. 5·1(h) because punctuated equilibria suggests that they are functionally linked);
(4) consistent correlation of the first appearance of new morphologies with environmental changes (e.g. in the Miocene-Recent African fossil record see below);
(5) large-scale contraction and expansion of distribution ranges of putative species in the fossil record (as under a punctuated model environmental changes are not expected to be tracked to any appreciable extent);

(6) the demonstration of phenotypic equilibrium *within* rapidly diver-
sifying monophyletic groups.

One must note here that all these tests are weak. Outcomes which do or
do not seem to support punctuated equilibria are usually open to alternative
interpretations in each case.

Causes of macroevolution

Some currently debated hypotheses of macroevolution were classified and
discussed at some length in Vrba and Eldredge (1984). A shorter version of
some of the discussion from that paper is reproduced here. Table 5·1 (after
op. cit.: Table 3) is a very rough scheme. These are not the only hypotheses
that can or have been proposed, and this is obviously not the only way to
classify them. Broadly speaking they apply to all kinds of
macroevolutionary patterns, although trends will feature more prominently
in the ensuing discussion than other kinds of pattern. The proponents of
such models all acknowledge that more than one of the factors here
classified may be acting simultaneously to produce a given macroevolutionary pattern.

One basic distinction is whether the highest level, at which the definitive
processes responsible for long-term evolutionary direction are seen to
occur, is that of organisms (Table 5·1 A, B) or the level of species (Table 5·1
C-F). The four models which do regard differential introduction or sorting
of variation among species as crucial were all originally discussed within a
framework of punctuated equilibria. The causes of differential patterns of
species diversity *per se* were hardly ever explicitly hypothesized in the neo-
Darwinian tradition. As a palaeontologist I am amazed at the vista of
systematically correlated speciation and extinction rates, and of significant
differentials in diversity patterns between and within phylogenies (cf.
Stanley, 1979; *and below*), and even more amazed that so few attempts at
explicitly explaining them have been made. In articulating the notion of
punctuated equilibria Eldredge and Gould (1972) broached the notion of
independant sorting among species, of hierarchy above the level of
organisms in populations.

Another distinction between the hypotheses in Table 5·1 is whether
intrinsic factors are seen as prime determinants of long-term evolutionary
directions (Table 5·1: A, C, D) versus random (E) or selective sorting (B, F)
at the focal level. Some of the models imply any trend direction *per se* to be
adaptive (B, F) while others do not. I conclude this section with a brief run-
through of these models (after Vrba and Eldredge, 1984).

TABLE 5.1 Classification of some hypotheses of macroevolution (after Table 3 in Vrba and Eldrege, 1983). The hypotheses directional phenotypic change in monophyletic groups. They vary according to whether they stress: (a) the organism versus the species level as the highest focal level; (b) introduction versus sorting of variation at that focal level; (c) causation from lower levels versus at the focal level. Numbers in brackets refer to some authors who have either formulated the hypothesis, or explicitly explored it. (1) Riedl, 1978; (2) Zuckerkandl, 1983; (3) Simpson, 1953; (4) Dobzhansky *et al.*, 1977; (5) Bock, 1979; (6) Ho and Saunders, 1982; (7) Stanley, 1979; (8) Thomson and Rachootin, 1981; (9) Alberch, 1980; (10) Vrba, 1980; (11) Raup *et al.*, 1973; (12) Eldredge and Gould, 1972.

		Source level of cause of process	
Focal level	Process	From lower levels	At focal level
Organism	Introduction	A. Directed organismal phenotype (1) (2)	
	Sorting by birth/death		B. Synthetic (3)(4) (5)
Species	Introduction	C. Directed speciation (6) (7) (8) (9)	
	Sorting by birth/death	D. Effect (10)	E. Random (11) F. Species selection (12) (7)

The hypothesis that organismal variation is intrinsically directed

Riedl (1978), Zuckerkandl (1983) and others have discussed organizational features of genetic and epigenetic system which may be expected to direct intrinsically the introduction of phenotypic variation at the organism level. Thus parallel evolution and trends, in this view, are largely intrinsic to genomic and epigenetic organization and function. The cited authors seem to be thinking of directional changes in lineages which are more or less gradually evolving and not necessarily branching. Hierarchical organization within organisms is stressed, but upward causation to among-species sorting is not discussed. In this respect only, the hypothesis resembles the synthetic one, just as the two kinds of patterns in the fossil record may look similar as far as distribution of rates of change along lineages is concerned. The suggestions of others that phenotypic evolution has a large intrinsic component will be mentioned below under the hypothesis of directed speciation.

The synthetic hypothesis

Recombination and mutations of genes and chromosomes provide the input of new variation at the organism level. Sorting among organismal phenotypes by natural selection is the main force of evolution. Many other kinds of processes are explored and acknowledged as possibly occurring (e.g. Dobzhansky *et al.*, 1977). The most important point about a synthetic approach to long-term evolution seems to be this: no matter how many diverse processes are recognized to occur within species and during speciation, the directions of eventual species evolution *and* transpecific evolution are seen as determined overwhelmingly by intrapopulation selection of small-step variations. Thus in this view 'macroevolution' is nothing more than extrapolated, additive microevolution.

The hypothesis of directed speciation

The hierarchical organization of organisms places constraints on the introduction of phenotypic variation, and has been suggested to be responsible for canalization and the presence of unexpressed stores of epigenes. From such considerations one could postulate, (1) long-enduring evolutionary conservatism at the phenotype level (species equilibrium); but (2) that, when changes occur, they may come in relatively large 'parcels' and be strongly directed. This is precisely what some biologists have suggested (e.g. Rachootin and Thomson, 1981; Alberch, 1980; Oster and Alberch, 1982).

The effect hypothesis of macroevolution

The basic statement (Vrba, 1980, 1983) is this: differences between lineages in characters of organisms and their genomes, may incidentally determine a pattern of among-species evolution in a monophyletic group. Processes at the among-organism level (directly by upward causation) cause speciation, characteristic speciation rate (S) and species extinction rate (E) in a lineage. Thus any characteristically different probabilities of S and E in different subclades of a monophyletic group will lead to a trend towards higher intrinsic rate of species increase (R; R = S–E, as in Stanley, 1975, 1979). If a pattern of punctuated equilibria predominates in such a phylogeny, then a trend in directional phenotypic evolution will result as well, which may not be adaptive (Vrba, 1980, 1983; *see* Fig. 5·1(h), page 117). The effect hypothesis embodies a controversial new idea: that life's diversity patterns, and through them some long-term directional tendencies, may result as incidental, non-adaptive consequences. Why the term effect hypothesis? Many agree with Williams (1966) that the adaptations of organisms, shaped by natural selection to perform particular functions, may have incidental *effects* which are not the direct consequence of natural selection. It has been explicitly argued (Paterson, 1978, 1982a) and is implicitly widely accepted (e.g. Charlesworth *et al.*, 1982) that speciation usually results as an incidental consequence of the accumulation of genotypic and phenotypic differences between populations. Such species *per se* are not adaptations (although the component organisms may be adapted). Rather speciation and species can be said to result as *effects*. For instance, where selection causes phenotypic divergence of an isolated population to result in allopatric speciation (*suggested by many e.g.* Paterson, 1978; Futuyma and Mayer, 1980, *to be the usual case*), selection acted for adaptations that promote reproductive success, certainly not *for* isolation or speciation. Relative to the selection regime speciation is an effect. I extend the concept to suggest that speciation may also be effected by processes containing large non-selective components. Certain characters of genomes and organisms may confer on their lineages a characteristic probability of speciation. If organisms in related species and lineages differ in such characters, then differential S will result. Combinations of S and E may vary across a monophyletic group to cause trends towards higher R.

Which characters of genomes and organisms could possibly effect differential R and trends? Suggestions of potential candidates, adaptations or other, abound in the literature. Some focus on differential, intrinsic susceptibility to the introduction of new genomic and phenotypic variation (e.g. Dover *et al.*, 1982; Rachootin and Thomson, 1981; Zuckerkandl, 1983). Others stress different susceptibilities of organisms in related species to directional versus stabilizing selection. For instance the breadth of resource utilization of organisms may be expected to confer both characteristic S and E on the species and lineages they occur in. A correlation between S and E and organismic adaptation for resource use has been noted by many (e.g. Stebbins, 1950; Rensch, 1959; Eldredge, 1979b). I have suggested (Vrba, 1980; 1983a) that features which confer breadth of resource use may be *characters of organisms*, with incidental but deterministic effects which translate to patterns of *among-species evolution*. Organisms whose adaptations allow use of resources in alternative environments (such as a lineage may encounter through time) are subject to less directional selection, their lineages to lower S and E. Specialist organisms, whose resources disappear as environments change, should more often be subject to directional selection pressure, range fragmentation, and their lineages to high S and E.

In its most general form the effect hypothesis does not require a pattern of punctuated equilibria: differential R may occur as an effect in phylogenies in which rate of phenotypic change is independent of lineage splitting. To the extent that a punctuated pattern predominates, long-term phenotype divergence may result as an effect (*see* Vrba, 1980: e.g. Fig. 23). In the original statement (Vrba, 1980), Wright's (1967) suggestion that the direction of speciation events may be random with respect to trend direction was used. But this is not a necessary condition, as effect evolution may superimpose on biases in speciation direction.

The random hypothesis was proposed by Raup et al. (1973)

It adheres to notions of punctuated equilibria and species as individuals in spirit. Briefly, the authors show that trends can be simulated if both direction and frequency of speciation events, and frequency of extinction, are allowed to vary randomly. It is obvious that this possibility needs to be tested, in assessing causes of among-species (as of among-organism) evolution, and discounted before proposing deterministic explanations of pattern.

The hypothesis of species selection

The several long discussions of this model which have appeared to date (Eldredge and Gould, 1972; Stanley, 1975; Gould and Eldredge, 1977; Stanley, 1979; Eldredge and Cracraft, 1980; Gould, 1982) all agree on four basic points: (1) monophyletic groups in which species selection operates are assumed to show a predominant pattern of punctuated equilibria; (2) directions of speciation events which are random in relation to any long-term trend among species (see Wright, 1967); (3) species are seen as individuals which have births, durations and deaths and which give birth to other species; (4) species selection is deterministic sorting among species individuals, a sorting which is irreducible to sorting among organisms within species. My own retrospective view of what this model entails in addition to requirements 1–4 is this: species selection is deterministic sorting among species by differential birth and/or death which involves causation *at the focal level of species* (Table 5·1). This means that species should vary in characters that enter at the species level (i.e. are more than simple sums of characters of organisms); that such species characters are the cause of species sorting; and that characters and processes at the lower levels of genotypes and phenotypes should *not* be sufficient on their own to explain among-species pattern.

The units of selection in this model are explicitly stated to be the species. Hence species selected macroevolution has variously been claimed to be irreducible to, or 'decoupled' from microevolution (e.g. Stanley, 1979). But how does this unambiguous identification of an irreducible process of selection at the species level relate to character variation and adaptation? On this topic all arguments of species selection up to 1980 were vague. Thus, if one argues that equilibrium may result as a consequence of stabilizing selection, the units of selection are alleles and organisms. But a view of species as 'homeostatic systems [which] resist change by self-regulation' (Eldredge and Gould, 1972) implies species adaptation. Similarly Stanley (1979) emphasizes the gene flow characteristics of species as *mechanisms* which determine S and E. To put it simply: a separation of characters in terms of the levels at which they enter, and the possibility of simple upward causation from character variation and processes among organisms to species sorting, was simply not addressed by species selectionists.

Both the hypothesis of effect macroevolution and that of species selection were originally developed in the framework of assumptions 1–4 above. Throughout the several long discussions of species selection which appeared up to 1980 differential *survival* (or E) is stressed as of paramount importance in producing trends. In contrast the initial statement of the effect hypothesis concentrated on trends driven by intrinsically different S. However the crux of the distinction does not lie in whether E or S is focussed on as determining trends; but in the level(s) at which adaptation and selection are identified. Effect macroevolution is explicitly a model which accounts for species characters and sorting among species by direct upward causation from combinations of phenomena at lower levels.

In sum, there is a theoretical separation of two processes (*see* Vrba, 1983): one in which characters which enter at the level of species result in their selection, called species selection; the other in which Darwinian selection of organisms and other lower level processes simply determine among-species patterns, called effect macroevolution.

SOME EVIDENCE FROM AFRICAN MAMMALS

Many mammals include a strong visual component in their communication systems. This is particularly true of the Bovidae, i.e. antelopes (e.g. Walther, 1974), to which I will mainly refer. In particular I have suggested (Vrba, 1983a) that certain aspects of horn and facial bone structure are important in specific-mate recognition in some antelope monophyletic groups, and may be expected to vary between bovid species but not within. This expectation is borne out by evidence on how cranial variation distributes within and between extant taxa which are recognized as species on other grounds. The study of several groups of extant Bovidae indicates that closely related species are distinguishable even on cranial fragments and certainly on horns—in other words that there are no sibling species among the modern survivors. All this lends support to hypotheses of species recognition in the relevant fossil patterns, and to hypothetical phylogenetic trees like that in Fig. 5·2 (based on cladistic analysis, Vrba, 1979). This tree is plotted as accurately as possible against the time axis (the horizontal axis is blank and does *not* represent phenotype variation). At the top of Fig. 5·2 are sketches drawn to scale of some of the diverse horn and face morphologies within this monophyletic group.

The Pliocene-Recent African fossil record is relatively well sampled, especially in the African savanna areas, and includes several well dated stratigraphic successions. It is thus reasonable, on the basis of these data, to venture tentative comments on the distribution of rate of phenotypic change

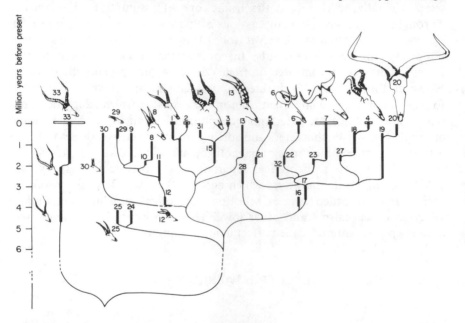

FIG. 5·2 Durations and phylogenetic tree of the antelope sister-group Alcelaphini (species 1–32, including extant hartebeests, wildebeests, blesbuck) — Aepycerotini (species 33, the extant impala *Aepyceros melampus*). Solid lines give estimated durations of species. Numbers refer to species names in Table 1, Vrba (1983a). Horizontal bars drawn at 0 million years represent estimates of relative abundance of extant species. Sources for duration dates as well as cladistic arguments, are given in Vrba (1979).

along lineages in relation to lineage splitting, i.e. speciation events. Extensive details for the phylogeny in Fig. 5·2 are given in Vrba (1983a). In a nutshell, many monophyletic groups of Bovidae (e.g. Fig. 5·2) and of other Miocene-Recent mammals, show (1) a strong positive correlation between morphological divergence along lineages and net rate of increase in species; (2) an equally strong positive correlation between speciation and extinction rates (*see also similar evidence for diverse groups of organisms in* Stanley, 1979); and (3) synchronous waves of first records, 'sudden appearances' of distinct morphologies which subsequently persist for relatively long periods of time. There are particularly many such first records of species of Bovidae and other large mammals near 2 million years ago (Vrba, 1983b). In fact, close to 2 million years ago there occurred a peak, unprecedented in African bovid evolution, in both extinction and origination of species in all African bovid groups. Figure 5·3 shows that the peak in first records of Bovidae coincides with the earliest records of the genus *Homo* and of the 'robust' australopithecines, as it does with earliest

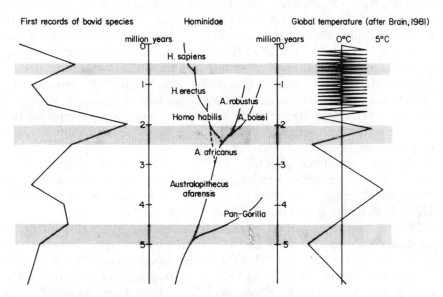

FIG. 5·3 Rates of origination of all species of Bovidae known from subsaharan Africa, Miocene-Recent, and of Hominidae in relation to the timing of global temperature fluctuation (temperature data from Brain, 1981; 1983). Stippled bands represent approximate timing of major evolutionary events in bovid and hominid radiation. See Vrba (1983b) for data on which first records of Bovidae are based.

records of hypothesized species in many other mammalian phylogenies (*see contributions in* Maglio and Cooke, 1978). The fact that numerous African fossil assemblages span the 3–1 million year period, notably from the Omo in Ethiopia, East Turkana in Kenya and the Transvaal cave breccia successions in South Africa, gives special significance to such sudden appearances.

Furthermore it is of interest that global temperature fluctuations can be shown to have predictably affected rainfall and vegetation changes in many parts of Africa (review in Brain, 1981; 1983) and to be correlated with waves of speciations and extinctions (Vrba, 1983a,b). The evidence suggests that the evolutionary events were caused by the climatic events. For instance during the end of the Miocene when temperatures plunged to a low point unprecedented during the entire preceding Tertiary, a number of mammalian monophyletic groups of relatively high taxonomic rank appeared for the first time. Among Bovidae these include the essentially African tribes Alcelaphini, Aepycerotini (see Fig. 5·2), Reduncini, Hippotragini and Tragelaphini. Thus a spectacular bovid radiation into new 'adaptive zones' occurred at the same time as the origin of other groups including the Hominidae. Around 2 to 3 million years ago, when temperatures again plunge to a lowpoint (at 2,6 my corresponding to the first formation of the Arctic Icecap, Brain, 1981), bovid diversification and extinction both rise dramatically. Again bovid evolutionary activity correlates with a major event in the hominid and other phylogenies. A part of this evidence for coincident climatic and mammalian evolutionary events, and for a probable causal relationship between them, is shown in Fig. 5·3. The African stratigraphic successions certainly offer evidence for the contraction and expansion of distribution ranges of species and lineages in predictable synchrony with temperature and environmental changes. Temperature correlated gaps in a species distribution, in the record of a single relatively complete stratigraphic succession, can be documented in many bovid groups, in other African mammals like primates (Vrba, 1983a, b) as well as in invertebrates (e.g. Coope, 1979).

On the whole the data I have studied seem to support punctuated equilibria if it is moderately stated as a *predominant* pattern of distribution of differential rate of morphological change along lineages in association with splitting versus non-splitting evolution. The patterns I have studied are not rigidly 'rectangular' or 'catastrophic'. They are quite compatible with an hypothesis of strong directional selection, enduring generation by generation for tens of thousands of years or longer, responsible for 'punctuation', and stabilizing selection as a cause of equilibrium. There is also evidence of gradual evolution in non-splitting lineages, i.e. gradual average modification through time. But the gradual average divergence seems to occur within species, generally slow enough to be what one may call 'virtual equilibrium' (Vrba, 1980; for example Fig. 2: species 33, the impala, as well as several alcelaphine species of longer duration). I certainly do not claim that I have adequately tested and supported punctuated equilibria. It is obvious that more precise testing, via quantification of rate of change in particular characters, is required.

I have found no support for the much cited notion of gene flow as the preserver of species integrity (e.g. Mayr, 1963; Stanley, 1979). Take the example of the monophyletic group in Fig. 5·2. The more mobile migratory Alcelaphini, which today potentially cover large distances annually, have split more rapidly. The impala lineage, for which good evidence exists of low gene flow between populations today and probably also in the past has not (Vrba, 1983a). Thus the data support others which suggest that not gene flow, but another factor such as stabilizing selection (e.g. Ehrlich and Raven, 1969; Charlesworth *et al.*, 1982) or ontogenetic determinism (e.g. Riedl, 1978; Alberch, 1980) may be the primary cause of stasis.

I have argued above (*see also* Vrba, 1980; 1983) that effect macroevolution may result if the organisms within different species of a monophyletic group vary along a eurytopy-stenotopy axis (i.e. an axis representing generalist–specialist resource utilization). To test the hypothesis of the influence of resource utilization on speciation and extinction rates I compared speciation and extinction rates with feeding behavior of extant survivors of monophyletic mammalian groups (Fig.5·4). I found speciation and extinction rates positively correlated; the correlation between breadth of feeding behaviour and evolutionary rate (Fig. 5·4) was also as predicted in the hypothesis.

My results caution against a facile concept linking generalists with low rates, specialists with high ones. Instead the degree of *independence of the feeding niche* of such *alternative environments* as a lineage encounters through time appears to be of paramount importance. The data I analyzed accorded well with an hypothesis that climate/vegetation oscillations provided recurrent alternative environments during Miocene-Recent African large mammal evolution. Herbivores were all predictably affected. Pure grazers (e.g. Alcelaphini in Fig. 5·2) or browsers evolved more rapidly than related clades which both graze and browse (e.g. *Aepyceros* in Fig. 5·2). The analysis suggests that *distribution of resource patches* in alternative environments is of primary importance. Thus the exclusively myrmecophagous specialist lineage of the extant aardvark, *Orycteropus*, today finds its feeding patches in a large range of vegetational physiognomies, has a trans-African distribution, and a fossil record indicating very slow evolution. A similar observation can be made in relation to carnivore specialists, and substrate specialists like *Pedetes*, the springhare (specialized on sandy substrates along talus slopes, M. Coe, personal communication): Their resource patches are distributed across vegetational physiognomies, and so are they, while their evolutionary rates are low.

A subject of basic interest is how one might use monophyletic patterns including extinct and extant species to test whether directional evolution in lineages is mainly caused by intrinsic directionality of introduced variation (e.g. models A, C in Table 5·1), versus whether it is predominantly accounted for by natural selection among randomly introduced variants. I suggest that such testing might be possible via cladistic analysis. In the groups I have studied, parallel evolution of certain characters in different limbs of the same phylogeny seems to be rampant (as noted by systematists

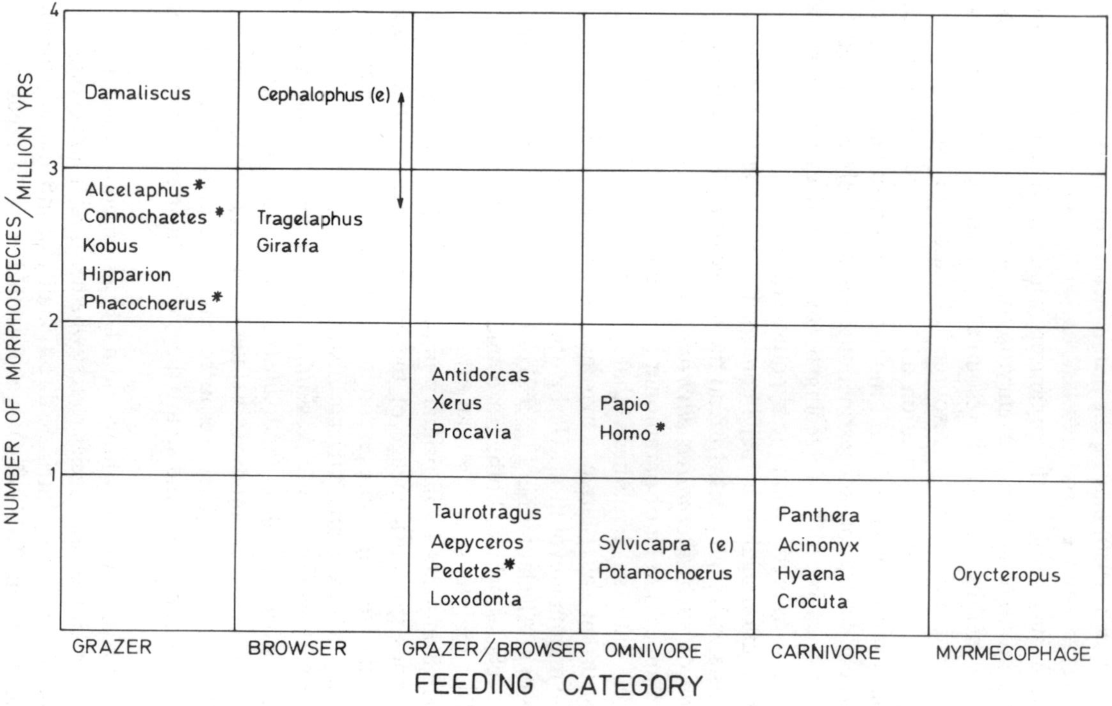

FIG. 5·4 Speciation rate in some monophyletic mammal groups, contrasted with feeding behaviour of extant survivors. Each generic name marked with an asterisk refers to a monophyletic group larger than, and including the stated genus; thus *Homo** refers to the Hominidae. The *Hipparion* data refer only to an African monophyletic subgroup of that genus. The letter (e) connotes estimates based on extant diversity, the fossil records in these cases being poorly known.

working on diverse other groups of organisms); and predictably associated with similar habitat and function. It is clear that the *kind* of parallelism, occurring in a particular part of the phenotype in response to similar environments, is very much 'phylogeny-dependant.' For instance, many mammalian herbivore clades exhibit recurrent responses to hard vegetation diets in xeric environments. Thus is some clades the evolution of thickened tooth enamel tends to recur in parallel (e.g. primates), in others extreme hypsodonty is evolved again and again on separated limbs of the same monophyletic group (e.g. within perissodactyls and artiodactyls). Among grazing clades some exhibit repeated parallel reduction of premolars (e.g. many bovid groups including the Alcelaphini in Fig. 5·2), others premolar expansion and molarization (e.g. Perissodactyla). The cladistic distribution of many features thus strongly suggests intrinsic determinism of the variation that is expressed in the phenotype and made available for any subsequent selection. This subject will not be further pursued here, but one can note this: the erection of hypotheses and predictions for fossil patterns, in terms of the distribution of variation within and between monophyletic groups, to test for intrinsic directionality versus selection, should surely be a high priority in macroevolutionary study.

To end this section I want to return briefly to rival hypotheses of why there should be differential net rates of increase in species. I have presented some support from patterns of African mammal lineages for the hypothesis of effect macroevolution (Vrba, 1980, 1983, 1983 a, b). *Characters of organisms* which confer breadth of resource use do seem to correlate with, and therefore perhaps incidentally but deterministically effect, differential patterns of *species diversity* (Figs 5·2, 5·4). I have also tested for possible instances of species selection (*op. cit.*). Thus one could perhaps argue that characteristic population size and separation, and gene flow frequency between populations, amount to characters which enter at the species level (although originally produced via sorting at the level of organismal phenotypes). Insofar as such species variation is the cause of differential speciation and extinction rates, one could make a case for species selection. But, as mentioned above, I see no evidence for this in the data I have examined. Similarly one could argue that eurytopy, resulting from *between-phenotype niche width* (that is from among–organisms, genetically based differences in resource utilization within a species; *see* Roughgarden, 1972; and Van Valen, 1965), may result in a process of differential species sorting that is validly terms species selection. But in all cases I have looked at in detail niche width is a *within-phenotype* phenomenon. For instance each 'impala organism' in the lineage of low S, E and R (Fig. 5·2: species 33) is a herbivore eurytope (decisively demonstrated by many separate studies; see Vrba, 1983a), while each extant alcelaphine organism in the speciose sister-group is a grazing specialist. Each species only bears the relevant character in the sum-of-the-parts sense. It seems to me that species selection is an onerous concept. I have difficulty in circumscribing what a set of emergent species characters might include (*see also* Vrba and Eldredge, 1984). From the perspective of the evidence I have examined, I see no need to postulate that species selection has occurred.

CONCLUSION

In this paper the biota is viewed as a genealogical hierarchy of nested sets of individuals, genomic constituents to species in monophyletic groups (*see also* Vrba and Eldredge, 1984). At each level the individuals are subject to processes of birth and death, of introduction and sorting of variation. A review of hypotheses of macroevolution shows that fundamentally different explanations of evolutionary patterns result depending on which processes of introduction and sorting at particular levels are stressed. The synthesis has traditionally addressed two kinds of basic processes occurring within populations as of preeminent importance in evolution. (1) The introduction of genic and phenotypic variation is held to come from random gene and chromosome mutations and recombination. (2) Such variation is seen as sorted mainly by the deterministic process of natural selection or by the random one of genetic drift. As a consequence patterns of variation, from those among genes to trends among species, are generally interpreted in the adaptive mode. An hierarchical approach includes wider possibilities of causation within additional levels, as well as upward and downward causation between levels; and an acknowledgement that many patterns of variation may not be well described as adaptive.

I have argued that sexual reproduction itself, particular hypotheses of species, speciation, microevolution and of processes of introduction and sorting at various levels of the genealogical hierarchy, may lead to predictions for phylogenetic patterns. Thus diverse hypotheses of micro- and macro-evolution can be tested by examining character variation in time and space, in the fossil record and among extant organisms. Patterns in the fossil record allow only an extremely restricted view, 'through a glass darkly', of phenotypic variation. Nonetheless they afford opportunities for tests that are complementary to those on extant organisms. Indeed, in attempting to falsify or corroborate certain hypotheses of evolutionary rate distribution and species diversification, fossil patterns are all we have to proceed with. I illustrated such arguments with evidence from Miocene-Recent mammalian evolution in Africa. I conclude that most of these data support, some less and some more strongly, the notion of sexually reproducing species as individuals subject to processes that are more than smooth, additive extrapolations from evolution within populations. It seems to me, as it does to others from diverse biological subdisciplines, that many patterns within and among species, including species *per se*, are not appropriately interpreted in the adaptive mode. My main conclusion is that concepts additional to those of the synthesis (of potential processes of introduction and sorting of variation at all levels of the genealogical hierarchy) be incorporated and tested, alongside of what we know of population evolution, in a more complete evolutionary theory.

ACKNOWLEDGEMENTS

I thank Niles Eldredge for allowing me to repeat here some of the discussion from our unpublished manuscript (Vrba and Eldredge, 1984).

REFERENCES

Alberch, P. (1980). Ontogenesis and morphological diversification. *Am. Zool.* **20**, 653–667.

Arnold, A. J. and Fristrup, K. (1982). The theory of evolution by natural selection: an hierarchical expansion. *Palaeobiology* **8**, 113–129.

Avise, J. C. and Ayala, F. J. (1976). Genetic differentiation in speciose versus depauperate phylads: evidence from the California minnows. *Evolution* **30**, 46–58.

Bock, W. J. (1979). The synthetic explanation of macroevolutionary change—a reductionist approach. *Bull. Carn. Mus. Nat. Hist.* **13**, 20–69.

Bonde, N. (1981). Problems of species concepts in paleontology. *Inter. Symp. Concpt. Meth. Paleo. Barcelona*, 19–34.

Brain, C. K. (1981). Hominid evolution and climate change. *S. Afr. J. Sci.* **77**, 104–105.

Brain, C. K. (1983). The evolution of man in Africa: was it a consequence of Cainozoic cooling? 17th Alex. L. du Toit Memorial Lecture. *Geol. Soc. S. Afr.* **84**, 1–19.

Campbell, T. (1974). 'Downward causation' in hierarchically organized biological systems. *In* 'Studies in the Philosophy of Biology' (F. J. Ayala and T. Dobzhansky, *Eds*), pp.179–186. University of California Press, San Francisco, USA.

Carson, H. L. (1976). Inference on the time of origin of some *Drosophila* species. *Nature* **259**, 395–396.

Charlesworth, B., Lande, R. and Slatkin, M. (1982). A neo-Darwinian commentary on macroevolution. *Evolution* **36**, 474–498.

Coope, C. R. (1979). Late Cenozoic fossil Coleoptera; evolution, biogeography, ecology. *Annu. Rev. Ecol. Syst.* **10**, 247–267.

Dobzhansky, T. (1951). Mendelian populations and their evolution. *In* 'Genetics in the 20th Century' (L. C. Dunn, *Ed.*), pp.573–589. Macmillan, New York, USA.

Dobzhansky, T. (1976). Organismic and molecular aspects of species formation. *In* 'Molecular Evolution' (F. J. Ayala, *Ed.*), pp.95–105. Sinauer Associates, Sunderland, Massachusetts, USA.

Dobzhansky, T., Ayala, F. J., Stebbins, R. L. and Valentine, J. W. (1977). 'Evolution'. W. H. Freeman, San Francisco, USA.

Doolittle, W. F. and Sapienza, C. (1980). Selfish genes, the phenotype paradigm and genome evolution. *Nature* **284**, 601–603.

Dover, G. A., Brown, S., Coen, E., Dallas, J., Strachan, T. and Trick, M. (1982). The dynamics of genome evolution and species differentiation. *In* 'Genome Evolution' (G. A. Dover and R. B. Flavell, *Eds*), pp.343–372. Academic Press, London, UK.

Dover, G. A. and Flavell, R. B. *Eds* (1982). 'Genome Evolution'. Academic Press, London, UK.

Ehrlich, P. R. and Raven, P. H. (1969). Differentiation of populations. *Science* **165**, 1228–1232.

Eldredge, N. (1979a). Cladism and common sense. *In* 'Phylogenetic Analysis and Paleontology' (J. Cracraft and N. Eldredge, *Eds*), pp.165–198. Columbia University Press, New York, USA.

Eldredge, N. (1979b). Alternative approaches to evolutionary theory. *Bull. Carn. Mus. Nat. Hist.* **13**, 7–19.

Eldredge, N. and Cracraft, J., *Eds* (1980). 'Phylogenetic Patterns and the Evolutionary Process'. Columbia University Press, New York, USA.

Eldredge, N. and Gould, S. J. (1972). Punctuated equilibria: an alternative to phyletic gradualism. *In* 'Models in Paleobiology' (T. J. M. Schopf, *Ed.*), pp.82–115. W. H. Freeman, San Francisco, USA.

Fryer, G. and Iles, T. D. (1969). Alternative routes to evolutionary success as exhibited by African cichlid fishes of the genus *Tilapia* and the species flocks of the Great Lakes. *Evolution* **23**, 359–369.

Futuyma, D. J. and Mayer, G. C. (1980). Non-allopatric speciation in animals. *Syst. Zool.* **29**, 254–271.

Gaffney, E. S. (1979). An introduction to the logic of phylogeny reconstruction. *In* 'Phylogenetic Analysis and Paleontology' (J. Cracraft and N. Eldredge, *Eds*), pp.79–112. Columbia University Press, New York, USA.

Ghiselin, M. T. (1975). A radical solution to the species problem. *Syst. Zool.* **23**, 536–544.

Gordon, D. H. and Dennet, N. (1979). Ultrasonic calls and courtship behaviour of sibling species of multimammate mice, *Praomys (Mastomys) natalensis*, and *Praomys (Mastomys) coucha*. Address to 1979 Zoological Society of Southern Africa Symposium on Animal Communication, Cape Town, South Africa.

Gould, S. J. (1982). Darwinism and the expansion of evolutionary theory. *Science* **216**, 380–387.

Gould, S. J. and Eldredge, N. (1977). Punctuated equilibria: the tempo and mode of evolution reconsidered. *Paleobiology* **3**, 115–151.

Hennig, W. (1966). Phylogenetic Systematics. University of Illinois Press, Urbana, USA.

Ho, M. W. and Saunders, P. T. (1982). The epigenetic approach to the evolution of organisms—with notes on its relevance to social and cultural evolution. *In* 'Essays in Evolutionary Epistemology' (H. C. Plotkin, *Ed.*), pp.343–361. Wiley, London, UK.

Hull, D. L. (1976). Are species really individuals? *Syst. Zool.* **25**, 174–191.

Hull, D. L. (1980). Individuality and selection. *Annu. Rev. Ecol. Syst.* **11**, 311–332.

King, M. C. and Wilson, A. C. (1975). Evolution at two levels in humans and chimpanzees. *Science* **118**, 107–116.

Lewontin, R. C. (1970). The units of selection. *Annu. Rev. Ecol. Syst.* **1**, 1–16.

Løvtrup, S. (1981). The epigenetic utilization of the genomic message. *In* 'Evolution Today' (G. G. E. Scudder and J. L. Reveal, *Eds*), pp.145–162. Hunt Institute, Pittsburgh, USA.

Maglio, V. J. and Cooke, H. B. S., *Eds* (1978). 'Evolution of African Mammals'. Harvard University Press, Cambridge, Massachusetts, USA.

Maynard Smith, J. (1976). Group Selection. *Quar. Rev. Biol.* **51**, 277–283.

Mayr, E. (1963). 'Animal Species and Evolution'. Harvard University Press, Cambridge, Massachusetts, USA.

Mayr, E. (1969). 'Principles of Systematic Zoology'. McGraw-Hill, New York, USA.

Nelson, G. J. (1971). Paraphyly and polyphyly: redefinitions. *Syst. Zool.* **20**, 471–472.

Oster, G. and Alberch, P. (1982). Evolution and bifurcation of developmental programs. *Evolution* **36**, 444–459.

Paterson, H. E. H. (1978). More evidence against speciation by reinforcement. *S. Afr. J. Sci.* **74**, 369–371.

Paterson, H. E. H. (1982a). Perspective on speciation by reinforcement. *S. Afr. J. Sci.* **78**, 53–57.

Paterson, H. E. H. (1982b). Darwin and the origin of species. *S. Afr. J. Sci.* **78**, 272–275.

Paterson, H. E. H. and James, S. H. (1973). Animal and plant speciation studies in Western Australia. *J. Roy. Soc. West. Aust.* **56**, 31–43.

Rachootin, S. P. and Thomson, K. S. (1981). Epigenetics, paleontology, and evolution. *In* 'Evolution Today' (G. G. E. Scudder and J. L. Reveal, *Eds*), pp.181–194. Hunt Institute, Pittsburgh, USA.

Raup, D. M., Gould, S. J., Schopf., T. J. M. and Simberloff, D. (1973). Stochastic models of phylogeny and the evolution of diversity. *J. Geol.* **81**, 525–542.

Rensch, B. (1959). 'Evolution above the Species Level'. Methuen, London, USA.

Riedl, R. (1978). 'Order in Living Organisms'. John Wiley, New York, USA.

Roughgarden, J. (1973). Possibilities for palaeontology. *Science* **179**, 1225.

Stanley, S. M. (1975). A theory of evolution above the species level. *Proc. Natl. Acad. Sci. U.S.A.* **72**, 646–650.

Stanley, S. M. (1979). 'Macroevolution: Pattern and Process'. W. H. Freeman, San Francisco, USA.

Stebbins, G. L. (1950). 'Variation and Evolution in Plants'. Columbia University Press, New York, USA.

Vrba, E. S. (1979). Phylogenetic analysis and classification of fossil and recent Alcelaphini (Family Bovidae, Mammalia). *J. Linn. Soc. (Zool).* **11**, 207–228.

Vrba, E. S. (1980). Evolution, species and fossils: how does life evolve? *S. Afr. J. Sci.* **76**, 61–84.

Vrba, E. S. (1983a). Evolutionary pattern and process in the sister-group Alcelaphini–Aepycerotini (Mammalia: Bovidae). *In* 'Living Fossils'

(N. Eldredge and S. M. Stanley, *Eds*). Springer-Verlag, New York, USA.

Vrba, E. S. (1983b). Palaeoecology of early Hominidae, with special reference to Sterkfontein, Swartkrans and Kromdraai. *In* 'l'Environment des Hominidés' (Y. Coppens, *Ed.*). Proceedings of Colloque International held at Fondation Singer–Polignac, Paris, June 1981.

Vrba, E. S. (1983). Macroevolutionary trends: new perspectives on the roles of adaptation and incidental effect. *Science* **221**, 387–389.

Vrba, E. S. and Eldredge, N. (1984). Individuals, hierarchies and processes: towards a more complete evolutionary theory (Submitted for publication).

Walther, F. R. (1974). Some reflections on expressive behaviour in combats and courtships of certain horned ungulates. *In* 'The Behaviour of Ungulates in its Relation to Management'. (V. Geist and F. Walther, *Eds*), pp.56–106. International Union for Conservation of Nature and Natural Resources, Morges, Switzerland.

Wiley, E. O. (1978). The evolutionary species concept reconsidered. *Syst. Zool.* **27**, 17–26.

Williams, G. C. (1966). 'Adaptation and Natural Selection'. Princeton University Press, Princeton, New Jersey, USA.

Wright, S. (1967). Comments on the preliminary working papers of Eden and Waddington. *In* 'Mathematical Challenges to the Neo-Darwinian Theory of Evolution'. (P. S. Moorehead and M. M. Kaplan, *Eds*). *Wistar Inst. Symp.* **5**, 117–120.

Zuckerkandl, E. (1983). Molecular basis for directional evolution. *In* 'Modalités, Rytmes et Mécanismes de l'Evolution' (J. Chaline, *Ed.*), Centre National Recherche Scientifique, Dijon, France. Proceedings of Colloque International, Dijon, May, 1982. (In press)

6

Systematics and Evolution

GARETH NELSON and NORMAN PLATNICK

Abstract

Biological data integrate at the level of classification (cladogram), not at the level of phylogeny (phylogenetic tree). As commonly conceived, phylogeny — evolution from ancestral to descendant taxon — is an artifact of classification. Improvement in classification amounts to elimination of ancestral (paraphyletic) taxa and ancestral geographic areas (centers of origin).

INTRODUCTION

A systematist need attach no particular meaning to the term 'Darwinism' (Darwinism + neo-Darwinism) in order to proceed with his everyday work of discovering what species exist, what their biological characteristics are, where they occur, how they are interrelated, how they are to be named, and so on. A systematist may pursue all of these activities, and achieve unlimited success in their objectives, while giving no thought to notions of struggle for existence, natural selection, gene recombination, punctuated equilibria, or for that matter evolution. Systematics and Darwinism might, therefore, seem entirely unrelated, or at most related in a trivial or obscure way, the meaning of which remains open to argument.

We believe that such is not the case. We believe that Darwinism has an identity within the area of biological systematics, that it has a history within that discipline, that it is, in short, a theory that has been put to the test and found false. We believe further that the process of test and falsification has

BEYOND NEO-DARWINISM
ISBN: 0-12-350080-X

changed the nature of systematics, so that the discipline is already 'beyond Darwinism' in a significant sense. By this we mean that the process has not been merely negative. Rather, something novel and unexpected has been gained in our knowledge of the world. In other words, some positive progress has been achieved.

The elements of our thesis are not novel. Systematic literature abounds in argument about the meaning and significance of two key concepts: paraphyletic group and center of origin. If our discussion contains anything novel it is partly the perception that these two concepts characterize a Darwinian theory of systematics; it is partly in the perception that this theory has been falsified; and it is partly in the perception of what has been achieved in the process.

Charles Darwin accepted as a fact that the many different kinds of organisms, both living and extinct, are ordered in a hierarchical system. The main outlines of this system had already been discovered by systematists, who had referred to it as the natural system of classification. In arguing that evolution is the true explanation of the natural hierarchy, or system, and that descent with modification is its cause, Darwin offered nothing novel. Yet he implied that the history of the hierarchy could be documented by the discovery of taxa that stand in relation to one another as ancestors and descendants. He implied further that there is a geographical dimension to the history of the hierarchy, which could also be documented. These implications were followed up early by Darwinian systematists who sought, in fact, to document the evolution from ancestral taxon to descendant taxon, and from the area of origin of a taxon to areas into which the taxon spread as it evolved and diversified. These implications seemed real and reasonable enough, given the gradualistic and competitive nature of natural selection — Darwin's novel mechanism of evolution. These implications seemed most real and most reasonable to paleontologists, who in general came to believe that through the fossil record they have direct access to the documentary evidence of evolution both in time and in space. As a result chiefly of their activities the notions of ancestral group and area became standard in the scientific and educational literature of the late nineteenth and twentieth centuries.

These implications are nowhere better illustrated than in the widely publicized story of human evolution. The story is of successive finds of fossil ancestral taxa, and a narrowing focus on a particular geographical area, lately eastern Africa, as the center of origin — that area wherein the ancestral taxa have been found. We argue not that such stories are demonstrably false, but that their theoretical framework is demonstrably false. The particular stories are mere artifacts of that framework. As artifacts the stories themselves have a significance that is indeterminate.

One may dispute that these two notions—of ancestral taxa and areas—are necessary implications of Darwin's theory. They are, after all, present in rudimentary form in the biblical account of creation. This account was incorporated early into the systematic literature by Linnaeus. Historically, however, these two notions were accepted by Darwinians as necessary implications. Without them there is no Darwinian theory of systematics, beyond the idea that evolution is the explanation, and descent with modification the cause, of the natural hierarchy. These ideas are not novel with Darwin, nor are they a theory of systematics, for they place no constraints upon the nature of the hierarchy. The implications, however, do impose constraints: they specify that the hierarchy in its detail will show taxa and areas ancestral one to another. After 1859 such taxa and areas were soon 'discovered' and the literature is now replete with them. We argue that such taxa and areas were discovered for no other reason than because the theory, as conceived or misconceived by Darwinians, demanded them. We argue that the taxa and areas are not phenomena, but rather artifacts. We argue, finally, that the theory that demands them is false, and that the falsifying evidence presently available is decisive.

If this thesis is true, what is the significance of that truth? For the falsification itself we see two areas of significance: within systematics and beyond. The significance within systematics is an improved theoretical structure—and hopefully a better assessment of the nature of the world and a better understanding of the processes of its investigation and discovery. The significance beyond systematics lies in what systematics means to biology and to humanity in general. We are certain that biology will survive and prosper without the notions of ancestral taxa and ancestral areas as empirical entities documented by fossil evidence. We suppose that if systematists stop presenting to other biologists artifacts in the guise of phenomena, science in general will be improved. And if science stops presenting to humanity as large artifacts in the guise of phenomena, everyone's understanding will be improved. With respect to the overall status of Darwinism, which has been much debated in recent times, there is little to say. The intrusion of Darwinism into systematics was at worst a misadventure, but it was not without value. Science learns through mistakes, and science may count as progress the falsification of a theory, even if in retrospect the theory seems to have been wrongheaded from the beginning. In this connection, we do not imply that developments within systematics falsify the theory of evolution as that word is normally understood—the appearance of life in all of its diversity through historical process of long duration. Rather the data of systematics are a massive corroboration of evolution in that sense, which is distinct from any notion of theory properly termed Darwinian.

Apart from the falsification, which is basically a negative aspect, there is a positive side to recent developments in systematics. The corroboration stems from life's geographical dimension, and the apparent congruence of pattern between relationships that are, on the one hand, geological and, on the other, biological. So far, congruence of pattern is the bottom line on the positive side. The significance of such congruence, both for science and for humanity at large, remains to be determined. Its occurrence was unexpected and unpredicted by Darwinian theory or, for that matter, any other theory of systematics.

ANCESTORS AND DESCENDANTS

Darwinian systematics is the theory that taxa stand in relation, one to another, as ancestors and descendants. An example is Romer's diagram of vertebrate phylogeny (Fig. 6·1(a)), wherein reptiles are shown as ancestors of birds and mammals; amphibians, of reptiles; bony fishes, of amphibians; placoderms, of bony fishes and cartilaginous fishes; and agnathans, of placoderms.

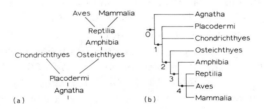

FIG. 6·1 (a) Phylogeny of vertebrates according to Romer (1962). (b) Classification implied by phylogeny in A: 0, Vertebrata; 1, Gnathostomata; 2, Teleostomi; 3, Tetrapoda; 4, Amniota.

Always inherent in such notions of phylogeny are more general relationships that are hierarchical and that become evident when inclusive taxa are specified (Fig. 6·1(b)). Analysis of this example shows that there are five inclusive taxa, each at a different hierarchical level: 0, Vertebrata; 1, Gnathostomata; 2, Teleostomi; 3, Tetrapoda; 4, Amniota. These inclusive taxa were all recognized and named (sometimes by different names) in pre-Darwinian times.

The inclusive taxa offer a basis for understanding the nature of ancestral taxa, two series of which are considered here. The first series (Fig. 6·2(a)) is shown in Romer's phylogeny. *Reptiles*, for example, are those amniotes that are not birds or mammals; *amphibians*, those tetrapods that are not amniotes; *bony fishes*, those teleostomes that are not tetrapods; *placoderms*, those gnathostomes that are not teleostomes or cartilaginous fishes; *agnathans*, those vertebrates that are not gnathostomes. The second series (Fig. 6·2(b)) is the complement of Romer's: *amniotes* are those tetrapods that are not amphibians; *tetrapods*, those teleostomes that are not bony fishes; *teleostomes*, those gnathostomes that are not placoderms or

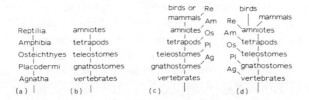

FIG. 6·2 (a) Series of ancestral taxa of Romer's phylogeny (cf. Fig. 6·1(a)). (b) Complementary series of ancestral taxa (cf. Fig. 6·1(b)). (c) Alternative phylogeny with reversed ancestor-descendant relationships of complementary taxa. (d) Corrected version of phylogeny in C (cf. Fig. 6·1(b)).

cartilaginous fishes; *gnathostomes*, those vertebrates that are not agnathans. The taxa of each series are logically equivalent in the sense that the ones are complements of the others: *amphibians* are those tetrapods that are not *amniotes*, and *amniotes*, those tetrapods that are not amphibians; *bony fishes* are those teleostomes that are not *tetrapods*, and *vice versa*, etc.

Either series (Fig. 6·2(a)–(b)) may seem credible in itself as a depiction of certain aspects of evolution. Yet the two series are different in one respect. In the first series the taxa are mutually exclusive: reptiles are not amphibians; amphibians are not bony fishes, etc. In the second series the taxa are partly or wholly inclusive: amniotes are tetrapods, and some tetrapods are amniotes; tetrapods are teleostomes, etc. Because of this difference the first series specifies a phylogeny, whereas the second series specifies a classification. In view of the logical equivalence of the taxa of the two series (the ones being complements of the others), it might seem strange that the two series differ in so fundamental a way.

Consider the second series in more detail. Attempt to add to it the taxa of the first series, by reversing the ancestor-descendant relationships of each pair of complementary taxa, so as to achieve a different phylogeny (Fig. 6·2(c)). Let birds or mammals stand as ancestors of reptiles; amniotes as ancestors of amphibians; etc. The result is peculiarly defective, but its defect—beyond its mere novelty as a phylogeny of vertebrates—is not obvious. Its defect may be remedied, however, by a slight adjustment in the position of the taxa of the first series (Fig. 6·2(d)). The result is a classification that is defect-free.

Achieving the final result permits understanding the nature of the defect that at first was obscure—the inconsistency in the inclusive relationships of the taxa. The different phylogeny implies a different classification that conflicts with the original (Fig. 6·1(b)). According to the original, for example, amphibians are not amniotes, yet both groups are tetrapods. The defective result (Fig. 6·2(c)) implies that amphibians are amniotes, and therein is the defect—a conflict between one classification (and its related phylogeny) and another classification (and its related phylogeny).

In itself the above analysis is no more than a logical exercise. Its purpose is to show the relation between two kinds of concepts, which for convenience here are termed phylogeny (Fig. 6·2(a)) and classification (Figs. 6·1(b), 6·2(d)). This is a special usage of these terms (according to another usage Fig. 6·2(a) is a phylogenetic tree and Figs. 6·1(b) and 6·2(d) are cladograms). And in a less precise sense either concept may be viewed in either way (the phylogeny involves classification insofar as it includes taxa; and the classification involves phylogeny insofar as vertebrates may be said to have evolved, or differentiated, into agnathans and gnathostomes). We argue, however, that the concept of phylogeny inherent in Darwinian systematics is that of Fig. 6·2(a), wherein ancestral taxa are exclusive rather than partly or wholly inclusive in their mutual relations. This concept of phylogeny implies that ancestral taxa have an objective identity independent of their descendants. This implication means that ancestral taxa can be discovered; and if discovered can be identified as such. In brief, it means that ancestral taxa are under the constraints of direct empirical investigation rather than mere logical necessity.

We argue in contrast that such phylogeny is really an artifact of classification—that classification imposes the only constraint relevant to the resolution ('discovery') of ancestral taxa. Within a certain classification (Fig. 6·3(a)) certain taxa are in effect already designated as ancestors: taxon A1 (agnathans) is the ancestor of inclusive taxon 1 (gnathostomes); taxon A2a (placoderms) or A2b (cartilaginous fishes) is the ancestor of inclusive taxon 2 (teleostomes); taxon A3 (bony fishes) is the ancestor of inclusive taxon 3 (tetrapods); taxon A4 (amphibians) is the ancestor of inclusive taxon 4 (amniotes). Given the classification, no other resolution ('discovery') of ancestral taxa is possible without violation of the classification, as is evident from the above analysis (Fig. 6·2(c)). In other words, the only taxa acceptable as ancestors are exclusive taxa; inclusive taxa cannot serve as ancestors. Whether a taxon is exclusive or inclusive is evident only in the context of a classification.

If all of the formal problems of phylogeny and classification were immediately reducible to the above formulation, such would have been evident years ago. There would have been no need for the extended argument that has taken place over the last 25 years in the literature of systematic biology. Alas, such is not completely the case. The above formulation is oversimplified. Yet it does permit entry into series of problems that have been extensively discussed in recent times.

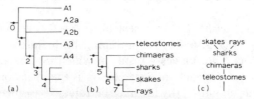

FIG. 6·3 (a) Classification showing the relation between exclusive (ancestral) taxa (A1, A2a, A2b, A3, A4) and inclusive (descendant) taxa (1, 2, 3, 4). A1 is the ancestor of 1; A2a or A2b is the ancestor of 2; etc. (b) Classification of gnathostomes (1), with particular reference to cartilaginous fishes (5). (c) Phylogeny of gnathostomes implied by classification in (b).

One series is briefly considered here, for it is exemplified by Romer's phylogeny of vertebrates, wherein a group of fossil fishes, the placoderms, is ancestral to teleostomes and cartilaginous fishes. Given the above classification of vertebrates (Figs. 6·1(b), 6·3(a)), either placoderms or cartilaginous fishes are the taxon ancestral to teleostomes. Cartilaginous fishes traditionally were so considered in the late nineteenth and early twentieth centuries. Why, then, are placoderms now recognized as ancestral rather than cartilaginous fishes?

We approach this question through a consideration of cartilaginous fishes, which include a variety of living forms, such as chimaeras, sharks, skates, and rays, classified as in Fig. 6·3(b). Skates and rays together form the inclusive taxon of batoids (7); these together with sharks are elasmobranchs (6); these together with chimaeras are chondrichthyans (5). This classification implies a certain phylogeny (Fig. 6·3(c)), wherein teleostomes appear as ancestors of chondrichthyans. This result is paradoxical, inasmuch as chondrichthyans appear as a possible ancestor of teleostomes in what really is the same classification. This paradox is the reason why the above formulation is oversimplified, for the formulation is dependent on a particular starting point. With mammals or birds (or both) as a starting point, one sees chondrichthyans as possible ancestors at a certain level in phylogeny (Fig. 6·1(b)). With skates or rays (or both) as a starting point, one sees teleostomes as ancestors at the same level in phylogeny (Fig. 6·3(b)). The notion that chondrichthyans are ancestral is logically absurd if chondrichthyans are specified as an inclusive group of the classification.

Inclusive taxa were termed *monophyletic* by Hennig (1950, 1966)—a term with an older, but ambiguous, usage. He defined the term to mean a taxon that includes all of the descendants of a common ancestral species. He argued that the members of a monophyletic taxon can be recognized only through their sharing advanced characters (*synapomorphies*)—in effect, evolutionary novelties unique to each monophyletic group. From this perspective, therefore, ancestral taxa that are monophyletic are logically absurd.

In more general terms, CD is ancestral to A + B (Fig. 6·4(a)), and AB is ancestral to C + D (Fig. 6·4(b)). Combining both statements into a single classification (Fig. 6·4(c)) gives a result in which no one ancestral taxon is evident. In other words each inclusive taxon is the ancestor of the other. If on these grounds chondrichthyans may be rejected as ancestors, what sort of ancestral group might be acceptable?

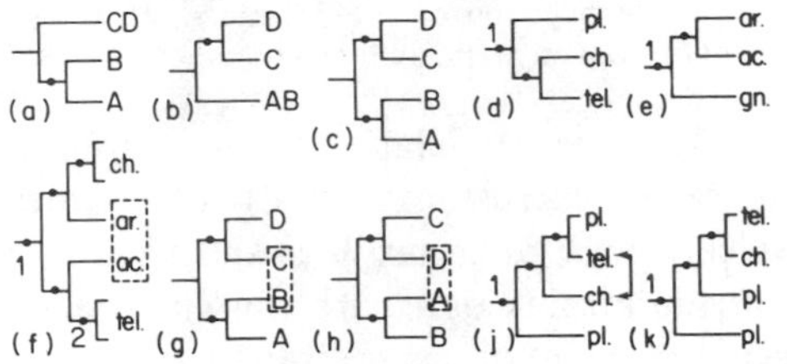

FIG. 6·4 (a) Classification implying that taxon CD is ancestral to taxa A and B. (b) Classification implying that taxon AB is ancestral to taxa C and D. (c) Classification implying that taxa AB and CD are ancestral to each other. (d) One possible classification of gnathostomes (1), wherein chondrichthyans (ch) and teleostomes (tel) form a group that excludes placoderms (pl). (e) Same classification, with particular reference to placoderms, which include arthrodires (ar) and acanthodians (ac), and which form a group that excludes recent gnathostomes (gn = chondrichthyans and teleostomes). (f) Second possible classification of gnathostomes (1), wherein some placoderms, arthrodires (ar), are related to chondrichthyans (ch); and other placoderms, acanthodians (ac), are related to teleostomes (tel); and the limits of the group of placoderms are indicated by the dashed lines. (g) Classification wherein the paraphyletic taxon BC is ancestral to taxa A and D. (h) Classification wherein the paraphyletic taxon AD is ancestral to taxa B and C. (j) Third possible classification of gnathostomes (1), wherein some placoderms (pl) are related to teleostomes (tel) or to chondrichthyans (ch), and other placoderms (pl) are more distantly related; arrows indicate that teleostomes and chondrichthyans may be interchanged in position. (k) Fourth possible classification of gnathostomes (1), wherein placoderms (pl) are distantly related to teleostomes (tel) and chondrichthyans (ch).

Romer did not make clear his reasons for portraying placoderms as ancestral. The possibilities, in terms of alternative classifications, are reviewed here. One possible classification (Fig. 6·4(d)) implies that placoderms are ancestral if teleostomes or chondrichthyans (or both) are the starting point (this allows chondrichthyans to be ancestral to teleostomes, and *vice versa*). However, Romer included among placoderms two major groups, acanthodians and arthrodires. Given these two groups as a starting point, complete analysis again leads to paradox if placoderms are an inclusive taxon (Fig. 6·4(e)): gnathostomes are ancestral to placoderms (and *vice versa*). With respect to the first possible classification, placoderms are rejectable on the same grounds as chondrichthyans.

Another possible classification, possibly the one Romer actually had in mind, specifies that some placoderms (acanthodians) are related to teleostomes, and others (arthrodires) are related to chondrichthyans (Fig. 6·4(f)). This classification has an interesting property: it eliminates one paradox and leads to another. Given teleostomes as a starting point, one sees placoderms (acanthodians) as ancestors of teleostomes; placoderms (arthrodires) again appear as part of an older ancestral taxon that also includes chondrichthyans. Given chondrichthyans as a starting point, one sees placoderms (arthrodires) as ancestors of chondrichthyans; placoderms (acanthodians) appear again as part of an older ancestral taxon that also includes teleostomes. The analysis may be summarized thus:

Fig. 6·4(f)	ancestor (tel):	pl (ac)·
	ancestor (tel + ac):	pl (ar) + ch
	ancestor (ch):	pl (ar)
	ancestor (ch + ar):	pl (ac) + tel
taxon common to all ancestors:		pl

Placoderms emerge as a taxon that is formally ancestral both to teleostomes and to chondrichthyans, and that figures in the ancestry of the two other inclusive taxa of the classification. On these grounds placoderms emerge with good credentials as ancestors. It would seem, therefore, that one kind of paradox has been eliminated. But where is the new one?

In more general terms, if taxon B + C is allowed by the classification to be ancestral to taxa A and D (Fig. 6·4(g)), then taxon A + D is allowed to be ancestral to taxa B and C (Fig.6·4(h)). Therein is the paradox. Why does it arise? The classification specifies as inclusive taxa only A + B and C + D. Neither taxon B + C nor taxon A + D is specified. If taxon B + C (or A + D) meets the formal requirements of an ancestral taxon, it does so at the expense of a classification in which it has no existence. If its purpose is to meet those requirements, its creation is mere artifice, achieved by ignoring the constraints of classification.

There are two other classifications, each specifies placoderms as an artificial group, and in each placoderms may be said to meet the formal requirements of an ancestral taxon. In one (Fig. 6·4(j)) some placoderms are related to teleostomes (or, alternatively, chondrichthyans) and other placoderms are more distantly related. In the other (Fig. 6·4(k)), both groups of placoderms are more distantly related to teleostomes and chondrichthyans. The analyses may be summarized thus:

Fig. 6·4(j)	ancestor (tel):	pl
	ancestor (tel + pl):	ch
	ancestor (tel + pl + ch):	pl
taxon common to most ancestors (2 of 3):		pl
Fig. 6·4(k)	ancestor (tel + ch)	pl
	ancestor (tel + ch + pl):	pl
taxon common to all ancestors:		pl

The three classifications (Figs. 6·4(f), (j) and (k)) in which placoderms appear as an artificial group specify that group as *paraphyletic*—a term introduced by Hennig (1950, 1966). He defined the term to mean a taxon that includes some but not all of the descendants of a common ancestral species. He argued that the members of paraphyletic groups can be united only on the basis of shared primitive characters (*symplesiomorphies*)— in effect, by the lack of characters unique to other groups. This means that paraphyletic groups can have no characters of their own. If placoderms are paraphyletic, for example in the sense of Fig. 6·4(f), they are nothing more than gnathostomes that are neither teleostomes nor chondrichthyans. Whereas gnathostomes, teleostomes, and chondrichthyans can be characterized by biological properties (synapomorphies), placoderms logically cannot be so characterized. Gnathostomes, for example, are currently understood to have jaws, paired fins, three semicircular canals, etc; teleostomes, to have dermal jaw bones, lungs, etc; chondrichthyans, to have claspers, placoid scales, ampullae of Lorenzini, etc. How then are placoderms characterized, if they are paraphyletic in the sense of Fig. 6·4(f)? As gnathostomes that lack the characters unique to teleostomes and chondrichthyans.

Hennig argued that paraphyletic groups are artifacts of classification, and further that to improve existing classification paraphyletic groups, if discovered, must be eliminated. If placoderms are paraphyletic in the sense of Figure 6·4(f), for example, acanthodians should be classed with teleostomes, and arthrodires with chondrichthyans. This change in classification would eliminate the taxon of placoderms; it would also eliminate an ancestral group.

The above analysis indicates that paraphyletic groups might satisfy certain formal requirements of ancestors. Further analysis, omitted here, shows that only paraphyletic groups can logically satisfy these formal requirements. If Hennig is correct—that paraphyletic groups are artifacts and must be eliminated from classification—all ancestral groups will disappear and with them all phylogenies as the term is used here.

Developments since Hennig include much discussion and argument, pro and con, about these fundamental matters. Much of the published discussion can be found in the pages of the journal *Systematic Zoology*. Developments also include systematic investigations of particular groups, which are scattered through the literature of systematic biology. They have not been the subject of comprehensive review and evaluation, and their import is thus open to varied judgment. We judge that in their aggregate they confirm the argument of Hennig with respect to the nature and fate of paraphyletic groups—with one exception. The exception is the occasional author of Darwinian persuasion who sees ancestral groups as desirable (demanded by theory) and who refuses to give them up while admitting that they are artificial (e.g. Charig, 1982). Nevertheless, we expect that in time all paraphyletic groups will disappear and with them all ancestral taxa of the Darwinian tradition.

Of the ancestral groups listed in Romer's phylogeny, all are generally admitted to be paraphyletic (Janvier, 1981; Gardiner, 1982; Rosen *et al.*, 1981). These groups survive in Romer's phylogeny because they are paraphyletic. That chondrichthyans have usually been considered a monophyletic group explains why this taxon did not survive as the ancestor of teleostomes, and why it was replaced by placoderms. The tendency of Darwinian phylogenies to evolve in the direction of ancestral groups that are extinct and paraphyletic has been noted by Patterson (1982:64):

> Is it not strange that the justification of phylogeny, as something beyond systematics, resides in extinct paraphyletic groups? For those groups are the inventions of evolutionists, those who appeal to them as demonstrating the path of descent. So far as I know, such groups did not exist in pre-Darwinian taxonomy, for paleontologists were then preoccupied with the real problem of allocating fossils to Recent groups (Patterson 1977b, p.596). Nor do I find any extinct paraphyletic groups in Haeckel's (1866) trees. Such groups are therefore a later invention, imagined by evolutionists, those most committed to the confirmation of Darwin's views.

If phylogenies of one sort (Fig. 6·2(a)) are to pass away, is the notion of phylogeny doomed also? We judge not, for there is an alternative notion, here simply termed classification (Fig. 6·2(d)). Notions of this kind can be looked upon as phylogenies—as historical statements of ancestry and descent. But they are different in character. They include no ancestral taxa. They deny the postulates of Darwinian systematics: that ancestral taxa have an objective identity independent of their descendants; that ancestral taxa can be discovered and identified as such; that ancestral taxa are under the constraints of empirical investigation.

This shift in meaning of the term *phylogeny* from a Darwinian to a cladistic sense marks a revolution in biological systematics. Nominally it is associated with Hennig (Dupuis, 1979; Eldredge and Cracraft, 1980; Wiley, 1981), but actually its roots extend back to the origin of systematics as a rational endeavor (Nelson and Platnick, 1981). The negative aspect of cladistics is the denial of the Darwinian postulates. The positive aspect is the affirmation that the data of biology integrate empirically at the level of classification. Integration, or congruence, of biological data has become the chief concern of cladistic theory and practice (Funk and Brooks, 1981; Platnick and Funk, 1983).

ANCESTRAL AREAS AND DISPERSAL HISTORIES

Within biogeography, centers of origin are the conceptual analog of paraphyletic groups. In the Darwinian tradition, each species is envisaged as originating in some (relatively small) area separated from other areas by one or more barriers. Subsequently, individuals cross the barrier(s) by whatever means of dispersal are available to them. Selection for new conditions in the newly colonized areas may result in the origin of new species, which subsequently disperse outward across new barriers, etc. The literature abounds with 'explanations' of the geographic history of taxa in which centers of origin are 'discovered' and routes of dispersal across barriers are 'traced'.

These dispersal histories have the formal structure of phylogenies (like Fig. 6·2(a)) with different areas, rather than different taxa, portrayed as ancestral to each other. Like their taxonomic analogs, the phylogenies can be interpreted instead as classifications. And here again, that shift in interpretation transforms the Darwinian view: the ancestral areas become inclusive rather than partly or wholly exclusive in their mutual relations. In other words, the ranges of ancestral taxa, instead of being separate from those of their descendants, include the distributions of all their descendants.

Viewing ancestral distributions as originally inclusive, and as having subsequently undergone vicariance (division by the appearance of new barriers) was one contribution of Croizat (1958, 1964). Like Hennig's, Croizat's contribution marks something of a revolution. Beyond its negative aspect (the denial of Darwinian explanation in space as well as time) is a positive one allowing an integration of data at higher levels.

Darwinian biogeography stressed the uniqueness of each group's geographic history, involving unique dispersal events made possible by the unique capabilities and opportunities of that group's ancestors. Croizatian, or vicariant, biogeography allows a different, more general mode of investigation.

Given, for example, a group containing eight species occurring in eight different areas and cladistically interrelated in some fashion (such as in Fig. 6·3(a)), one need not assume that the group's history is unique. Instead, one can ask whether other groups of organisms include species endemic (restricted to) the same areas (Fig. 6·3(a): areas A1, A2a, A2b, etc.). If they do, one can ask whether those endemic species are interrelated in the same way. That is, are species endemic to, say, area A2a always more closely related to species endemic to areas A2b, A3, etc., than they are to species endemic to area A1, as Fig. 6·3(a) specifies?

The few empirical studies conducted so far (such as Brundin, 1966; Rosen, 1976, 1978) indicate that substantial congruence of this sort exists, even among groups of organisms whose capabilities and opportunities for dispersal are widely disparate. As Croizat stressed, if the geographic relationships displayed by a group of birds are congruent with those displayed by groups of cacti, earthworms, and spiders, then those distributions can have a common cause despite the different dispersal capabilities of those organisms.

The discovery of such general biogeographic patterns raises the possibility of congruence of data on an even higher level. It is possible also to construct cladograms of areas on the basis of geological data, uniting areas that share unique features of stratigraphy, geochemistry, paleomagnetism, etc. One can thus compare a cladogram of areas based on the interrelationships of endemic species of various groups of organisms with a similar cladogram based on independent geological information. If the biological and geological area cladograms agree in the interrelationships they postulate among areas (i.e. if they are congruent), then the historical events reflected in them can be associated with the origins of the areas and their biotas. The events may be large scale (such as separations of continental plates) or small (such as local climatic changes), and can often be dated fairly accurately by radiometric and other techniques.

Developments since Croizat include much discussion and argument, pro and con, about these matters. Relatively few particular groups and areas have been the subject of comprehensive studies. We judge that in their aggregate, however, these studies confirm the argument of Croizat with respect to the nature and fate of ancestral areas and dispersal histories — most, if not all, of the dispersal histories in the literature are merely artifacts of a theory that required them, and will not survive critical examination. If histories of one sort (Fig. 6·2(a)) are to pass away, is the notion of biogeographic history and evolution doomed also? We judge not, for there is an alternative notion, here termed area cladograms or classifications. Because the alternative notion allows data to be integrated at higher levels, first as general biological area cladograms and then as geological area cladograms, we judge that progress beyond Darwinism in biogeography, as in systematics, already exists and is likely to continue.

SOCIOLOGY OF SCIENTIFIC CHANGE

By 1970 there had developed within biological systematics enough ferment to permit a relatively unbiased observer to perceive three schools of systematics: evolutionary, phenetic, and cladistic (Hull, 1970). At the time these appeared as independent, competitive, and mutually antagonistic, yet in the main they developed historically in the order listed. What we now see as Darwinian systematics includes the phenetic and evolutionary schools, which agree at least in principle on the necessity of paraphyletic groups (from the phenetic standpoint, mistakenly we believe, such groups are necessary accurately to depict nature in its static aspect; from the evolutionary standpoint, again mistakenly we believe, such groups are necessary to depict nature in its dynamic, or transformational, aspect).

Although it has never been so described, cladistics to a large degree began as a reform of paleontology (Brundin, 1968). The point at issue was the role fossils play in deciding questions of systematics, biogeography, and evolution. The issue would never have been raised unless it was made necessary by the over-zealous advocation of a purely paleontological approach to all such questions. In recent times the paleontological approach is best represented in the writings of a single person, George Gaylord Simpson, one of the founders of the neo-Darwinian synthetic theory of evolution.

Most contemporary contributors to cladistics are professional biologists who were trained within the tradition, and represent first- or second-generation academic by-products, primarily of the synthetic theory and secondarily of one of the older schools of systematics. Thus it was inevitable that cladistics be viewed, and viewed critically by older or more conservative colleagues, through the many different lenses of pre-existing points of view. The critical commentary arising from these older and more conservative sources — much of it severely and obviously flawed — is virtually unanimous in its negative appraisal of cladistics. The most bitter denunciations have arisen from paleontologists, to some of whom cladistics has the aspect of a

Marxist 'spectre haunting palaeontology'. (Campbell, 1975:87 and Halstead in various issues of *Nature* during 1978-1982).

We view these developments with mixed feelings. We agree with Wilde that condemnation is preferable to indifference, but we are still dismayed to observe the scientific process descend—however temporarily—into demagoguery. The general reader is forewarned.

REFERENCES

Brundin, L. (1966). Transantarctic relationships and their significance. *Kungliga Svenska Vetenskapsakademiens Handlingar* (4) **11(1)**, 1-472.

Brundin, L. (1968). Application of phylogenetic principles in systematics and evolutionary theory. *In* 'Current Problems in Lower Vertebrate Phylogeny: Nobel Symposium 4' (T. Ørvig, *Ed.*), pp.473-495. Almqvist and Wiksell, Stockholm, Sweden.

Campbell, K. S. W. (1975). Cladism and phacopid trilobites. *Alcheringa* **1(1)**, 87-96.

Charig, A. J. (1982). Systematics in biology: a fundamental comparison of some major schools of thought. *In* 'Problems of Phylogenetic Reconstruction' (K. A. Joysey and A. E. Friday, *Eds*), pp.363-440. Academic Press, London and New York.

Croizat, L. (1964). 'Space, Time, Form: The Biological Synthesis'. Published by the author, Caracas, Venezuela.

Croizat, L. (1958). 'Panbiogeography'. Published by the author, Caracas, Venezuela.

Dupuis, C. (1979). La systématique phylogénétique de W. Hennig (historique, discussion, choix de références). *Cahiers des Naturalistes* n.s. **34(1)**, 1-69.

Eldredge, N., and J. Cracraft. (1980). 'Phylogenetic Patterns and the Evolutionary Process'. Columbia University Press, New York, USA.

Funk, V. A., and Brooks, D. R. *Eds* (1981). 'Advances in Cladistics: Proceedings of the First Meeting of the Willi Hennig Society'. New York Botanical Garden, The Bronx, USA.

Gardiner, B. G. (1982). Tetrapod classification. *Zool. J. Linnean Soc.* **74(3)**, 207-232.

Hennig, W. (1950). 'Grundzüge einer Theorie der phylogenetischen Systematik'. Deutscher Zentralverlag, Berlin.

Hennig, W. (1966). 'Phylogenetic Systematics'. University of Illinois Press, Urbana, USA.

Hull, D. L. (1970). Contemporary systematic philosophies. *Annu. Rev. Ecol. Systematics* **1**, 19-54.

Janvier, P. (1981). The phylogeny of the Craniata, with particular reference to the significance of fossil 'agnathans'. *J. Vertebrate Paleontol.* **1(2)**, 121-159.

Nelson, G., and Platnick, N. (1981). 'Systematics and Biogeography: Cladistics and Vicariance'. Columbia University Press, New York and London.

Patterson, C. (1982). Morphological characters and homology. *In* 'Problems of Phylogenetic Reconstruction' (K. A. Joysey and A. E. Friday, *Eds*), pp.21-74. Academic Press, London and New York.

Platnick, N. I., and Funk, V. A. *Eds* (1983). 'Advances in Cladistics, Volume 2: Proceedings of the Second Meeting of the Willi Hennig Society'. Columbia University Press, New York, USA.

Romer, A. S. (1962). 'The Vertebrate Body'. W. B. Saunders, Philadelphia, USA.

Rosen, D. E. (1976). A vicariance model of Caribbean biogeography. *Systematic Zool.* **24(4)**, 431–464.

Rosen, D. E. (1978). Vicariant patterns and historical explanation in biogeography. *Systematic Zool.* **27(2)**, 159–188.

Rosen, D. E., Forey, P. L., Gardiner, B. G. and Patterson, C. (1981). Lungfishes, tetrapods, paleontology, and plesiomorphy. *Bull. Am. Mus. Nat. Hist.* **167(4)**, 159–276.

Wiley, E. O. (1981). 'Phylogenetics: The Theory and Practice of Phylogenetic Systematics'. John Wiley and Sons, New York, USA.

7

Ontogeny and Phylogeny

SØREN LØVTRUP

Abstract

Ontogeny concerns the creation of adult organisms from fertilized eggs, phylogeny concerns the changes to which ontogeny has been subject in the course of time.

Phylogeny thus primarily deals with the history of evolution. We can learn important details about this phenomenon through the fossil record, however, the knowledge gained is fragmentary and imprecise. But if we concentrate on living beings we may, through phylogenetic classification, get a much more correct, if still incomplete idea about the course of evolution. Modern phylogenetic classification shows that, in contrast to Linnean systematics, living organisms are naturally classified in dendrograms which are to a very large extent dichotomous.

If ontogeny is mainly the descriptive aspect of the creation of adult organisms, then epigenetics is the group of sciences which study the mechanisms through which transformation from egg to adult takes place. Epigenesis is to a very large extent a matter of morphogenesis, the creation of form. This is in the last analysis a geometrical transmutation, accomplished by physical forces.

The morphogenetic aspect of ontogeny can subdivided into three distinct phases with respect to the nature of the epigenetic mechanisms involved. The first one is the phase of *form creation*, during which the basic body plan is formed through the activity of various kinds of cells and cellular differentiation products. The latter may either accumulate inside the cell or be secreted to the extracellular matrix. Once the body plan has been created, the next phase of morphogenesis consist of *differential growth*, a process which primarily concerns the skeleton. During this phase the body plan is modified such that the peculiarities of the body distinguishing the lower taxonomic classes are formed. The process of differential growth continues after birth or hatching, but at a much lower relative rate. This phase of *allometric growth* ends when the organisms reach sexual maturity.

Some of the epigenetic mechanisms involved in the earliest developmental stages are illustrated with observations made on amphibian embryos.

BEYOND NEO-DARWINISM
ISBN: 0-12-350080-X

Die Phylogenese ist die mechanische Ursache der Ontogenese.

Ernst Haeckel (1874, p.5)

L'univers organisé ne présenterait donc bientôt que confusion et désordre, si des lois fixes et immuables ne présidaient à la formation des êtres, et ne maintenaient chacun d'eux dans des limites qui lui sont assignés.

A.E.R.A. Serres (1860, p.ix)

Die Entwickelung der Organismen ist ein *Physiologischer Prozess*, welcher als solcher auf mechanischen 'wirkenden Ursachen', d.h. auf physikalisch-chemischen Bewegungen beruht.

Ernst Haeckel (1920, p.342)

Ontogeny is 'the entire sequence of events involved in the development of an individual organism', whereas phylogeny is 'the sequence of events involved in the evolution of a species, genus, etc.' (Collins Dictionary of the English Language). For reasons that will become apparent later I would replace 'species, genus, etc' by 'kingdom, phylum, etc.', but apart from that these definitions constitute an excellent point of departure for the present discussion.

The two concepts deal with the origin of living beings, partly as it occurs during the development of single organisms, and partly as the consequence of the changes in the developmental processes which have occurred in the course of evolution. This suggests that there must be an intimate correlation between ontogeny and phylogeny. The aim of the present survey is to elucidate the nature of this correlation.

We shall begin with a brief discussion of the theories of evolution, and the historical and the creative aspects of the relations between the two concepts. Living organisms are created in epigenetic processes, and we shall devote the main body of this chapter to a discussion of the mechanisms involved in these events, from a theoretical and an experimental point of view.

THE FOUR THEORIES OF EVOLUTION

The notion of phylogeny implies that evolution has occurred, and thus, the validity of *the theory on the reality of evolution*. This theory is usually known under the name of 'Darwinism'. In fact, many savants, from Charles Lyell to Gavin de Beer, have publicly affirmed that this theory was advanced by Lamarck; some have even claimed that it rightly ought to be called *Lamarckism*, as we shall see, a better name would be *Lamarck's first theory of evolution*.

Once it is accepted that evolution has occurred, it follows that a unique classification of living beings must exist, which truly depicts the course of evolution. Such, at least, was Lamarck's conviction. This principle, according to which the phylogenetic relationships between living organisms may be determined through classification, is generally accepted by present-day biologists. Consequently, *the theory on the history of evolution* is *Lamarck's second theory of evolution*.

Prima facie it may seem that the result of such classificatory endeavours deals with the temporal aspect of evolution, and thus merely with phylogeny. It is a remarkable fact, recognized only by a minority in the biological community, that the classifications also outline the course of ontogeny for the members of the various terminal taxa in the phylogenies (*see below*).

However, the classifications thus established are only descriptive. They may account for the temporal course of evolution and development, but they do not indicate the mechanisms through which these events were or are brought about. In order to understand the latter, we must first understand how individual organisms are created, that is, we must study the epigenetic mechanisms which are responsible for ontogenetic development. Once we understand the mechanisms involved in the creation of the members of a particular lineage in the phylogenetic classification, we may be able to account for the changes in the epigenetic mechanisms which have been necessary for the transition from one taxon to another.

These creative aspects of ontogeny and phylogeny, dealt with by *the epigenetic theory on the mechanism of evolution*, are the subject of the second section below.

The mechanism of evolution is not only a matter of creation of new forms, but also of their survival. This aspect is accounted for by the fourth and last theory of evolution, the *ecological theory on the mechanism of evolution*. The implications of this theory will be briefly discussed in the following section. A detailed discussion of the four theories of evolution has been published recently (Løvtrup, 1982).

THE HISTORICAL ASPECT

For more than two centuries the classification of living organisms has been based on the Linnaean system. Without contemplating any evolutionary implications, Linnaeus envisaged the existence of a natural system, which we may assume coincides with the one quested by Lamarck for outlining the course of evolution.

Lamarck introduced an important innovation, namely, that classification should be based on anatomical and physiological features, as well as on the traditional external morphology. This led to his discovery of two kinds of evolution, which may be called *divergent* and *progressive* respectively.

The evolution of life is to a considerable extent constituted by fortuitous events which cannot be accommodated within a rigorous conventional system. However, since all classifications are based upon similarities and dissimilarities, it is unavoidable that Linnaean classification, at least that of animals, outlines the major features of the history of evolution. For a very long time biologists were satisfied with this crude simile, but the situation has changed radically during the last decades, owing notably to the contributions of Hennig (1966), the founder of cladism and of Sokal and Sneath (1963), the inventors of numerical taxonomy. The fundamental idea in Hennig's cladistic method is that the various features exhibited by living organisms are not of equal value in phylogenetic classification. Rather, some features, characterized as being primitive (plesiomorphous), are detrimental in phylogenetic classification, because they are misleading. Only derived (apomorphous) features lead to correct results.

An example illustrating this principle are the craniate animals, which are classed together in the Linnaean class Cyclostomata. This taxon comprises two orders of jawless fish, Myxinoidea (hagfishes) and Petromyzontidae (lampreys). The particular feature uniting these animals is the absence of jaws and teeth, and the consequently rounded shape of the mouth, as indicated by the class name. To a Linnaean systematist, this is an excellent taxonomic character, but the cladist will be suspicious, for the absence of jaws is surely a primitive feature in the phylum Craniata.

This does not necessarily imply that the Linnaean classification is wrong, but it urges the cladist to search for apomorphous features which are

TABLE 7·1 Characters shared by Petromyzontidae and Gnathostomata (from Løvtrup, 1977 and Janvier, 1981)

(1) Arcualia
(2) Radial muscles
(3) Spiral intestine (or spiral fold in intestine)
(4) No persistent pronephros in adult
(5) Collecting tubules in kidneys
(6) No accessory heart
(7) More than one semicircular canal
(8) Synaptic ribbons in retinal receptors
(9) Histology of adenohypophysis
(10) Blood volume below 10 per cent
(11) Hyperosmoregulation
(12) Properties of insulin
(13) Perfected immune response
(14) Nervous regulation of heart
(15) Chondroitin 6-sulphate in cartilage
(16) Absence of a unique dermatan sulphate
(17) Amino acid composition of collagen
(18) Peptide pattern and amino acid composition of hemoglobin (foetal hemoglobin)
(19) Photosensory pineal organ
(20) Structure and function of pancreas
(21) True lymphocytes
(22) No preanal median fold
(23) Braincase with larger cartilaginous walls
(24) Atrium and ventricle of heart close together
(25) Extrinsic eye musculature
(26) No naso-pharyngeal duct
(27) Perfected sensory line system (with true neuromasts)

common to one of the cyclostomatous orders and to the jaw-possessing vertebrates (Gnathostomata). The outcome of an attempt to follow this precept is given in Table 7·1, which shows that twenty-seven characters are shared between Petromyzontidae and Gnathostomata (Løvtrup, 1977; Janvier, 1981). * Consequently, the three groups should be classed as shown in Fig.7·1.

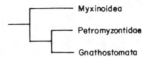

FIG. 7·1 Phylogenetic classification of the taxon Craniatia. It is seen that according to this classification the taxon Cyclostomata (i.e. Myxinoidea + Petromyzontidae) is no monophyletic taxon, and therefore ought to be suppressed. Instead we observe the existence of the taxon Vertebrata (Petromyzontidae + Gnathostomata). (From Løvtrup, 1977b.)

The result of modern phylogenetic classifications are dendrograms, overwhelmingly dichotomous, although occasionally three or more branches may occur (Fig. 7·2). We shall later see that this state of affairs is a logical consequence of the ecological theory on the mechanism of evolution.

Thus, the phylogenetic hierarchy which outlines the course or the history of evolution is represented by a largely bifurcated dendrogram. It is important to notice that *all* the vertical lines represent taxa, which increase in numerical rank as we move towards the base of the hierarchy, i.e. the top of the dendrogram. The frequency of bifurcation varies considerably from one region to another in the dendrogram, and consequently the terminal taxa, that is, generally those which are called 'species', may have greatly varying numerical rank. This implies that Linnaean concept of taxonomic categories is invalid, or at least deceptive.

The taxa arising at a branching, usually two in number, are called sister groups or twin taxa (Hennig, 1966; Løvtrup, 1977). The horizontal lines uniting such taxa represent an event of isolation.

There is one important inference which can be made from a phylogenetic classification, presuming that it truly represents the course of evolution. It is seen that in the hierarchy the superior taxa, kingdoms, phyla, etc., are located at the apex, and the inferior taxa, genera, species and subspecific taxa, at the base. From this we may conclude that the course of evolution has been from the superior to the inferior taxa. The species are found among the terminal or quasi-terminal taxa, showing that the Linnaean notion that the species stands out among the taxa is something unique, 'the evolutionary unit' in modern terminology, is a mistake. The species are not a category of taxa, but simply taxa among other taxa, to wit, relatively

* This list has been further extended by Hardisty (1982).

recent ones. Darwin's book, if correctly named, should have been called *On the Origin of Taxa*.

If we want to establish the evolutionary history of some taxon, for instance Mammalia (mammals), within the phylum Craniata, then we must follow the sequence of steps which limit the lower edge of the dendrogram (Fig. 7·2). Recalling that each vertical line represents a taxon, we find that

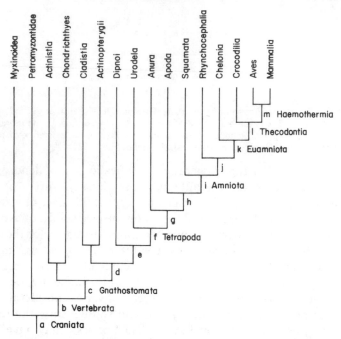

Fig. 7·2 In *The Phylogeny of VERTEBRATA* I listed several characters uniting the taxa Crocodilia and Mammalia (p.185 and p.204). Upon reading the typescript Dr. Colin Patterson pointed out that if I had consistently followed my own precepts I should have made the two taxa twins — or sister groups. If that had been done it would almost certainly have led to the union of the taxa Aves and Mammalia as a pair of twins. I did not do that. The reason was that the books on paleontology which I consulted all so unhesitatingly asserted the early origin of the mammal-like reptiles and hence also of the mammalian lineage. The fact that the specialists generally were willing to admit the reptiles having crossed the boundary to the mammals several times should have raised my suspicion; yet, I did not dare to challenge the authorities.

However, recently it has happened that a paleontologist (Gardiner, 1982) has come out in support of the classification which I should have proposed. In the present classification of the phylum Craniata I have therefore introduced the new classification of Mammalia, a classification which, I believe, deserves to be considered seriously, notably on the background of the characters listed by Gardiner. Relative to my previous classification I have made one more change, that is, I have separated Urodela and Anura, thus suppressing the taxon Amphibia. The reason for this is that nothing but plesiomorphous characters seem to unite these two taxa. I do not contend that the taxon cannot be a monophyletic taxon, but only that evidence in favour of this classification is missing for the time being. (From Løvtrup, 1985.)

the mammals have passed from Craniata (a) to Vertebrata (b), to Gnathostomata (c), through two unnamed taxa (d and e) to Tetrapoda (f), through two further unnamed taxa (g) and (h) to Amniota (i), through one more unnamed taxa (j) and finally through three for which I have adopted the names suggested by Gardiner (1982), *viz.* Euamniota (k), Thecodontia (l) and Haemothermia (m). If we disposed of a phylogenetic classification of the taxon Mammalia, resolved to the species level, we would be able to follow the history of an individual species through a number of inferior mammalian taxa, some of which would coincide with Linnaean taxa, while many would be unnamed. By tracking the phylogenetic dendrograms we may thus establish the evolutionary history of any taxon. It will be shown in the following section that at the same time we have outlined the course of ontogenetic development of the taxon in question.

The ecological theory on the mechanism of evolution is based on the *Axiom of Competitive Exclusion* (Hardin, 1960), which states that two taxa cannot subsist in the same ecological niche. If they happen to share a niche, one of the taxa, the dominant one, will cause the extinction of the other one. Living organisms do in fact live side-by-side, and it therefore follows that they do not share niches, consequently they subsist in isolation.

As the axiom is supposed to be generally valid, it must hold also for twin taxa, and it therefore follows that an act of isolation is a prerequisite for taxonomic divergence. There are various kinds of isolation, but the most important distinction is between random and non-random isolation. The former usually, but not always, depends on properties in the environment such as its geography, and may occasionally be broken. When that occurs, it may have fatal consequences for a previously isolated taxon.

Non-random isolation is based on properties possessed by the organisms themselves, involving ecological specialization of various kinds, feeding habits, protective devices, etc. Non-randomly isolated organisms are less likely to become extinguished through competition, but, of course, dominant forms may arise even in isolation. Whenever this kind of isolation happens to involve the colonization of a new environment (territory, niche, etc.) there is usually a very rapid dispersal of forms associated with taxonomic divergence. This 'adaptive radiation' is generally ascribed to an excessive 'selection pressure' prevailing under the new conditions. In my opinion, it is the abundant possibilities for isolation and thus for avoidance of competitive selection, which ensures the survival of new forms, and thus the occurrence of divergent evolution.

The extinction of one of the taxa sharing the same ecological niche exemplifies *interspecific natural selection*, resulting in an increase in dominance in the prevailing organisms. The outcome of this kind of selection is nothing but Lamarck's *progressive evolution*. Although he did not believe in progressive evolution himself, Darwin thus devised a mechanism to account for it.

We may now return to the phylogenetic dendrogram, and observe that if isolation is the prerequisite for divergent evolution, then we should indeed expect that, as a general rule, the hierarchy is dichotomous. But we can even account for progressive evolution, for when we face a pair of twin taxa, it will hold almost without exception that one of them is distinguished by the primitivity of its members, by a small number of subordinate taxa, by some kind of specialization or by some combination of these features. This is the isolated taxon. If we introduce the rule that the dominant twin is always turned to the right, then the lower side of the dendrogram will be bounded by a step-like line which represents a progressive phylogenetic lineage, with the most dominant taxon being situated furthest to the right (Fig. 7·2).

THE CREATIVE ASPECT

In discussing the historical aspect of evolution we had to begin with phylogeny; when treating the creative aspect we must deal first with ontogeny. And the subject which is of primary interest in this context is the so-called 'embryonic recapitulation'. In the beginning of the last century it was generally believed that during their development animal embryos pass through a succession of stages corresponding to the *adults* of more primitive forms. Thus, for instance, it was thought that the mammalian embryo at some stage is a fish, later on an amphibian, a reptile, etc.

This hypothesis, called 'Meckel–Serres law' (cf. Russell, 1916), is manifestly false. Yet, there is a reality behind the law; it is indeed possible to distinguish a succession of different forms in the course of embryonic development, but they never display adult features.

One may readily imagine that the sequence in question represents *embryonic* forms, but the great embryologist Karl Ernst von Baer (1828) came up with another suggestion, namely, that the succession outlines a passage 'from the general to the particular'. This expression implies that, say, a mammalian embryo in turn represents a number of superior taxa, among which Animalia, Craniata, Vertebrata, Gnathostomata, Tetrapoda, Amniota, Euamniota, Thecodontia and Haemothermia (Fig. 7·2), besides a further number of inferior taxa.

Since von Baer did not believe in evolution when he wrote the mentioned work, he did not realize that the generalization submitted by him represented a recapitulation of the epigenetic or creative aspect of evolution. But it does, and it is easy to understand why. For if the phylogenetic classification of living organisms depicts the historical aspect of evolution, it must also outline the creative aspect, for *the existence of an epigenetic mechanism capable of creating a particular organism is a necessary and sufficient precondition for the existence of the organism, provided, of course, that the external conditions are not prohibitive.*

Therefore, once upon the time it happened that an agnathous animal arose possessing the craniate body plan, which consists of a notochord, surrounded above by a spinal cord ending in a brain at the anterior end and by a row of metameric somites at each side, and below by the gut. *This body plan has been preserved throughout the evolution of phylum Craniata* and, as envisaged by von Baer, *its realization is the first step in the morphogenesis of every craniate embryo.*

This body plan prevails unchanged in the taxon Myxinoidea. If cartilage-producing cells engage in the formation of neural arches around the spinal cord, the vertebrate body plan prevails, as it occurs in the taxon Petromyzontidae. This feature, together with the other characters in Table 7·1, has the consequence that Petromyzontidae are united with Gnathostomata in the taxon Vertebrata. If, during the early development of the head, some cartilage producing cells, otherwise forming the first gill arch, accumulate to form the primordia of the jaw bones, the gnathostome body plan is realized. It is not possible to specify all of the following steps in the formation of the embryonic body plan, but it may be observed that if limb buds are formed, the embryo will develop into a tetrapod, otherwise, it will become a fish.

Whether the adult is going to weigh ten grams or ten tons, the embryo is never more than about ten millimeters long when it is settled into the major craniate taxa where it belongs. This is because the body plan is formed by cells, assisted by certain extracellular differentiation products. Cells are very small, and can operate only over distances a fraction of a millimeter long. For purely mechanical reasons, therefore, the embryos must have 'cellular dimensions' when the initial morphogenetic events take place. This early ontogenetic period may be called the phase of *form creation*.

The subsequent morphogenetic development is based on another mechanism, namely *differential growth*. Growth processes determine the absolute size of an animal, but the fact that these processes are differential means that the basic body plan is modified to conform to the inferior taxa to which the animal belongs. This modification advances so far that the specific, or at least the generic affinity is established at the time of hatching or birth. In general, the process of differential growth continues until the organism reaches sexual maturity, but no growth, i.e. increase in size is the preponderant aspect. It may be useful to call this third period of ontogenesis the phase of *allometric growth*, to distinguish it from the preceding one (Fig. 7·3).

Neither Meckel nor Serres were evolutionists, so their law does not imply any evolutionary recapitulation. However, when Ernst Haeckel, under the influence of Darwin, became an evolutionist, he adopted Meckel–Serres law, renamed it 'the biogenetic law', and claimed that it asserts phylogenetic evolution of adult forms to be recapitulated in the course of ontogenetic

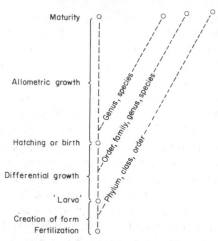

FIG. 7·3 An attempt to outline the ideas stated on von Baer's laws. The stippled vertical line represents a particular case of ontogenesis, say, of a reptile. This event has been subdivided into three phases (1) creation of form, stretching from fertilization to the onset of larval development, (2) differential growth, lasting until hatching and (3) allometric growth, distinguishing the period from hatching to maturity.

It is suggested that the body plans of the phylum, the class and, possibly, the order are determined during the first phase. It therefore follows that an evolutionary event giving rise to a new taxon of his high rank must involve changes in the very early stages of development. The affiliation to the lower ranking taxa, possibly order, surely family and most often even genus and species, is determined during the second phase. The evolutionary transfer from one of these taxa to another of same rank must therefore occur during this second phase of ontogenesis. Exceptionally, the genus and species characteristics may arise only after birth or hatching; this situation is typical in all changes involving terminal modifications, and thus the particular form of recapitulation called *Haeckelian recapitulation*.

development. This step did not make the law less ridiculous; as affirmed above, embryos are never miniatures of adults. Yet, in some particular cases a recapitulation occurs of the kind implied by Haeckel; namely, when evolutionary change involves terminal modification. This phenomenon always occurs during the phase of allometric growth, as for instance, when a species evolves to another one distinguished by a larger size. In that case the members of the latter species may pass through a stage corresponding to those of the former species. The phenomenon of neoteny may also belong here.

This 'Haeckelian recapitulation' is of relatively slight evolutionary importance. However, as we have seen, in the course of embryonic development there is a recapitulation of the course of evolution from 'the general to the special' as asserted for by von Baer. This phenomenon, much the most important one, I have called 'von Baerian recapitulation'.

It thus appears that there is a very intimate association between ontogeny and phylogeny, in fact, *ontogeny recapitulates the historical as well as the creative aspects of phylogeny*. I do not claim that the resolution is such that it is possible to make phylogenetic classifications on the sole basis of embryological studies. Nevertheless, I venture to make the following assertion: *in the course of their ontogeny the members of a set of twin (sister) taxa follow the same course of von Baerian recapitulation up to the stage of their divergence into separate taxa*. This idea was first, I believe, stated by Herbert Spencer, and I have therefore named the theorem after him (Løvtrup, 1981).

EPIGENETICS

Epigenesis is traditionally the antonym of preformation, and implies that living organisms are created through epigenetic processes during development. Consequently, and in accordance with the preceding discussion, the creative aspect of evolution must be covered by an epigenetic theory.

The current theory of evolution, the synthetic theory or neo-Darwinism, is based upon Mendelian genetics. For this reason the advocacy of an epigenetic theory of evolution may seem to imply that genetics and epigenetics are somehow antagonistic notions or disciplines. This is not so, for epigenetics represents a synthesis of numerous disciplines, concerning widely separated levels of organization, which deal with the creation of living organisms. Epigenetics thus comprises biophysics, biochemistry, molecular, cellular and developmental biology, anatomy, morphology, physiology, genetics, etc.

Epigenesis and information

It is an incontestable fact that the fertilized egg contains the information required to transform itself into an organism possessing all the features distinguishing a number of taxa of varying rank. Some will object to this statement that the environment plays a decisive role in the ontogenetic transformations, but so long as the parameters of the environment are kept within certain limits, their effects are rather slight.

For centuries the nature of this information was an enigma, but with the decipherment of the genetic code the situation has radically changed. For now it is claimed by many that *all* biological information resides in the genome. That this statement is made at all is astounding, because it is so easily refuted. (I have no space to rehearse the arguments here, but see, for example, Løvtrup (1974; 1983b) and Stent (1978).)

I will stress that the structural information 'to be a cell' is contained in the fertilized egg, and is transmitted to other cells through cell division. The information required for cell division presupposed the existence of a cell, but that is not enough, further necessary information is embodied in the centriole, which usually is supplied by the fertilizing sperm. Cell and centriole are indispensible elements in the process of cell division, but they are assisted by other elements at lower levels of organization, like microtubules and actin filaments.

The phenomenon of cell differentiation presupposes the existence of cells differing with respect to some parameter. If the fertilized egg were completely homogeneous, the cells arising through the early cell divisions would be identical, and no cell differentiation could take place. The circumstance that cell differentiation does occur is a consequence of the existence of inhomogeneties (polarities) in the fertilized egg. The polarities control the initial patterns of differentiation, the later ones result from interactions of various kinds, mainly between the products of earlier cell differentiations. So whereas it is correct to claim that the genome contains the information required for the synthesis of the various proteins which characterize each of the several patterns of cell differentiation observed in adult animals, it nevertheless holds that extragenomic, presumably extranuclear and extracellular factors determine which part of the genome is going to be transcribed. This fact underscores that the genome does not contain all the information required for the epigenetic processes. In passing, it may be mentioned that sometimes it is attempted to resolve this difficulty by invoking regulatory genes. This expedient may, indeed, be involved, but it must be presumed that the regulatory genes are the same in all cells.

Finally, it should be mentioned that the nature of the morphogenetic processes which follow upon the onset of cell differentiation depends, among other things, upon the spatial location of the various cell types. This, again, involves a kind of information which cannot, and does not, reside in the genome.

Epigenetic Mechanisms

Three different kinds of epigenetic mechanisms may be distinguished: cell division, cell differentiation and morphogenetic events. The sequence in which they are listed here represents a causal succession. That is, cell division is a prerequisite for cell differentiation and it is also responsible for the earliest and simplest morphogenetic event, the conversion of the egg, a solid sphere, into a hollow sphere, the blastula, which is the consequence of the cleavage divisions in the early embryo. All subsequent morphogenetic activities are consequences of the combination of cell divisions and cell differentiations occurring in the course of development.

Cell division

Cell divisions begin soon after fertilization. It has been found, notably in vertebrate embryos (fish and amphibia) that the first ten to fifteen cell divisions are extremely fast and almost synchronous (metachronous). The high rate may, in part, be referred to the fact that the G_1 and G_2 phases of the cell cycle are missing, but even the S phase is of very brief duration. In fact, in many embryos, the S and the M phases are of approximately equal length, so that the cells are in interphase half the time, and undergoing mitosis the other half (*for details, see* Graham and Morgan, 1966; Signoret and Lefresne, 1974).

The synchronous cell divisions seem to depend upon an exhaustible cytoplasmic resource involved in the synthesis of DNA. Various results support this suggestion (1) in haploid embryos the synchronous divisions go on for one round more than in diploid embryos, (2) substances that inhibit *de novo* synthesis of nucleotides are without effect during the synchronous cell divisions, but inhibit the subsequent development and (3) there is a store of deoxyribonucleotides in the fertilized eggs of various amphibia, large enough to sustain the synthesis of DNA during the period of synchrony (*see* Løvtrup *et al.* (1978) for detailed discussion).

The importance of cell division for cell differentiation will be dealt with in the next section.

Growth, that is, increase in size, may be included under the heading Morphogenetic events; but it may also belong to the present section. The reason is that within many superior taxa, like classes or orders, cell size is remarkably constant, and that means that variation in body size must be correlated with the number of cells. This evidently implies that the number of successive cell divisions is controlled. If this control is general, it seems to imply that evolutionary variation in size can occur only in steps that are multiples of two (cf. Løvtrup, Rahemtulla and Höglund, 1974; Roff, 1977; Løvtrup, 1977a).

Cell differentiation

Cell differentiation implies the presence, in the same organism, of cells which utilize different parts of the genome and which, consequently, are distinct with respect to the chemical activities into which they engage.

From this definition it follows that cell division is one precondition for cell differentiation, another is that the cells cohere to form a multicellular organism; cell differentiation is known only in metazoa.

It is a generally accepted dogma that, barring mutations, the genes are faithfully reproduced during mitosis, and that consequently the genomes are identical in all nuclei in any individual organism. On this premise it follows that the phenomenon of cell differentiation must be governed by extragenomic, and most likely by extranuclear factors. This circumstance permits the formulation of the third precondition for the occurrence of cell differentiation, namely, the existence of unequal cell divisions. As pointed out already, unequal cell divisions requires the existence of polarities, cytoplasmic or cortical, as the case may be, in the unfertilized egg, or the creation of such polarities in the course of development. Such polarities occur in all embryos, but as a general rule it may be asserted that their number and their sophistication increase as we ascend the classificatory hierarchy of the animal kingdom. Figure 7·4 outlines the polarities which can be demonstrated in the amphibian egg. There are six altogether, of which four control processes of cell differentiation and two control morphogenetic events.

Cell differentiation occurs continuously in living organisms, but it is most important in the early embryo, because at this stage the cell differentiation patterns originate *de novo*, so to speak. Many attempts have been made to study the mechanisms of cell differentiation in the early embryo, and thus the influence of the various polarities. One favourite method has been to study the differentiation *in vitro* of explants, isolated notably from chicken and amphibian embryos. There are two potential sources of error in this approach, which in many cases have hampered the proper interpretation of the results obtained. One is to work with too large explants, in which the process of cell differentiation may be blurred by the simultaneous occurrence of morphogenetic events of various kinds. The second is to work with explants taken from relatively late embryos, in which an unknown number of interactions may have occurred, which may affect the differentiation of the isolated cells.

In my laboratory (Løvtrup *et al.*, 1978; Løvtrup, 1983a) these potential sources of error were avoided by working with very small explants isolated from various regions of the axolotl blastula (Fig.7·5). When such explants are kept in culture for some days, it turns out that three *spontaneous* patterns of differentiation occur. (i) The ectodermal cells (regions (2), (4), (5), (6) and (8) differentiate into epidermal cells which, undergoing a simple morphogenetic process, form spherical ciliated vesicles, which move around

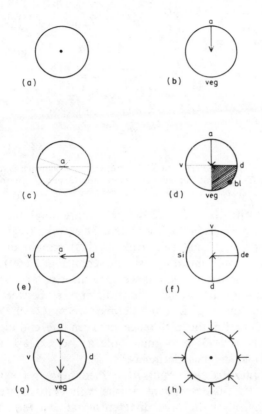

FIG. 7·4 Symmetries and polarities in a sphere. (a) Spherical symmetry: any plane including the centre divides the sphere into two equal halves. (b) An apical polarity vector defines two points (poles) on the surface. (c) b seen from above. Radial symmetry: any plane including the polarity vector divides the sphere into two equal halves. (d) A median polarity defines two further points on the surface. (e) d seen from above. Bilateral symmetry: only the plane including both polarity vectors divides the sphere into two equal halves. (f) Same as e, but rotated 90°. A dextro-sinistral polarity defines two further points on the surface. Bilateral asymmetry: no plane any longer divides the sphere into equal halves. (g) The animal-apical and vegetal-apical polarities. (h) The normal polarity. The small letters refer to the terminology used for the amphibian embryo: (a), animal pole; (bl), blastopore; (d), dorsal side; (de), right side; (si), left side; (v), ventral side; (veg), vegetal pole. The dorso-vegetal quadrant is cross-hatched. This region holds most of the grey crescent and it is here the incipient blastopore, and thus the first Ruffini cells are formed. (From Løvtrup, 1981.)

in the culture dishes (Fig. 7·6(a)). (ii) The endodermal cells from regions (1)
and (9) in the vegetal hemisphere form fibroblast-like cells which migrate
on the bottom of the dishes by means of slender filopodia (Fig.7·6(b)).
From their location and their behaviour it may be inferred that these cells

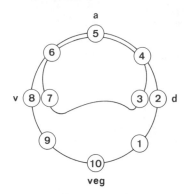

FIG. 7·5 Diagram outlining the regions of the blastula from which explants were taken. (4), (5)
and (6), animal ectoderm; (2) and (8), equatorial ectoderm; (3) and (7), mesoderm; (1) and (9),
endoderm; (10) circumpolar vegetal cells. (From Landström and Løvtrup, 1979.)

are the so-called 'Ruffini cells' known to initiate amphibian gastrulation.
(iii) The mesodermal cells in the interior of the embryo (regions (3) and (7),
and the circumpolar vegetal cells (region (10)) remain undifferentiated,
forming aggregates surrounded by a surface coat (Fig. 7·6(c)).

To account for these observations we need three polarities; (1) the apical
polarity is required to explain the difference between the animal
(ectodermal) and the vegetal (endodermal) cells, (2) the vegetal-apical
polarity to account for the difference between the endodermal and the
circumpolar cells and (3) a normal polarity to explain the difference
between ectodermal and mesodermal cells.

In order to interpret the results described here, we assume that the
fertilized egg and all its descendants in the early embryo are programmed
for one and only one pattern of differentiation and that, therefore, the
origin of the other patterns requires inductive processes. The original
pattern may be either the one represented by the undifferentiated cells, or it
may be one of the other differentiated patterns. In the former case, two
induction processes are required to account for our observations, in the
latter case, one induction suffices.

The principle of parsimony dictates the adoption of the second alternative.
In order to settle this question, we must refer to the phenomenon of primary
induction. During gastrulation endodermal Ruffini cells invaginate the
blastocoele in a process that starts at the dorsal side and gradually spreads
right and left until the ventral side is reached. As a consequence of this

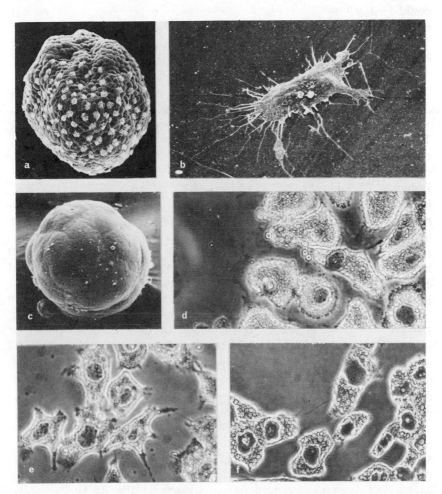

FIG. 7·6 (a) Vesicle consisting of ciliated and non-ciliated epidermal cells (75X); (b) endodermal sf-cell (Ruffini cell) (750X); (c) aggregate consisting of large undifferentiated endodermal cells, kept together by a surface coat (50X); (d) endodermal sf-cells, distinguished by their large size and their content of yolk platelets (300X); (e) ectodermal sf-cell, smaller than the endodermal ones, and loaded with pigment granules confined primarily to the filopodia (300X); (f) mesodermal sf-cells, intermediate in appearance between the other two types (300X). All three kinds have filopodia, but these are only distinct in the ectodermal cells.

activity the epibolic movements begin which ultimately lead to the formation of a gastrula. However, when contact is established between the invaginating cells and the ectodermal cells residing in the surface of the embryo, the latter are 'induced' to differentiate into nerve cells, which later contribute to the formation of the central nervous system. Owing to the fact that invagination begins at the dorsal side, the cells determined to become nerve cells are mainly located at this side (Fig.7·7).

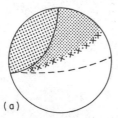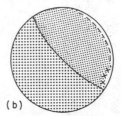

FIG. 7·7 (a) Fate map of the urodele blastula. The four regions outlined represent, from north-west to south-east' epidermis, neural plate, notochord and endoderm. x·x·x, limit of invagination. (b) Diagram showing the topological transformations occurring during gastrulation. The radical deformation of the notochordal primordium is indicated, but it should be recalled that it is situated below the presumptive neural plate; it is thus not visible from the outside. (From Løvtrup, 1983a.)

This induction phenomenon has been studied *in vitro* for more than half a century (cf. Saxen and Toivonen, 1962). It was observed that a number of quite unspecific agents cause the ectodermal cells to differentiate into nerve cells, a fact which does not accord with the notion that we are dealing with an instructive event. Indeed, at an early stage, it was suggested that the ectodermal cells are actually programmed to become nerve cells, but that they need some kind of activation in order to realize this destination. In the absence of such a stimulation the cells will become epidermis (Woerdeman, 1936). This proposition has recently been corroborated by two sets of experimental observations: (1) it has been found that unspecific substances like Li^+ and diverse sulphated glycosaminoglycans, as well as activating agents like ionophores, cyclic nucleotides and hormones, mostly in exceedingly low concentrations, may ensure the transformation of ectodermal cells into nerve cells; (2) it has been found that before the ectodermal and mesodermal cells differentiate into the patterns described below, they first assume a shape reminiscent of the Ruffini cells (cf. Fig. 7·6(d) (e) and (f)). Although the three cell forms remain morphologically distinguishable owing to differences in the content of pigment granules and yolk platelets (Fig. 7·6(f)-(h)), it nevertheless seems probable that these cells represent the fundamental cell type which is coded for in the genome of the fertilized egg (Løvtrup and Perris, 1983).

However this may be, our studies have shown that the outcome of this *activation* is not only nerve cells; rather, in a specific temporal sequence we observe the formation of mesenchyme cells, nerve cells and melanophores. Thus it appears that activation does not give rise to one pattern of cell differentiation, but to a spectrum which represent the *neural crest*. The circumstance that one activating stimulation can provoke a repertoire of different patterns reveals a mechanism of differentiation, *auto-differentiation*, the nature of which is unknown at the present time.

It should be observed that the neural crest cells arise only in aggregates taken from the regions (4), (5) and (6). Cells from the equatorial ectoderm (regions (2) and (8)) give rise to notochordal cells and collagen-producing fibroblasts, and mesodermal cells (regions (3) and (7)) to muscle cells. We here observe effects of the animal–apical and the normal polarities.

If the ectodermal cells are actually programmed to differentiate the way we observe in our cultures, then the formation of epidermis must involve an *induction*. We do not know anything about this phenomenon except that it requires that cells are exposed to a dilute medium (Landström, 1977).

There are some other induction phenomena observable in amphibian embryonic cells. Thus, in the presence of tyrosine many cultured cells, including mesenchyme cells and nerve cells, become black, demonstrating that they have begun to produce melanine. In the presence of guanosine triphosphate (GTP) many cells become yellow, showing that they produce pteridine. Tyrosine and GTP are substrates for the synthesis of melanine and pteridine, respectively, but as they exert their effect a long time before the synthesis begins, it may be concluded that they function as inductors rather than as substrates.

The last mechanism of cell differentiation we will mention is *inhibition*. This phenomenon is observable in the epidermis where it appears that as a general rule a ciliated cell inhibits its neighbours from becoming ciliated.

Very few cell differentiations occur during the phase of form creation, the great majority of new cell types originate in association with the histodifferentiation occurring during the initial stages of the phase of differential growth. All kinds of interactions must be invoked to explain the appearance of all these patterns of differentiation, but nothing is known about the mechanism involved.

The succession from the single cell represented by the fertilized egg to the repertoire of patterns found in the fully differentiated organism presumably may be represented by a bifurcated dendrogram. Each branching requires at least one instructive induction. At the terminal ends we may expect to find the various patterns found in the adult body, but the way they come into existence suggests that it may be possible to classify cells in some hierarchical system.

Following this idea it has been suggested (cf. Løvtrup, 1983a) that there are two main classes of cells, *solocytes*, distinguished by their motility and by the fact that they do not readily form stable aggregates, and epitheliocytes — or *colligocytes*, which through their adhesiveness tend to form stable aggregates. Since the formation of embryos to a large extent may be considered as topological transformation of epithelial aggregates, it is evident that the epitheliocytes are of great morphogenetic importance.

The contribution of the solocytes is due to their motility, which implies that they are able to carry out dynamic activities of various kinds. Many of the morphogenetic transformations which take place in the embryo are performed by solocytes. We may distinguish between two kinds of solocytes, *solo-lobocytes* (*sl*-cells), which form lobopodia and short actin filament-containing filopodia and *solo-filocytes* (*sf*-cells) which, because they contain cytoplasmic microtubules, may form very long filopodia.

Morphogenetic events

The morphogenetic elements are partly cells, partly extracellular elements, macromolecules produced and secreted by various cells. The topics dealt with here have been treated in greater detail in a recent review (Løvtrup, 1983a).

Colligocytes

Cells appear to conform to the physical law of minimum surface area: consequently, in the absence of any deforming forces cells will be spherical, as exemplified by the undifferentiated embryonic cells. Deviations from this shape require the application of deforming processes, involving either adhesion residing in the cell surface, or a cytoskeleton in the interior of the cell. Both kinds of forces are involved, but the former are of particular importance in colligocytes, and the latter in solocytes.

When ectodermal cells differentiate into epidermis, they form epithelial vesicular aggregates. Epithelial cells are known to be kept together by specialized junctions of various kinds which can only be established after the aggregates have been formed. So the initial aggregation must depend upon adhesive properties in the cell surface. It is known that cells are covered by a glycocalyx consisting of glycoproteins and proteoglycans, and it is likely that these determine cellular adhesivity. The circumstance that Ca^{2+} may be required for aggregate formation suggests that some sulphated substance, probably chondroitin sulphate, is involved.

When epithelial cells engage in aggregate formation, the adhesive forces residing in the surface will deform the cells, making them assume the shape of equilateral hexagonal prisms (Fig 7·8). This appears to be a compromise between the adhesive forces striving for maximum surface contact and the

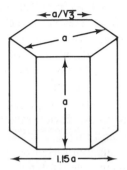

FIG. 7·8 The equilateral hexagonal prism with minimum surface: volume ratio. (From Løvtrup, 1974.)

cortical forces which aim to establish minimum surface area. The situation described here corresponds to the so-called 'cuboidal' epithelia; 'columnar' epithelia are made up of elongated cells, a fact which can be explained by the circumstance that these cells contain a cytoskeleton of microtubules.

In vitro, two kinds of aggregates are observed with explanted ectodermal cells, *two-dimensional* and *three-dimensional*, respectively. In the former case the cells are attached to the surface of the culture dish, in the second they mostly form spherical aggregates; in either case the cells are ciliated. The ratio between cell–substrate and cell–cell adhesion appears to determine the nature of the aggregate. The formation of spherical aggregates suggest that even in this case the law of minimum surface area prevails. Whenever epithelial aggregates are found in the embryo, whose shape deviate from a sphere, the involvement of extrinsic formative forces may be presumed.

The various epithelial aggregates formed in the interior of the vertebrate embryo, notochord, central nervous system, myotomes, etc. as well as the inside of the epidermis, are covered by sheaths of reticular collagen. In the amphibian embryo the interspaces between these aggregates is occupied by this collagenous network, through which the neural crest cells are observed to migrate. In later developmental stages this layer of collagen is in many cases covered by a basement lamella, consisting of very stout collagen fibrils arranged in orthogonal plies. Both the reticulous and the fibrillar collagen are deposited in a matrix in which sulphated proteoglycans are an essential component.

Solocytes

It has already been mentioned that there are two kinds of solocytes, the sl-cell represented by the cells in the early embryo and the sf-cell, a fibroblast-like cell type. Here the difficulties involved in cell classification become obvious, for in the embryo, as well as in cultures the sl-cells actually form aggregates. As shown by Holtfreter (1943), the cells in the early embryo are kept together by a surface coat, which serves to unite the cells into a supracellular unit. The surface coat, which probably is a glycoprotein, is dissolved at extreme pH values and in the absence of Ca^{2+}.

I do not know if this is a general sl-cell feature, but in any case there are several ways to distinguish the sl-cell aggregates from those made by colligocytes. Thus, the sl-cells do not form vesicular, but solid aggregates, they do not produce collagen, and they are not kept together by the typical specialized epitheliocyte junctions, but rather by short filopodia which form tight junctions with neighbouring cells. Isolated embryonic sl-cells typically produce lobopodia which circulate around the circumference of the cell. *In vivo* they seem to migrate by means of filopodia, which are known to contain actin filaments.

The solo-lobocyte has no other cytoskeleton than the cell cortex, and therefore it is principally a spherical cell, although it may become elongated under certain conditions. In contrast, the solo–filocyte has a cytoskeleton consisting of cytoplasmic microtubules.

There are many patterns of cell differentiation which form subgroups (families) within the order of sf-cells, for instance, mesenchyme cells, nerve cells, glia cells, muscle cells, cartilage cells, melanophores, etc.

It is therefore difficult to give an unambiguous characterization of an sf-cell, but probably the cell characterized above as a Ruffini cell is close to the original type. If this is true, then the sf-cell is a slightly elongated cell with filopodia extended in all directions and attached to appropriate surfaces. Evidently it moves by means of these filopodia, and we may presume that they are also the tools through which the cells may carry out the work involved in diverse morphogenetic processes.

Certain sf-cells flatten out on the substrate and move by means of a leading lamella. Since this kind of structure is found even in epitheliocytes, it may demonstrate that some sf-cells have surfaces with adhesive properties resembling, but not necessarily identical to those found in the colligocytes.

Other sf-cells, as exemplified by the secondary mesenchyme cells in the sea urchin embryo and by nerve cells in the amphibian embryo, are able to extend extremely long and rather rigid filopodia. In the former case these filopodia are engaged in the mechanical work required to bring about gastrulation. In the second case the long filopodia establish contact with distant parts of the body.

Like the colligocytes the solo-filocytes produce proteoglycans. It seems highly probable that the production of heparan sulphate is a characteristic of the whole order of cells; indeed, it is possible that the typical behaviour of these cells depends upon their content of heparan sulphate. Another proteoglycan produced by sf-cells is hyaluronate. In the higher animals this substance is produced by the cells representing the pattern of differentiation called mesenchyme cells.

Hyaluronate characteristically allows for swelling through the oncotic uptake of water. In the cavities thus formed sf-cells have free access, they may enter such 'morphogenetic spaces' and participate in various morphogenetic activites among which one of the most important is the formation of cartilaginous skeletal elements. Chondrogenesis begins with the production of hyaluronidase by the sf-cells, subsequently they form agglomerations and begin to synthesize chondroitin sulphate and collagen, thus contributing to cartilage formation. In most vertebrates the cartilage subsequently is transformed into bone, and the ossification processes through which this occur are of great morphogenetic importance.

Epigenesis

We are now ready to use our knowledge in an attempt to account for at least certain aspects of morphogenesis in a specific instance of embryogenesis. The case selected here concerns amphibian development.

Amphibian morphogenesis

As the result of a number of synchronous cell divisions in the early amphibian embryo, a blastocoele is formed. The swelling involved is either oncotic or osmotic; certain observations actually suggest that potassium ions are pumped into the blastocoele. The embryonic cells are kept together by an extracellular membrane, a surface coat. If the latter is dissolved the cells dissociate.

With the formation of the blastocoele the topological preconditions for gastrulation prevails. The mechanical one consists of cell differentiation leading to the formation of a new type of cell, solo-filocytes, which in the amphibian embryo are called 'Ruffini cells'. The function of Ruffini cells in the amphibian embryo seems to be confined to pulling the embryonic surface into the blastocoele, thereby ensuring contact between two layers of membrane-bound cells. Once this has happened, the two cell layers undergo displacement. The interior layer migrates towards the animal pole, the only direction available for advance, while the exterior cell layer is shifted in the opposite direction. It has been found that the interior cells move by means of filopodia which they attach to the cells of the outer layer (Nakatsuji, 1976). The latter may be displayed passively, but it is also possible that the cells are actively engaged in these epibolic movements.

The formation of the Ruffini cells begins at the dorsal side of the embryo and spreads from there in both directions around the circumference of the egg. There is thus a dorso-ventral polarity, prevailing in the form of a temporal gradient, imposed upon the differentiation of the endodermal sf-cells. It is easily visualized that this gradient is a topological precondition for the process of gastrulation; if the invagination began simultaneously everywhere around the circumference of the blastula, extrogastrulation would take place, leading to the separation of the germ layers. It is seen that in contrast to the polarities discussed earlier, the dorso-ventral polarity does not determine the origin of a particular pattern of differentiation, but rather the spatio-temporal distribution of the cells representing a particular pattern. In this way the polarity controls a morphogenetic event, the process of gastrulation.

In this context it may be mentioned that the last of the polarities discussed in Fig. 7·4, the dextro-simistral polarity, also controls a morphogenetic event, the bilateral asymmetry of the vertebrate body.

Gastrulation comes to an end when the blastopore is closed. At this stage, the endoderm and the notochordal primordium are found in the interior of the embryo, the outside being taken up by the epidermal and the neural plate primordia.

Although invagination occurs all the way around the embryo, cell migration is confined to a narrow strip around the dorsal median. The reason for this is that elsewhere the outer ectodermal and the inner endodermal cell layer are separated by a layer of free mesodermal cells, which prevents them from establishing the contact necessary for the migratory activity. As a consequence, the invaginated ectodermal cells in the dorsal median undergo a very extensive topological transformation, as shown in Fig. 7·7.

The next step of morphogenetic importance concerns the differentiation of the notochord. At the outset the presumptive chordocytes have the appearance typical of embryonic cells, but subsequently they become adhesive like epitheliocytes; one might say that they form a 'one-dimensional epithelium' (Fig. 7·9). This suggestion is supported by the fact that the cells become covered by a sheath which in the electron microscope is found to consist of a reticulum of collagen, a typical epitheliocyte feature.

The further differentiation involves a substantial swelling of the cells through binding of water, a phenomenon which may be referred to their content of hyaluronic acid (Mathews, 1975). The pressure thus developed is balanced by the tension in the sheath, and together these forces impose a

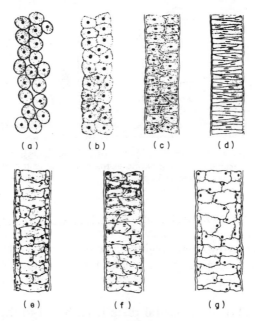

FIG. 7·9 Successive stages in the differentiation of the amphibian notochord. (a) Late gastrula, (b) early neurula, (c) neural plate, (d) early tail bud, (e) middle tail bud, (f) late tail bud, (g) larva. (From Mookerjee, Deuchar and Waddington, 1953.)

considerable rigidity on the notochord. Initially the rod shape of the latter is determined by the shape of the agglomeration of presumptive chordocytes in the dorsum, but the notochord continues to grow at the posterior end. The combined effect of these swelling and growth processes is a considerable elongation of the notochord and of the body.

At the same time the 'induced' ectodermal cells differentiate into nerve cells, a phenomenon which entails an elongation of the cell bodies. On the premises given above this elongation must be referred to the presence of a cytoskeleton of microtubules. This elongation implies by itself a shrinkage of the apical ends of the cells, a phenomenon which is further amplified by rings of contractile actin filaments situated at the outer ends of the neural plate cells (Wessels *et al.*, 1971). This cell differentiation causes a condensation and a curving up of the neural plate primordium (Fig. 7·10).

It is possible to account for the location and the direction of the actin filaments on two premises: (1) a microtubular activity similar to that suggested to be responsible for the location of the ring of actin filaments involved in cytokinesis and (2) the assumption that actin filaments cannot become attached to cell surfaces in close contact with another cell surface or

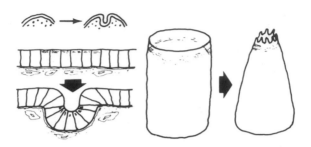

FIG. 7·10 Schematic illustration of the participation of actin filaments in the morphogenesis of the neural plate. Actin filaments forming a ring at the outer end of the neural plate cells are supposed to deform the cells as shown in the figure, and thereby cause the curvature of the neural plate. (From Wessels *et al.*, 1971.)

with the basement membrane. This implies that in epithelial cells actin filaments can become attached to the external surface, as ectoplasm or microvilli, or in narrow lateral region close to the outer surface, where the cells are separated by an external matrix (cf. Løvtrup, 1976). These two premises together explain the formation of the observed ring of actin filaments.

The posterior part of the neural plate is anchored to the notochord by means of cellular filopodia. Holtfreter (1943b) showed that this phenomenon is evidenced by the median neural groove. When the notochordal stretching takes place, the attached part of the neural plate becomes elongated at the same time. This physical association between two cell aggregates is thus the causal basis for the subdivision of the vertebrate nervous system into brain and spinal cord (cf. Jacobson, 1978).

The epidermis and the neural plate are separated by the neural crest. The latter is a source of sf-cells, representing, as we have seen, a variety of differentiation patterns; these cells enter the body during the subsequent development. The cells of the remaining ectodermal primordia, that is, notochord, neural plate and epidermis, all differentiate into colligocytes, a circumstance which may be a consequence of their exposure to the external dilute medium. If this is correct, it is curious that the neural crest cells avoid this differentiation.

At a time when the body is still only moderately elongated the sides of the neural plate are raised so much that the neural crests at some places touch and fuse. As this process spreads forwards and backwards the hollow central nervous system is formed, being simultaneously enclosed by the epidermis. We have already outlined a mechanism which may account for the rounding up of the neural plate. This mechanism does not, however, suffice to explain the closure of the central nervous system. This was shown by isolating the neural plate from amphibian embryos with and without the underlying notochord. In the former case neural tubes were formed, in the latter not (Jacobson, 1978). The elongation of the neural plate resulting from the stretching of the notochord is thus instrumental in the closure of the nervous system. One aspect of the mechanism is that the neural crests are brought closer to each other as a consequence of the elongation.

It is a distinguishing difference between the head and the trunk that extensive swelling occurs in the former, both inside and outside the brain. With respect to the nervous system the difference between brain and spinal cord may in part be explained by the fact that the latter is surrounded by a collagenous sheath which must resist any extensive swelling. This alone does not suffice to explain the difference in swelling between head and trunk. We know with great probability that swelling implies the presence of hyaluronate and in turn that of mesenchyme cells. There are two potential sources of this surplus of mesenchyme cells in the head—the prechordal plate mesoderm and the cranial neural crest; present knowledge does not allow to decide between these alternatives. But it is evident that the extensive swelling in the anterior part of the body must be referred to an antero-posterior gradient in the sequence of autodifferentiation—mesenchyme cell–nerve cell–pigment cell—discussed above.

It seems that the distinction between head and trunk induction made by many students of embryonic induction may be referred to the presence or absence, respectively, of hyaluronate-producing mesenchyme cells. In passing, it should be noticed that the particular shape of the anuran tadpole, involving an enormous swelling of the body cavity, must reflect a radical change in this gradient.

As the body is stretched, the mesodermal cells lying at both sides of the notochord become elongated accumulations of cells. Like the chordocytes they differentiate into colligocytes along an antero-posterior gradient. However, the dimensions of the cell agglomerations are now so large that two rows of epithelial vesicles are formed rather than a one-dimensional aggregate. This process is the mechanical precondition for the metamery of the vertebrate body. As may be expected, the vesicles become surrounded by reticulous collagen, which, among other things, prevent them from fusing together.

Between the vesicles cavities are formed which are accessible to sf-cells, mesenchyme cells, producing hyaluronate, chondrocytes, etc. The vertebrae and the ribs are formed in these cavities. This skeletal metamery is in turn the basis for the metamery of the spinal cord.

The present discussion could be continued, for we know still more about the epigenesis of the amphibian embryo, for instance the mechanisms involved in the formation of the tail, the median fins, the limbs and the gill clefts. But I believe that the points brought up here suffice to support my views concerning the relation between ontogeny and phylogeny which I shall present in the conclusion.

CONCLUSION AND SUMMARY

Of the four theories of evolution mentioned in this chapter, two are of particular importance for the topic of phylogeny and ontogeny: Lamarck's theory on the history of evolution and the epigenetic theory on the mechanism of evolution.

The salient point which issues from the combination of these two theories of evolution is that evolution has consisted in a gradual modification of a limited number of basic body plans, and that this course of events is *represented* by the phylogenetic classification, but it is *recapitulated* in the course of ontogenetic development. Inverting Haeckel's biogenetic law, we may assert that ontogeny is the mechanical cause of phylogeny. And it must be so, for ontogeny is a mechanical process, while phylogeny is a historical phenomenon. We may concur almost completely with Garstang when he writes (1922, p.98): 'The phyletic succession of adults is the product of successive ontogenies. Ontogeny does not [only] recapitulate Phylogeny: *it creates it*' (my italics).

For the understanding of this problem it is of great importance to distinguish between three ontogenetic phases, to which I have proposed the names of 'form creation', 'differential growth' and 'allometric growth'.

The phase of form creation is over before the embryo has reached a size of ten millimeters, and at this stage it is already possible to distinguish the superior taxa to which the organism is going to belong. The process of form creation in most animals is a spatial, to a certain extent a topological transformation of the spherical fertilized egg. It is also a mechanical process, in which physical forces of various kinds are deployed. Among these forces contractile and adhesive forces may be associated with certain cell types, whereas swelling and tension may be referred to specific intra- and extracellular elements.

The most typical feature of form creation is that very few patterns of cell differentiation arise at this early ontogenetic stage. Thus, the synthesis of informational RNA is limited during this phase, and this shows that only a very modest part of the genome is required for the creation of the basic body plan.

At a certain stage of development, presumably around the transition between the stages of form creation and differential growth, it may be impossible to say whether a particular embryo is going to be a mouse or an elephant. This is decided during the second ontogenetic phase, in which differential processes of growth, affecting primarily, but not exclusively the skeleton, determine both the absolute size and the several peculiar features characterizing the various interior taxa to which the organism belongs.

Physiologically the control of these differential growth processes is probably very simple; it may be presumed that hormones are the main regulatory agents. However, in many cases the rates of growth are spatially distributed in a way which cannot be explained on the basis of a humoral agent. No acceptable idea has yet been advanced to account for this phenomenon.

The process of allometric growth is well known to most people at least as concerns man and domesticated animals. Most often the characters developed during this phase are subspecific or even individual ones. The most significant event taking place is sexual maturation. It may be noticed that in some cases evolutionary innovation has been associated with neoteny, that is, acceleration of sexual maturation relative to morphological development.

The present survey is incomplete but, if I have been reasonably successful, I ought to have established the following points. First of all, the present reductionist approach, which claims that biology, including ontogeny, should be studied only at the macromolecular or cellular level, is doomed to fail. The creation and survival of the metazoans also involve processes going on at supracellular levels of organization, and these must be studied at the respective levels. Second, morphogenesis — and evolution is primarily a matter of morphogenesis — is accomplished by a number of relatively simple mechanisms. And changes in these simple mechanisms will generally involve considerable changes in the ontogenetic end product, that is, they will be macromutations. Such innovations may presumably, but not necessarily, be traced back to the genome, sometimes in the form of structural genes, sometimes as more extensive modifications.

As an example of a macromutation affecting the phase of form creation I may mention the origin of hyaluronate. We know that this substance is of great morphogenetic significance, and that in the animal kingdom it is found only in the higher taxa. And since the possession of the enzymes required for the synthesis of hyaluronate cannot arise through natural selection, it must be concluded that the first time hyaloronate was synthesized by an animal embryo, it was the result of the felicitous concurrence of several fortuitous events. Yet, once this has happened, the power to synthesize hyaluronate has no doubt been zealously preserved through selection.

We have seen that the growth processes taking place during the phase of differential growth are of great importance for the subsequent morphogenesis. Factors controlling growth, notably that of the skeleton, are involved in this phase, and changes in these must in many cases have entailed substantial morphological changes, amounting to macromutations. What are the nature of these changes? Is it possible that modifications of the chemical composition of various hormones are instrumental, or are we to go to the cellular level in order to find the proper explanation? We do not know, but one thing is sure, a purely humoral mechanism does not suffice.

Finally, as an example demonstrating evolutionary changes occurring during the phase of allometric growth we may mention the colour pattern of hairs and feathers in mammals and birds. These features, which constitute distinctions at the specific or generic level, arise after birth or hatching. Even here, we are most likely dealing with macromutations, for the colour of hairs and feathers is generally due to the presence or absence of particular chemical substance. Even if grey coloured animals exist, it is not likely that the change from melanic to white (albinotic) forms, or *vice versa*, usually follows scale of intermediate grey tones. As a matter of fact, all the known cases contradict this supposition.

The third point is that evolution is a question of using old materials to create new forms. This implies that the information prevailing in the genome is only distantly related to the processes of evolutionary change revealed by the phylogenetic classification. Consequently, in conclusion, we must accept that the study of evolution must now progress beyond 'neo-Darwinism', that is, beyond the micromutation theory which during almost two centuries has passed under a number of different names, without, therefore, transgressing the line of demarcation between metaphysical and empirical theories.

REFERENCES

Baer, K. E. von (1828). 'Ueber Entwickelungsgeschichte der Thiere Beobachtung und Reflexion'. Gebrüder Kornträger, Königsberg, Prussia.
Gardiner, B. G. (1982). Tetrapod classification. *Zool. J. Linn. Soc. (London)* **74**, 207–232.
Garstang, W. (1922). The theory of recapitulation: A critical restatement of the biogenetic law. *Zool. J. Linn. Soc. (London)* **35**, 81–101.

Graham, C. F. and Morgan, R. W. (1966). Changes in the cell cycle during early amphibian development. *Dev. Biol.* **14**, 439–460.

Haeckel, E. (1866). 'Generelle Morphologie der Organismen: Allgemeine Grundzüge der organischen Formen-Wissenschaft, mechanisch begründet durch die von Charles Darwin reformierte Descendenz-Theorie' (2 vols.) Georg Reimer, Berlin, Germany.

Haeckel, E. (1874). 'Anthropogenie: Keimes—und Stammes—Geschichte der Menschen'. W. Engelmann, Leipzig, Germany.

Haeckel, E. (1920). 'Natürliche Schöpfungs-Geschichte' (Zwölfe verbesserte Auflage). Vereinigung wissenschaftlicher Verleger, Berlin, West Germany.

Hardin, G. (1960). The competitive exclusion principle. *Science* **131**, 1292–1297.

Hardisty, M. W. (1982). Lampreys and hagfishes: An analysis of cyclostome relationships. *In* The Biology of Lampreys (M. W. Hardisty and I. C. Potter, *Eds.*). Academic Press, London.

Hennig, W. (1966). 'Phylogenetic Systematics'. University of Illinois Press, Urbana, USA.

Holtfreter, J. (1943a). Properties and functions of the surface coat in the amphibian embryo. *J. Exp. Zool.* **93**, 251-323.

Holtfreter, J. (1943b). A study of the mechanism of gastrulation. Part I. *J. Exp. Zool.* **94**, 261–318.

Jacobson, A. G. (1978). Some forces that shape the nervous system. *In* 'Form-shaping Movements in Neurogenesis' (C.-O. Jacobson and T. Ebendal, *Eds*), pp.13–21. Almqvist and Wiksell International, Stockholm, Sweden.

Janvier, P. (1981). The phylogeny of the Craniata, with particular reference to the significance of fossil 'Agnathans'. *J. Vert. Paleontol.* **1**, 121–159.

Landström, U. (1977). On the differentiation of prospective ectoderm to a ciliated cell pattern in embryos of *Ambystoma mexicanum*. *J. Embryol. Exp. Morphol.* **41**, 23–32.

Landström, U. and Løvtrup, S. (1979). Fate maps and cell differentiation in the amphibian embryo—an experimental study. *J. Embryol. Exp. Morphol.* **54**, 113–130.

Løvtrup, S. (1974). 'Epigenetics—A Treatise on Theoretical Biology'. John Wiley, London, UK.

Løvtrup, S. (1976). On the falsifiability of neo-Darwinism. *Evolutionary Theory* **1**, 267–283.

Løvtrup, S. (1977a). Derek Roff and the evolution of body size. *Evolutionary Theory* **3**, 155–157.

Løvtrup, S. (1977b). 'The Phylogeny of Vertebrata'. John Wiley, London, UK.

Løvtrup, S. (1978). On von Baerian and Haeckelian recapitulation. *Systematic Zool.* **27**, 349–352.

Løvtrup, S. (1981). The epigenetic utilization of the genomic message. *In* 'Evolution Today' (G. G. E. Scudder and J. L. Reveal, *Eds*), pp.145–161. Hunt Institute for Botanical Documentation, Pittsburgh, Pennsylvania, USA.

Løvtrup, S. (1982). The four theories of Evolution. *Rivista di Biologia* **75**, 53–66, 231–272, 385–409.

Løvtrup, S. (1983a). Epigenetic mechanisms in the early amphibian embryo Cell differentiation and morphogenetic elements. *Biological Reviews.* **58**, 91–130.

Løvtrup, S. (1983b). Reduction and Emergence. *Rivista di Biologia* **76**, 437–461.

Løvtrup, S. (1985). On the classification of the taxon Tetrapoda. *Systematic Zool.*

Løvtrup, S. and Perris, R. (1983). Instructive induction or permissive activation? Differentiation of ectodermal cells isolated from the axolotl blastula. *Cell Differentiation* **12**, 171–176.

Løvtrup, S., Landström, U. and Løvtrup-Rein, H. (1978). Polarities, cell differentiation and primary induction in the amphibian embryo. *Biol. Rev.* **53**, 1–42.

Løvtrup, S., Rahemtulla, F. and N.-G. Höglund (1974). Fisher's axiom and the body size of animals. *Zoologica Scripta* **3**, 53–58.

Mathews, M. B. (1975). 'Connective Tissue Macromolecular Structure and Evolution'. Springer, Berlin.

Mookerjee, S., Deuchar, E. M., and Waddington, C. H. (1953). An experimental study of the development of the notochordal sheath. *J. Embryol. Exp. Morphol.* **1**, 399–409.

Nakatsuju, N. (1976). Studies on the gastrulation of amphibian embryos: ultrastructure of the migrating cells of anurans. *Roux's Arch. Devel. Biol.* **180**, 229–240.

Popper, K. R. (1959). 'The Logic of Scientific Discovery'. Hutchinson, London.

Roff, D. (1977). Does body size evolve by quantum steps? *Evolutionary Theory* **3**, 149–153.

Russell, E. S. (1916). 'Form and Function A Contribution to the History of Animal Morphology'. John Murray, London.

Saxén, L. and Toivonen, S. (1962). 'Primary Embryonic Induction'. Logos, London.

Serres, A. E. R. A. (1860). Principes d'Embryologie, de zoogénie et de tératogénie. *Memoires de l'Academie des Sciences de l'Institut Impérial de France* **25**, i–xv, 1–943.

Signoret, J. and Lefresne, J. (1979). Détermination par incorporation de thymidine tritiée des phases du cycle cellulaires chez le germe d'Axolotlen période synchrone de segmentation. *C. R. Acad. Sci.* (Paris) **279**, 1189–1191.

Sokal, R. R. and Sneath, P. H. A. (1963). 'Numerical Taxonomy'. Freeman and Co, San Francisco.

Stent, G. S. (1978). 'Paradoxes of Progress'. Freeman and Co, San Francisco.

Wessels, N. K., Spooner, B. S., Ash, J. F., Bradley, M. O., Luduena, M. A., Taylor, E. L., Wrenn, J. T. and Yamada, K. M. (1971). Contractile microfilament machinery of many cell types is reversibly inhibited by cytochalasin B. *Science* **171**, 137–143.

Woerdeman, M. W. (1936). Embryonic 'induction' by chemical substances. *Proc. R. Dutch Acad. Sci., Section of Science* **39**, 306–314.

Section IV

STRUCTURE AND FORM

8

The Relations of Natural Forms

GERRY WEBSTER

Abstract

This chapter addresses the question of how taxonomy is possible; that is, what must the world be like (or what presuppositions must we make) in order that the relations of similarity and difference exhibited by natural forms can be arranged in a quasi-logical system.

The Darwinian theory of evolution explains similarity in terms of descent from a common ancestor and difference in terms of natural selection — on the necessity of functional adaptation — from a continuum of individuals.

In attempting to construct a taxonomy in which natural kinds are equated with empirical laws of coexistence, two particular problems are relevant. First, how are morphological elements to be identified as belonging to the same kind in order that connections can be established — a problem of sameness. Second, how is variation which appears to disrupt the integrity of natural groups to be conceptualized — a problem of difference.

In the case of sameness, the only legitimate criterion is the purely formal one proposed by St. Hilaire. But this was either misunderstood or ignored by the speculative Darwinism of the nineteenth century and later. Consequently, Darwinism proposed to explain sameness by material and temporal continuity. This position lacks correspondence with reality and results in inconsistency.

In the case of the problem of difference, Darwinism deals with observable variation in a typically empiricist fashion by multiplying kinds. Hence it arrives at the view that only individuals exist in nature and thus that there are no laws of form, i.e. no necessary relations of an internal kind. The systematic similarities and differences observed result not from internal but from external constraints — the necessity of functional adaptation. These conclusions are criticized from a realist perspective in which a distinction is made between observable patterns and the generative structures responsible for their production. From this perspective, empirical laws can have exceptions, and apparent differences can conceal real

BEYOND NEO-DARWINISM
ISBN: 0-12-350080-X

similarities. Natural kinds group together entities whose manifest properties have a common explanation in terms of common generative structures, hence it allows a revival of the original taxonomic project of a search for laws of form, albeit in different terms—relating to generative structures.

Finally, the problem of generative structures is addressed. Weismann's attempt to modify the Darwinist notions of material continuity has its value in introducing the realist dualism of generative structures and manifest patterns although the theory in which this distinction is couched is probably untenable. There is no adequate theory of generative structures at present and hence no theory of sameness, and at best there is only a partial theory of difference. The basis of taxonomy remains to be accounted for.

INTRODUCTION

In winter we pass the time drinking Ich'eng wine and discussing the teaching of the Masters. One maintains that the meaning of what can be seen is never clear and another that the truth of what must be imagined is always uncertain. I wait for the plum blossom to return.

(Li Ch'ing-wa, The Plum Blossom Spring)

In a previous paper Webster and Goodwin (1982) suggested that one of the central problems for a theoretical biology is how to account for the existence—or more exactly, the production and reproduction—of a diversity of natural forms which can be classified hierarchically. This is a variant of the question posed by Kant (*see* Cassirer, 1950): given that taxonomy is possible, that nature, in appearance at least, approximates to the form of a logical system, how is it possible? Whence comes the apparent harmony between the logical forms of species, genus and class and natural forms? This *essai* is organized around this question and, in particular, the nature and adequacy of a theory which purports to answer it, namely the theory of evolution by natural selection, as proposed by Darwin in *The Origin of Species* (1859). In order to explain the similarities and differences which are systematized in taxonomy, Darwin employs two distinct, though related, explanatory principles: similarity is explained in terms of 'propinquity of descent' from a common ancestor—essentially an explanation in terms of historical conservation or inertia; difference is explained as being the result of a historical and gradual process of differentiation caused by 'natural selection' from an original continuum of individuals. Both explanatory principles thus reflect the conceptual preoccupations of their time in their concern with 'origins' and the diachronic. I now wish to analyse them in some detail.

Whatever its achievements may have been in practice, the goal of

Linnean taxonomy was far more than the merely artificial codification of organisms, however practically useful that might be. Its goal was *knowledge* of the natural order, the 'Plan of Creation' as some called it, and in attempting to reduce the diversity of organisms to a unified and logical system of natural kinds it sought to reveal, in phenomena which at first sight may appear to be free and unconstrained, a kind of inherent necessity. In method it was empirical and in assumptions largely empiricist in that it granted ontological and epistemological primacy to observable objects and relations (in this respect, Webster and Goodwin, 1982, have overemphasized the distinction between 'Rational Morphology' and 'Darwinism'). It follows that natural kinds are conceived as empirical kinds and defined in terms of empirical regularities or 'laws of coexistence'; that is observable constant (and therefore necessary) conjunctions in space of morphological elements or parts. Hence Rousseau can define Linnaean botany as the 'science of combinations and relations'. Such an enterprise is, therefore, no more than the normal practice of a science conceived on empiricist lines—the search for empirical regularities—as conditioned by the particular nature of its object, the synchronic realm of organisms. It is from this perspective that Cuvier and his followers can regard the concept of a structural plan, which every creature is supposed to exhibit and to adhere to, as an explanatory concept—strictly speaking one which allows predictions (*see* Cassirer, 1950)—a view which seems to have mystified Darwin (1889) who dismisses it as 'merely metaphysical'. (I perhaps should add that, for the purposes of this argument, I am ignoring the assertion of Foucault (1970) that there is a profound epistemological discontinuity between the science of Linnaeus and that of Cuvier.)

Thus, in their practice, comparative anatomists and embryologists concentrated their attention on those similarities and differences in structure which were regular and, therefore, supposedly law-governed—the 'typical' or 'essential' features. The task as conceived for comparative embryology was well summarized by Reichert writing in 1838:

> Its aim is to distinguish during the formation of the organism, the originally given, the essence of the type and to classify what is added or altered . . . during the individual developments; they reach thus . . . the essential structure of the organism and demonstrate the laws that manifest themselves during embryogeny.
>
> (*quoted by Russell, 1916*)

Driesch (1914) concisely summarizes the project in comparison with that of Darwinism: it

> sought . . . to construct what was typical in the varieties of form into a system which should not be merely historically determined but which should be intelligible from a higher and more rational standpoint.

It should, perhaps, be noted here that 'type' is simply an expression denoting the constancy of relations (an invariance), and such an enterprise does not necessarily imply any naive realism as Mayr (1963) seems to suggest, in which the 'type' is treated as a distinct and independent thing. Reification could occur, however, and, ironically enough, did so in Darwinism with its concept of the 'common ancestor' (*see* Webster and Goodwin, 1982).

The attempt to construct a taxonomy on the basis of the empiricist equation of natural kinds with empirical laws of coexistence of morphological elements involves the assumption that particular elements in different organisms can be regarded as the same kind of individual thing, i.e. identified or related as members of a species (thus the vertebrate tibia would constitute a species). This raises two serious and related problems. First, what criteria are to be used for identifying elements and assigning them to a given species? — a problem of sameness, which in turn raises the problem of the legitimacy of the assumption itself. Secondly, how should one deal with variation which is so irregular as to appear to confuse the distinction between kinds? — essentially a problem of difference.

THE PROBLEM OF SAMENESS

At the lower levels of the taxonomic hierarchy it may be sufficient to rely on shared qualities to assign individual elements to a species but this rapidly becomes inadequate at higher levels — for example, the bones which are given the same name in the different vertebrate limbs look very different — and, in the absence of adequate criteria, arbitrariness and anthropocentricity rapidly set in. Bateson (1894) comments on this (in the context of a discussion of Darwinism and descent) and points out that in the case of parts in which there is no perceptible differentiation, no attempt is made to assign these parts to a particular species or, as he puts it, to establish individual homologies:

> no one considers that the individual segments in the intestinal region of the Earthworm have any fixed relations of this kind; no one has proposed to homologise single leaves on one tree with single leaves on another; it is not expected that the separate teeth of a Roach have definite homologies with the separate teeth of a Dace.

He adds, significantly, that:

> in series whose members are differentiated from each other the existence of such individuality is nevertheless assumed.

Bateson's arguments are still worth studying because he is concerned to reveal the unexamined nature of such assumptions and then to undermine their tacit acceptance on the basis of empirical studies of variation and transformation and theoretical considerations of the nature of biological organization. Thus, he points out that in certain cases of transformation, for example a change in the number of petals in a flower or the number of segments in an insect limb, there are no empirical grounds (i.e. in terms of observable qualities) for supposing that any particular element in the 'new' pattern is the same as any in the 'original'. The notion that this might, or must, be the case is based on the tacit assumption that elements have individuality and are members of determinate species which can come into being, endure, or become extinct; an assumption, moreover, which, as noted above, is not made consistently. It is at least as plausible to suppose that this is not the case and, therefore, that the 'new' pattern is new as a whole and not simply a piecemeal variant of the 'original'.

The problems raised by the variation of morphological elements with respect to their qualities led E. Geoffroy St. Hilaire to propose a new criterion for the comparison of organisms and the identification of their parts:

> the only general principle that can be applied is given by the position, the relations, and the dependencies of the parts, that is to say by what I name and include under the term *connections*.
>
> *(de Beer, 1971)*

This criterion makes it clear that Geoffroy conceives organisms not primarily as conjunctions of parts but as structures. 'Parts'—the elements of biological patterns—are entities which are identified or defined (for the purpose of comparison at any rate) in purely relational terms by the 'place' they occupy in a system of relations and not by any intrinsic qualities or properties they happen to possess as individuals—size, shape, function, etc. Two 'parts' are said to be the same if they occupy the same 'place' in the system—thus some 'parts' of the middle ear of mammals are deemed to be the same as some 'parts' of the reptile jaw even though they have virtually no significant qualities in common.

Geoffroy's criterion for identity in the study of organisms is similar to that proposed by Saussure (1916) for use in the study of language and pre-dates it by almost one hundred years. The latter's famous pedagogic example will serve to clarify what is at issue. We regard the 8·25 Geneva-to-Paris express as being the same train each day despite changes in engine, carriages, driver, guard and passengers. The identity of the train from one day to the next is neither a function of some unchanging, essential and intrinsic qualities which differentiate it from other trains, nor of material continuity in space and time, but of its position in the system of trains. The '8·25' is a 'form' not a substance. So, according to Geoffroy, with the elements of biological patterns whose sameness is not that of individual things or kinds of things (substances) which depends upon, for instance, shared qualities, but that of relational entities (forms) whose sameness is a function of their position in a system of relations.

The 'Principle of Connections' provides a sound but restrictive methodological criterion for relating individual organisms, hence constructing natural groups, in terms of an empirical invariance in the spatial relations of the elements comprising them; more exactly, an invariance in the *system* of relations. Relations of homology ('unity of type') are thus relations of relations and *not* relations of things (substances). The criterion, if strictly applied, means the relations of sameness cannot be established between organisms (or their parts, insofar as they can be considered as individuals, e.g. segments) which differ in the number of elements (Bateson's, 1894, meristic variation). A change in number changes relational structure so that the elements in the two organisms cannot be identified; it is meaningless to speak of 'duplicated elements' or 'missing elements' in this context. These considerations were ignored in the speculative Darwinism of the nineteenth century, a matter I return to below.

The science proposed by Geoffroy is a science of pure morphology, in effect an empirical formalism, in which empirical order—the systems of relations and relations of relations—can be studied in abstraction from the nature of the particular elements whose 'arrangement' allows the order to be characterized. This possibility was recognized by Bateson (1894) with his emphasis on the distinction between meristic (relational) and substantive variation and his attempts to formulate the general problems raised by all types of repetitive order. It was recognized even more clearly, and developed, by that virtuoso of the isomorph D'Arcy Thompson (1942) who demonstrated that even some individual and isolated elements can be identified (as transformations) by treating them as systems of relations. In these developments D'Arcy raises the possibility, still largely unexplored, of a systematic treatment of morphology and taxonomy in terms of the mathematical theory of groups.

In the previous paragraphs, I have attempted to spell out what seem to me to be the implications of Geoffroy's 'Principle of Connections' with respect to the establishment of relations of sameness between organisms. Whether all, or indeed any, nineteenth century anatomists conceived it in this way and employed it consistently is another matter, and it may be that the ambiguity and measure of confusion displayed over this issue in *The Origin of Species* is typical.

Despite numerous references to 'plans of organization', 'patterns', 'relative connections', etc. it is fairly clear that, in the final analysis, the concept of homology as a relation of relations is foreign to Darwin (1859, see e.g. pp.415–419) and he conceives it as a relation of individual things which happen to be juxtaposed in space in particular ways. Thus, all vertebrate limbs include the 'same bones, in the same relative positions . . . Hence the same names can be given to the homologous bones in widely different animals' (pp.415–416). Since elements are conceived to resemble each other as individuals it is natural for Darwin to propose an explanation of this sameness in terms of some form of material continuity of these individuals—direct or indirect. Darwin plumps for direct material continuity in time, i.e. historical continuity: 'On my theory, unity of type is explained by unity of descent' (p.233); 'the chief part of the organisation of every being is simply due to inheritance' (p.228) of parts, for example the 'several bones' in the limbs of a vertebrate, from a 'common progenitor'; descent, we are assured, is the 'hidden bond of connection' which naturalists have been seeking under the term of the natural system (p.427). Darwin later (1868) made explicit his specific theory of inheritance (an ancient one)—pangenesis—which confirms the above interpretation, since in it individual elements are seen as effectively self-reproducing, hence materially continuous.

Further, individuals (substances) related only in space are independent, they comprise an aggregate or a population not a system. That such is Darwin's view of organisms, at least as far as their typical form is concerned, can be seen from his use of the concepts of 'inheritance' and 'correlation', and by his concept of difference (*see below*). 'Inheritance' is conceived as a causal relation between the individual elements of two organisms, parent and offspring (cf. 'propinquity of descent,—the only known cause of the similarity of organic beings'), whereas 'correlation' is (at least sometimes) conceived as a causal relation between the elements of a single organism. Since, as we have seen, 'the chief part of the organisation . . . is simply due to inheritance', it follows that elements, at least as regards their bare existence, are conceived as independent individuals.

The ontological thesis implicit in the *Origin*, though not altogether coherent, is something like this. Organisms are irreducibly complex, 'given' populations of independent individual elements (substances). These actual individuals are the 'subjects' of history, they effectively endure and suffer changes in their qualities. There is a certain ambiguity in the thesis here since Darwin makes no distinction between numerical and specific (qualitative) identity; the tone of his remarks often suggests that he is employing, unconsciously no doubt, the former concept. This ambiguity is a consequence of the fact that, as Bateson (1894) notes, the whole

Darwinian concept of inheritance is based on an unexplicated analogy with the organization and transmission of items of property in human society.

With such an ontology, difference (variation) can be consistently conceived as difference in the qualities of individuals with regard to size, shape, function etc.—substantive variation in Bateson's (1894) terminology—and, as noted above, this is, on the whole, Darwin's view of the nature of historical change. However, he also wishes to speak of changes in the composition of populations of elements—'organization' as he conceives it—and, therefore, novelty resulting from 'the atrophy and ultimately . . . the complete abortion of certain parts . . . the soldering together of other parts . . . and the doubling or multiplying of others' (p.417)—a passage that makes clear the metaphor behind the thought (*see also* pp.416–419). While similar processes may well occur secondarily during the development of an individual organism this is not pertinent to Darwin's historical thesis which cannot be justified empirically. The only empirical criterion which can be used generally and consistently to establish a relation of sameness between elements—which is a prerequisite for statements of the type quoted—is Geoffroy's 'Principle of Connections'. Darwin provides no alternative criterion, or indeed any criterion at all. As I have pointed out, Geoffroy's criterion cannot be used to compare organisms which differ in their organization. Such a comparison is only possible—if it is possible at all—on condition that we abandon empiricist notions of the ontological (and epistemological) primacy of observable objects, qualities and relations and the concommitant equation of natural kinds with empirical regularities. It is necessary to get behind appearances to the real generative structures which are responsible for the production of the manifest patterns of organization.

The claims which Darwin makes for his theory of sameness are not modest:

'On this same view of descent with modification, all the great facts in morphology become intelligible' (1859, p.433).

The appropriate comment, if one is needed, is provided by the ever-reliable Bateson (1886) in a discussion of metameric segmentation:

This much alone is clear, that the meaning of cases of complex repetition will not be found in the search for an ancestral form . . . Such forms there may be, but in finding them the real problem is not even resolved a single stage; for from whence was their repetition derived? The answer to this question can only come in a fuller understanding of the laws of growth and variation, which are as yet merely terms.

The point Bateson is making is, of course, that there is no such phenomenon as 'inheritance' in the Darwinian sense of continuity between elements possessing individuality; rather in each generation the organism reproduces (usually, though not invariably) a specific pattern and it is in terms of this process that 'Unity of Type', insofar as it exists, must be explained. Chomsky (1968) has made a similar point about the inadequacy of historical explanations, involving common ancestry, of the supposed 'formal universals' of language. In the biological domain, the common ancestry of similar forms may be a fact, but it is a contingent fact, and one which is irrelevant to any theoretical explanation of the synchronic phenomena of natural forms and their relations (hence the tedium of the current 'evolution *versus* creation' debate), a point which was recognized by many writers in addition to Bateson earlier in the century (*see* Russell, 1916, D'Arcy Thompson, 1942). In passing, it might be added that a preoccupation with relations of descent results, as Bateson (1894) convincingly argues, in a completely artificial separation of relations of similarity *between* organisms (homology proper, so called) from relations of similarity *within* an individual organism (so-called serial homology).

The ontological assumptions concerning organismic structure implicit in the *Origin* became widely accepted in the latter half of the nineteenth century and subsequently despite their problematic nature and the availability of pertinent criticism. Thus, the palaeontologist Osborn (1915) can speak of a 'principle of *hereditary separability*, whereby the body is a colony, a mosaic, of single, individual and separable characters' and further: 'each individual is composed of a vast number of somewhat similar new or old characters, each character has its independent and separate history, each character is in a certain stage of evolution . . . Since the *Origin of Species* appeared, the terms variation and variability have always referred to single characters'. In a similar vein, Hardy (1946) quotes approvingly from Goodrich to the effect that organs are homologous when all their parts have been derived from corresponding parts in a common ancestor, and adds that throughout the tetrapods fore and hind limbs are homologous and 'can be traced back in an uninterrupted series' to some common ancestral form even though 'they are not necessarily made up of the same segments'. The circularity here is typically Darwinian and provides, no doubt, yet another example of that 'action of the imagination' which facilitates the transition from the perception of a succession of objects to the idea of 'uninterruptedness' and hence sameness. More recently, Kent (1969) in his textbook of comparative anatomy invites us to compare the hands of *Rana catesbiana* which has five digits with that of *Necturus* which has four and ponder 'which finger is missing in *Necturus*?' As a final illustration, de Beer (1971) observes that homology, conceived in

Darwinian terms, remains an unsolved problem since he can find no material basis for the supposed continuity — either direct or indirect — of the supposed individuals comprising morphological patterns. It may be that spurious problems are doomed to remain forever unsolved.

THE PROBLEM OF DIFFERENCE

A problem confronting any empirical taxonomy, where categories are constructed on the basis of observable relations of similarity and difference, is (as Leviticus knew) how to deal with differences (variation) which are so irregular as to appear to threaten the integrity of natural groups and obliterate the clear-cut distinctions between them. The typical empiricist solution to this problem is to multiply kinds. Thus at certain periods in the history of science it has seemed an obvious matter of fact (or common sense) that, for example, rest and motion are two different stages of being; that there are two kinds of motion, terrestrial and celestial; two kinds of embryonic development, mosaic and regulative; two kinds of human beings, homosexuals and heterosexuals, or the mad and the sane, or the savage and the civilized. The logical end of this approach is to posit as many kinds (essences) as there are individuals and hence to arrive at a conception of a population or continuum of unique individuals and a concomitant nominalism. Thus, Lamarck, following Buffon, criticized the Linnaean system as artificial (i.e. as not constituting real knowledge of a natural order) since only individuals exist in nature and the categories under which they are subsumed are merely products of thought.

Darwin (1859) expresses similar views and his nominalism is explicit:

> . . . no clear line of demarcation has as yet been drawn between species and sub-species . . . between sub-species and well marked varieties or between lesser varieties and individual differences. These differences blend in to each other in an insensible series; and a series impresses the mind with the idea of an actual passage (p.107).

Further:

> From these remarks it will be seen that I look at the term species, as one arbitrarily given for the sake of convenience to a set of individuals closely resembling each other, and that it does not essentially differ from the term variety, which is given to less distinct and more fluctuating forms. The term variety, again, in comparison with mere individual differences, is also applied arbitrarily, and for mere convenience sake (p.108).

Thus the idea of a gradual, historical, differentiation of a continuum is derived directly from the 'facts'. Essentially the same view is expressed by Mayr (1963):

> The populationist stresses the uniqueness of everything in the organic world. What is true for the human species, that no two individuals are alike, is equally true for all other species of animals and plants . . . All organisms and organic phenomena are composed of unique features and can be described collectively only in statistical terms. Individuals, or any kind of organic entities, form populations of which we can determine the arithmetic mean and the statistics of variation. Averages are merely statistical abstractions; only the individuals of which the population is composed have reality.

Since, if taken literally, these radically empiricist views would reduce biologists to silence or, at best, to pointing like the old men of Laputa, one must assume an element of hyperbole here.

It will be remembered that the goal of the taxonomic endeavour was to discover the 'laws of form' which manifest themselves as, or are equated with, each level in a hierarchy of natural kinds; specific laws, generic laws and so on. Darwinian empiricism inevitably leads to the conclusion that no such laws exist, that *fundamentally* the domain of natural forms is irreducibly diverse. If all that really exists are populations of individuals, then each individual is a law unto itself and—God being dead—*a priori* everything is possible (*see* e.g. Monod, 1972). Thus particular forms can no longer be related to each other in terms of an ascending or nested set of 'laws' (i.e. natures or essences) but can be connected only materially and historically; as Darwin puts it:

> Our classification will come to be, as far as they can be so made, genealogies; and will then truly give what may be called the plan of creation (p.456).

Following this empiricist line of argument, Maynard Smith (1981) suggests that the assumption that there may be 'intrinsic' constraints on organismic form of the type discussed is

> in some cases . . . demonstrably false. For example, some of the earliest vertebrates had more than two pairs of fins . . . Hence there is no general law forbidding such organisms.

Maynard Smith, therefore, prefers to follow Darwin in explaining such similarities of vertebrate form as currently exist in terms of descent. It is indeed possible that there are no 'laws of form', but their existence is not refuted by observation of variation in the number of vertebrate appendages any more than the existence of laws of motion is refuted by observing that not all moving bodies describe elliptical trajectories. I return to this point below.

Since in the Darwinist view existence precedes specific essence, it follows that any order of discontinuous difference (apparent essence) which can be observed and systematized in a taxonomy must be explained as an *a posteriori* consequence of the operation of *universal* laws—that is laws which govern every living being as the laws of mechanics govern every body—and these laws must be concerned with the *external* relations of already-existing individuals. Darwin formulated such a universal law in terms of the necessity of functional adaptation and the consequent natural selection of those individuals best fitted to their environment. Thus an *a posteriori* necessity results in the differentiation of a continuum to give the appearance of discrete natural groups. It is easy to see how such a picture might have psychological (and ideological) appeal to those, such as Monod, with existentialist leanings and who, as Nietzsche put it, 'would rather have the void for a purpose than be void of purpose'.

On this view, then, organisms do not have 'natures', all they have is history and the necessity of being adapted to some environment; any discussion of form will therefore be couched in terms of natural history and universal laws. Since *a priori* everything is possible, the fact that any viable organism possesses a 'harmonious arrangement' of parts such that they are functionally adapted to each other—Cuvier's 'Conditions of Existence' (*see* Russell, 1916)—is a contingent matter, the result of natural selection (*see* Darwin, 1859, p.233).

This is, broadly speaking, the position adopted by, among others, Hull (1978, also 1973) who maintains that species cannot be defined in terms of essential characteristics or properties (presumably taken to be observable) but only in terms of spatio-temporal continuity. Species are particular, contingent, historical entities or lineages so no specific law-like statements are applicable to them. Putting the same point differently: species names are not the names of natural kinds, which would be applicable to all entities of that kind no matter where and when they existed, like the names 'gold' or 'chlorine', but are more like the proper names of individuals, 'Charles Darwin' or 'Li Ch'ing-wa', concerning whom no specific scientific laws can be formulated. Thus if some new form were to appear, on earth or elsewhere, which was identical to one already or previously existing it could not, on this view, be given the same name; as Hull (1978) puts it: 'If a species evolved which was identical to a species of extinct pterodactyl save origin, it would still be a new, distinct species'. Even Darwinists might find this position slightly odd since it ignores the 'biological' criterion of ability to function as a member of an interbreeding population and produce fertile offspring. But, it may be remarked in passing, this criterion itself is far from adequate and cannot be 'primary' as Mayr (1976) maintains. Inter-breeding populations are not 'given', they must be constituted, by the organisms themselves if not by biologists, and this must presumably depend upon recognition or discrimination by means of specific perceivable characteristics, some of which may be morphological. Though these characters *may* be historically contingent, they are, *contra* Mayr, not 'secondary' but are prerequisite for the very existence of determinate populations.

At this point the reader (if I still have one) will protest impatiently, and correctly, that while a genetic account of Darwinist explanatory principles may serve to disabuse the few who believe that scientific theories are epiphenomena of nature, it has no relevance to the truth or falsity of the theory in question. Since this is a matter on which I have no competence to pronounce, the following section is concerned firstly with methodological issues and then with a rehearsal of some classical criticisms.

Firstly, the empiricist nature of Darwinian theory with its heavy dependence on the 'facts' might be thought to give grounds for concern, for there are compelling arguments that empiricism (empirical realism) cannot provide an adequate methodology for science (*see* Bhaskar, 1978). As Mepham (1973) puts it: 'coherence is of the order of theory and not of fact'. Thus, the formulation of a theory which serves to make intelligible the relations between entities, for example, by explaining how these entities are produced, may result in new criteria for sameness and difference and a need to identify the entities themselves in a different way, hence a revision of empirical taxonomies. It may be the case that seven out of ten consumers cannot tell 'Stork' from butter; nevertheless, the manufacturers know the difference. In other words, scientific theory may lead to a 'critique of the facts', and a 'repudiation of experience' (Mepham, 1973). The naive empiricist or 'common-sense' belief in the sacred and given nature of 'facts' and 'experience' can act as a major epistemological obstacle to the development of adequate scientific theories in that we may prefer to think of the 'world' as falsifying these theories rather than accepting that we may need to reorganise our experience. Needless to say, this does not mean that facts should be ignored.

A few brief examples chosen at random from the history of science will illustrate these points. Thus, Newtonian mechanics served to abolish the apparent distinction between terrestrial and celestial motion and Freud provided theoretical grounds for supposing that the obvious difference between the 'mad' and the 'sane' might be merely a matter of degree rather than one of kind. Conversely, theories of morphology maintain that whales are not fish even though they resemble them and that the petals of the garden clematis are really sepals.

Thus the Darwinist line of argument outlined above in which observable variation leads to multiplication of kinds and the consequent belief that 'everything is possible', that there are no 'laws of form' and all is 'chance' and contingency, is methodologically suspect because it simply takes the 'facts' at their face value. It ignores the possibility that behind the empirical diversity there may be a hidden unity that cannot be observed but must be imagined.

The above discussion is pertinent to the question of essentialism, which many biologists (and others) seem to regard as an antiquated philosophical notion of no relevance to science, or worse, a positive hindrance to the development of scientific theory: 'a major misconception that had to be eliminated before a sound theory of evolution could be proposed' (Mayr, 1963). As I have noted above, Darwinism does not, strictly speaking, eliminate essences, it merely multiplies their number; and it is far from being the case that essentialism is either irrelevant to science or a major misconception. There are recent philosophical justifications of essentialism by Putnam (1975) and Kripke (1980), who argue that knowledge of 'natures' and 'real essences' can be discovered empirically, and essentialist notions play a central role in Bhaskar's (1978) transcendental realist philosophy of science. Following Locke, Bhaskar distinguishes between nominal essences and real essences. The nominal essence of a thing consists of the behaviour, properties or qualities the manifestation or possession of which is necessary for the thing to be correctly *identified* as being of a certain type. Nominal essences are thus involved in the construction of empirical taxonomies. Real essences, however, are those underlying structures and generative mechanisms— imagined but not necessarily imaginary—by virtue of which a thing manifests or possesses particular behaviour or properties. The discovery of real essences is, therefore, an important goal of science for these provide *explanations*. Thus, according to current theory, gold has the manifest properties that it does because it has a particular atomic constitution which is its real essence. The real difference between gold and silver does not lie primarily in their manifest properties but in that which serves to explain these properties, i.e. in what they *are*. Knowledge of real essences permits the formulation of real definitions, so that natural kind terms, insofar as they are based on such definitions, group together those entities whose manifest properties have a common explanation, that is, whose real essence is the same. *Contra* Darwin, natural kind terms are not just a matter of convenience or convention. However, as noted above, empirical descriptions and taxonomies are subject to revision in the light of theories—themselves fallible; as Bhaskar puts it: 'science consists of a continuing dialectic between taxonomic and explanatory knowledge . . . It aims at real definitions of the things and structures of the world as well as statements of their normic behaviour'.

Things thus act, or rather tend to act, in particular ways because of the kinds of things they are, because of their intrinsic natures. It is important to note, however, that it is contingent whether any particular tendency will be actualized and manifest itself in terms of overt behaviour or properties. Thus, it is necessary that copper, given its intrinsic nature, be a good conductor, but it is contingent whether this is ever made manifest, i.e. whether a current is ever passed through a sample of it. Thus knowledge of what things are—of their real essence—enables us to predict how they will behave *only* if we know that certain conditions prevail. From a transcendental realist perspective, therefore, Hull's contention—if I have understood it correctly—that because the manifest characteristics of organisms are historically contingent they do not have natures and are merely historical entities, is untenable.

As Bhaskar observes, the common belief that essentialism is 'incompatible with evolutionary thinking' (Mayr, 1963) is unfounded. There is no reason to suppose that natures are necessarily 'fixed' and 'unchanging' (other than at the level of 'ultimate' entities, where they are so by definition); whether, when and how they change is a matter for scientific investigation.

It is now possible to see the taxonomic *project* of the 'Rational Morphologists' as a proper attempt to define natural kinds and thereby characterize the system of necessity—the 'laws of form'—present in the domain of organisms. From the perspective of modern philosophies of science, the partial failure of the original project and the consequent development of certain Darwinian notions can be seen to be a result of a faulty methodology stemming from a faulty ontology; a failure to recognize that reality must be conceived as ontologically stratified, that natural kinds are not primarily of the 'order of fact' and that empirical laws can have exceptions. Once this is recognized, the way is open, in principle at any rate, for a revival of the 'rational' project and such a revival is envisaged by Webster and Goodwin (1982) in their attempt to discuss morphology in structuralist terms. As Detienne (1979) has remarked in a very different context:

> In various sectors structural analysis can show how the unfolding of history is at times subject to certain constraints, even if these are in turn dependent on another history that is all-encompassing, void of events, consisting of slow epochs and lengthy traverses.

Even history cannot be reduced to that which actually happened.

I turn now, briefly, to certain classical critiques of Darwinism which are relevant to the argument above, in the sense that they are concerned to address the question of the 'logic of morphology', denied or ignored within the Darwinist scheme which conceives the only law operating within the organismic domain to be the universal law of the necessity of functional adaptation.

From Naegeli onwards (*see* Cassirer, 1950), several people have pointed out that the utility of many of the characteristic features of organisms used in the construction of taxonomies is problematic or obscure, a fact which raises difficulties for a theory which claims to have identified the cause of specific differences. As Bateson (1894) notes, 'it is precisely on the utility of specific differences that the students of adaptation are silent'. Darwin (1889) himself accepted this criticism and explained his excessive preoccupation with teleology as being a consequence of the lingering influence of natural theology. When the students of adaptation do speak, and the sociobiologists are particularly vocal, they all too often disembogue whole tempests of uncontrolled speculation upon their audience: 'Paley *redivivus* . . . with natural selection instead of a divine artificer as the *deus ex machina*' (Huxley, 1942). For whereas the study of adaptation was originally conceived as a test of the theory of natural selection, by its very nature it all too easily degenerates into Kuhnian 'normal science' where what is tested is the imaginative power of the scientist; as Bateson (1894) caustically remarks: 'on this class of speculation the only limitations are those of the ingenuity of the author'.

The fact that Darwinism restricts itself to universal laws is closely bound up with the idea that 'everything is possible', that forms are the product of 'chance' or 'random variation'. In popular expositions (and sometimes elsewhere) it is frequently forgotten that what the theory of natural selection requires ('requires' in the sense that if it were not the case, the theory would be partially or wholly redundant) is only that variation in a population of forms should be unconstrained with respect to the external, functional relationships of these forms. Even if the theory is correct, therefore, the possibility of a 'logic of morphology' remains. Needless to say, none of the above remarks has any bearing on the validity of the theory of natural selection *vis a vis* its explanation of the stability or otherwise of given states of adaptation in populations.

With regard to the view that 'everything is possible' there are some results from empirical studies which, if they have been correctly interpreted render it problematic. As Driesch (1914) has noted, much of the work of the late nineteenth century on the regulative and regenerative capacities of embryonic and adult organisms was carried out in an attempt to refute this Darwinist conception. The experimental demonstration of the reproduction of 'typical' forms follows unnatural perturbations suggesting to Webster and

Goodwin (1982), as to Driesch, the existence of internal constraints on form; some sort of necessary connections which remain to be explicated theoretically. In their discussion they follow Galton, Bateson and Waddington in considering biological forms in terms of 'positions of organic stability'. Along similar lines, Bateson (1894) draws attention to the importance of the phenomena of homeosis, that class of transformations particularly common in arthropods and flowering plants, in which one member of a meristic series assumes the form and characteristics proper to another. D'Arcy Thompson's (1942) demonstration that at least some similar forms are orderly and systematic deformations or transformations of each other has already been referred to above.

At the end of his massive empirical study of variation, Bateson (1894) feels able to draw the following conclusion:

> For the crude belief that living beings are plastic conglomerates of miscellaneous attributes, and that order of form or symmetry have been impressed upon this medley by selection alone; and that by variation any of these attributes may be subtracted or any other attribute added in indefinite proportion, is a fancy which the study of variation does not support.

THE PROBLEM OF GENERATIVE STRUCTURES

It has been repeatedly asserted during the course of this *essai* that a proper understanding of morphology, and therefore of the possibility of taxonomy, is dependent upon the development of a theory of morphology couched in terms of the generative structures or mechanisms which are responsible for producing the manifest patterns of organisation; that is, of identifying the real essence of organisms, their 'genetic constitution' in the most literal sense. This presupposes a rejection of empiricist notions and an acceptance of the idea that reality is ontologically stratified; that a distinction must be made between the real — the plane of generative structures — and the actual — the plane of events and manifest patterns (Bhaskar, 1978).

A decisive step in this direction was taken by Weismann (1883, 1885, 1904), but it is most important in any discussion of his ideas to make a distinction between his crucial ontological innovation and the specific theory in which this was embodied. Webster and Goodwin (1981, 1982), for example, are not very clear about this distinction, and in their no doubt justifiable eagerness to criticize Weismann's theory seem at times to reject the vital dualism which it embodies. In so doing they produce a critique which is to some extent incompatible with their own overall thesis. There is thus an element of justice in Maynard Smith's (1982) comment that they 'wish to throw away the baby with the bath water'; rough justice, however, since Maynard Smith's baby is a specific theory, not an ontology.

A rational reconstruction of Weismann's position is somewhat as follows. 'Inheritance' — the central concept in the Darwinist explanation of sameness — is not to be thought of in terms of some form of material continuity or causal connection between observable individual elements of the patterns of parent and offspring. Rather, it must be thought of in terms of the control of growth and development — the task is to 'trace heredity back to growth' (Weismann, 1883). Offspring resemble parents because they are both the 'effects' of identical processes of growth and development, and — a classical empiricist view — constant effects imply constant causes. The cause is to be found in a particular entity, the germ cell, which contains a specific substance, the germ plasm, which has the power of developing into an organism of specific form. The supposed constant nature of the cause, and hence the effect, that is the phenomenon of hereditary resemblance, is simply a reflection of the historical 'continuity of the substance of the germ cell or germ plasm' for the germ cells 'are not derived at all, as far as their essential and characteristic substance is concerned from the body of the individual, but they are derived from the parent germ cell'. This substance 'has remained in perpetual continuity from the first origin of life' and during the course of time successive changes have occurred in it such that there have been successive changes in the manifest patterns of organisms which are its effects. The crucial dualism is explicated in terms of these specific concepts: 'the germ plasm and the substance of the body, the somatoplasm, have always occupied different spheres'.

In an elaboration of the basic theory (*see* McSherry, 1975), the germ plasm was localized in the nucleus from which it actively directed the course of growth and development; 'the essence of heredity is the transmission of a nuclear substance of specific structure' (Weismann, 1885). This nuclear substance was supposed to consist of 'active' discrete units (determinants) each (or each group) of which stood in a specific causal relation to a particular part of the manifest pattern; there were to be as many of these units (or groups) as there were independently variable and heritable parts of the organism, and individual development was a consequence of the determinants being activated or liberated in a definite and regular temporal and spatial order.

It will be apparent that although Weismann's theoretical scheme is dualist, the generative structures are constituted by simply transferring the structure of the actual or manifest pattern to the level of the real, so that Darwin's overall scheme is retained — hence the new picture is subject to many of the criticisms outlined above—but clarified to some extent. Thus individual elements of manifest patterns which are supposedly the same are now considered to possess a specific identity of resemblance—there is no material continuity at this level—and what endures (or not) is the species of element which suffers change as a consequence of changes in the historically given, discrete and independent units of the germ plasm which stand in causal relation to its members. The units of the germ plasm possess a (quasi-) numerical identity as a result of the 'perpetual continuity' of this material and this ensures the persistence of the species of pattern element. Thus, an individual pattern element of some modern form is considered to be a member of the same species as an individual pattern element of some ancestral form, and hence is entitled to the same name e.g. tibia, because that which causes both of them to exist is the same individual determinant which has endured through time. It is clear that there are grave problems as to what is to count as a 'pattern element', but I will not pursue this here.

Since in the Weismannist scheme there is a one-to-one relation between 'cause' and 'effect', the system is closed in the sense that there must necessarily be a unique relationship between a given set of determinants and the manifest pattern. When Weismann is confronted by a situation in which the system appears to be open, his solution can only be to multiply determinants (i.e. natures) in an analogous manner to the empiricist taxonomist. Thus, in the case of seasonal and regional polymorphism in butterflies where the manifest pattern of a specific organism shows variation which can be correlated with environmental change, he postulates the existence of alternative sets of determinants, each set being activated by a value of an environmental variable, in this case temperature (*see* McSherry, 1975).

Weismann's theoretical position can be characterized as involving a holistic conception of the organism as an 'expressive totality'—the expression of the activity of a 'central directing agency', itself conceived as the material embodiment of the 'idea' or 'plan' of the organism, that is the 'type' reified (*see above*). This 'centre' is supposed to be the real essence of the organism. Such a conception, despite modifications, remains central to most contemporary thinking about organismic form (*see* Webster and Goodwin, 1982) in which the real essence is identified as DNA (*see, for example,* Monod, 1972).

A broadly similar ontology underlies the theoretical constructs of the early Mendelians and thereby facilitates the fusion between Weismannism and Mendelism in the chromosome theory of the gene. Thus de Vries (1900) expounds his theory of pangenesis which requires that the concept of species recede into the background in favor of the consideration of a species as a composite of independent factors . . . the units of species-specific traits are to be seen in this connection as sharply separate entities;

further,

the total character of a plant is built up of distinct units. These so-called elements of the species, or its elementary characters are conceived of as tied to bearers of matter, a special form of material bearer corresponding to each individual character.

Somewhat later, Fisher (1936) quotes with approval Roberts' characterization of 'the modern period in the study of heredity' as involving the 'development of the idea that a living organism is an aggregation of characters in the form of units of some description' and with this we already have the combinatorial atomism of Neo-Darwinism in which organismic form and transformation is to be studied in terms of the behaviour of populations of determinants (gene pools).

As Marx (1926) observes: 'the traditions of dead generations weigh like a nightmare on the minds of the living' and there appears to be a certain reluctance on the part of contemporary biology both to acknowledge that this whole position seems to have been completely undermined by the — largely empirical — discoveries of classical genetics and to draw the appropriate conclusions. It is no doubt pious to hope, in the Freudian manner, that dreams when completely analysed lose their affective force, especially when that affect is ideologically compounded.

For the purposes of the present argument, it suffices to point to two related findings of classical genetics (see Dobzhansky, 1951; Huxley, 1942; Waddington, 1957; Goldschmidt 1958; Webster and Goodwin, 1982). Firstly, genes do not appear to be discrete units standing in specific causal relation to discrete elements of manifest patterns, as proposed by Weismann and the early Mendelians, for there is neither a one-to-one correspondence between units and elements nor constant conjunction between genetic and morphological composition; organisms with a wild-type genetic composition can be of mutant morphology and, conversely, organisms of mutant genotype can be of normal morphology. Rather, genes (effectively chemical composition) must be understood as only one of the factors involved in the production of manifest patterns so that a change in the genome may cause a change in morphology but does not necessarily do so. As Goldschmidt (1958) has it:

If we wish to express this factual situation by saying that a phenotype trait is the product of the action of many or all genes, we must realise that this *façon de parler* is nothing but a circumscription, in terms of the atomistic theory of the gene, of the fact of the unity and integrity of the organism.

While one might quibble with this formulation—the concept of the gene, however confused, represents an addition to a system of knowledge, not merely to a vocabulary—the message is (fairly) clear. Secondly, organisms appear to be open systems in the sense that a distinction must be made between the set of alternative morphological states which is possible for them (their competence) and the actualization of one or more of these states which is dependent upon the prevailing conditions and hence requires a 'structural' or realist concept of causation and not a Humean—same cause, same effect—concept (*see* Bhaskar, 1978; Webster and Goodwin, 1982). The open nature of organisms is suggested by the facts which led to the first conclusion, by the phenomena of homeosis referred to above and even more clearly by that of phenocopying. Here 'practically any kind of shock' will result in a organism whose morphology is that normally associated with the presence of a mutant allele and 'every known type . . . [of Drosophila mutant] . . . can be copied as a phenocopy'. Further, the same type of shock can produce different phenocopies depending on the time of application. Goldschmidt (1958), from whom these observations are taken, concludes:

> If the quantitative element of mutation is so sensitive to external and internal environment, it might follow that all or most mutational effects take place within a spectrum of possible developmental changes the range of which alone determines the possibility of appearance of both mutational and environmental effects.

The distinction between possibility and actuality is clear and it would be unreasonable to attempt to save the phenomena by multiplication of determinants in the Weismannist manner discussed above. Implicit in Goldschmidt's remarks is the recognition that these facts are of the same kind as those produced by students of developmental regulation where, again, the organism appears to possess a capacity to produce a coherent 'response'—selected from its repertoire of possibilities—to a 'stimulus' never previously encountered. Such facts formed one strand of Driesch's (1914) critique of the 'vulgar' Darwinism of his time (*see above and* Webster and Goodwin, 1982).

If these sorts of events have been correctly described and interpreted they render Weismann's theory untenable and this is a conclusion drawn by many writers, for example Dobzhansky (1952) and Huxley (1942). Nevertheless, as Dobzhansky notes, Weismannism 'continues to influence the thinking of many geneticists'. And not only them; the idea of the discrete retains its charm for other members of the bourgeoisie, especially those lacking all discretion such as the sociobiologists. Even the more responsible appear to cling to the theory and attempt to discuss incompatible facts in terms of it, with mystifying results. Thus, Harland (1936) concludes:

> we are able to see how organs such as the eye, which are common to all vertebrate animals, preserve their essential similarity and structure or function, though the genes responsible for the organ must have become wholly altered during the evolutionary process.

The rejection of Weismann's theory means that the Darwinist quest for intelligibility in history has failed—at least for the time being—for the elements of morphologies cannot be conceived as members of an enduring species in his sense (let alone Darwin's); there is no historical continuity, either direct or indirect, between them. Thus we have no theory of sameness. Hence De Beer (1971) can speak of homology as 'an unsolved problem'—whose solution, like that of all problems in the Anglo-Saxon world, requires the acquisition of more facts. As I have remarked above, there has never been an adequate theory of difference, for in Darwinism continuous difference, at any rate, is simply taken as given, and classical genetics has never satisfactorily theorized its empirical productions, though Waddington (1957) and Goldschmidt (1958) have given pointers.

While this conclusion concerning Weismann's theory means, of course, that the genetic constitution of the organism cannot be equated with, or reduced to, an ensemble of genes, and hence that the real essence of the organism is not, or not only, its DNA, this does not involve a rejection of a dualist ontology which can be conceived in alternative terms to those introduced by Weismann. I have noted above that both classical geneticists and students of development have found it necessary to make a distinction between potentiality—the repertoire of alternative morphological states which the generative structures (whatever they may be) are competent to produce—and actuality—the manifest morphological patterns which require explanation. The actual must be explained as an instance of the possible (Piaget, 1971; *see* Webster and Goodwin, 1982). This demand, general enough, awaits specific theoretical realization.

At present there is a poverty of theory. Since both Nature and Science abhorr a vacuum, it is no doubt to be expected that the void will be filled by a proliferation of metaphors, and such is the case. Depending upon taste, one can choose between the cosy and culinary (Dawkins, 1981), the literary cum semiotic (Monod, 1972) or the severe and hi-tech (Jacob, 1974). All, with the exception of Monod whose views are untenable (*see* Webster and Goodwin, 1982), remain suitably vague concerning the causal relations involved in the production of manifest patterns. No doubt Webster and

Goodwin's (1982) account, in which the generative structures are identified with that dynamic system of relations—the so-called developmental field— in which manifest patterns are elaborated could also justly be characterized as largely metaphorical (and the reader can provide suitable epithets) but it is at least capable of elaboration, albeit of a speculative and, as yet, purely formal kind (*see* Goodwin, this volume).

If we abandon, at least for the time being, unification in terms of history we are not necessarily thrown into an unintelligible chaos of sheer diversity. Already, D'Arcy Thompson (1942) has envisaged a move away from the Darwinist preoccupation with kin and a return to the taxonomists concern with kind—what Bergson calls the 'logical affiliation between forms'. With regard to the Foraminifera, D'Arcy opines that: 'we can trace in the most complete and beautiful manner the passage of one form into another among these little shells' just as 'the mathematician can trace one conic section into another, and "evolve", for example, through innumerable graded ellipses, the circle from the straight line; which tracing of continuous steps is a true "evolution" though time has no part therein'. D'Arcy's analogy with classical mechanics and his view of organisms as permutations and deformations of each other, that is as orderly and law-governed transformations, raises the possibility of a knowledge of the world of natural forms and their relations in terms of a theory of generative transformations which would embody Bateson's 'laws of growth and variation'.

Possibly. But it is salutary to remember, as Merleau-Ponty (1962) reminds us, that thought and the world are not one; that the world, 'of which knowledge always *speaks*', precedes knowledge, and the terms of discourse have there origin elsewhere than in a rational contemplation of it.

REFERENCES

Bateson, W. (1886). The Ancestry of the Chordata. *Quart. J. Micros. Sci. (N.S.)* **26**, 535–571.

Bateson, W. (1894). 'Materials for the Study of Variation'. Macmillan, London, UK.

Bhaskar, R. (1978). 'A Realist Theory of Science'. Harvester, Brighton, UK.

Cassirer, E. (1950). 'The Problem of Knowledge'. Yale University Press, New Haven, USA.

Chomsky, N. (1968). 'Language and Mind'. Harcourt, Brace, New York, USA.

Darwin, C. (1859). 'The Origin of Species'. 1st edition Penguin, Harmondsworth, UK.

Darwin, C. (1868). 'Variation of Animals and Plants Under Domestication'. John Murray, London, UK.

Darwin, C. (1889). 'The Descent of Man'. 2nd edition John Murray, London, UK.
Dawkins, R. (1981). In Defence of Selfish Genes. *Philosophy* **56**, 567–580.
De Beer, G. (1971). 'Homology: An Unsolved Problem'. Oxford University Press, Oxford, UK.
Detienne, M. (1979). 'Dionysos Slain'. Johns Hopkins, Baltimore, USA.
De Vries (1900). The Law of Segregation of Hybrids. Reprinted (1966) *In* 'The Origin of Genetics' (C. Stern and E. R. Sherwood, *Eds*). Freeman, San Francisco, USA.
Driesch, H. (1914). 'The History and Theory of Vitalism'. Macmillian, London, UK.
Fisher, R. A. (1936). Has Mendel's work been rediscovered? Reprinted (1966) *In* 'The Origin of Genetics' (C. Stern and E. R. Sherwood, *Eds*). Freeman, San Francisco, USA.
Foucault, M. (1970). 'The Order of Things'. Tavistock, London, UK.
Goldschmidt, R. B. (1958). 'Theoretical Genetics'. University of California Press, Berkeley, USA.
Hardy, A. (1946). Edwin Stephen Goodrich 1868–1946. Reprinted (1958) *In* 'Studies on the Structure and Development of vertebrates'. Dover, New York, USA.
Harland, S. C. (1936). The genetical concept of the species. *Biol. Rev.* **11**, 83–112.
Hull, D. (1973). 'Darwin and his Critics'. Harvard University Press, Cambridge, Massachusetts, USA.
Hull, D. (1978). A matter of individuality. *Philos. Sci.* **45**, 349–360.
Huxley, J. (1942). 'Evolution: The Modern Synthesis'. Allen and Unwin, London, UK.
Jacob, F. (1974). 'The Logic of Living Systems'. Allen Lane, London, UK.
Kent, G. C. (1969). 'Comparative Anatomy of Vertebrates'. Mosby, St. Louis, USA.
Kripke, S. (1980). 'Naming and Necessity'. Oxford University Press, Oxford, UK.
McSherry, G. M. (1975). August Weismann: The methodology of a research programme. MSc Thesis, University of Sussex, UK.
Marx, K. (trans. 1926). 'The Eighteenth Brumaire of Louis Bonaparte'. Allen and Unwin, London, UK.
Maynard Smith, J. (1981). Did Darwin get it right? *London Rev. Books.* **3(15)**, 21–22.
Maynard Smith, J. (1982). Commentary on Webster and Goodwin's 'The Origin of Species: a structuralist approach'. *J. Soc. Biol. Struct.* **5**, 49–51.
Mayr, E. (1963). 'Animal Species and Evolution'. Oxford University Press, London, UK.
Mayr, E. (1976). The Biological Meaning of Species. *In* 'Concepts of Species' (C. N. Slobodchikoff, *Ed.*). Strondsberg, Pennsylvania, USA.
Mepham, J. (1973). The Structuralist Sciences and Philosophy. *In* 'Structuralism' (D. Robey, *Ed.*). Clarendon Press, Oxford, UK.
Merleau-Ponty, M. (1962). 'Phenomenology of Perception'. Routledge and Kegan Paul, London, UK.
Monod, J. (1972). 'Chance and Necessity'. Collins, London, UK.
Osborn, H. F. (1915). On the Origin of Single Characters as Observed in Fossil and Living Animals and Plants. *Am. Nat.* **49**, 193–239.
Piaget, J. (1971). 'Structuralism'. Routledge and Kegan Paul, London, UK.
Putnam, H. (1975). 'Mind Language and Reality: Philosophical Papers 2'. Cambridge University Press, Cambridge, UK.

Russell, E. (1916). 'Form and Function'. John Murray, London, UK.

Saussure, F. (1974 reprint). 'Course in General Linguistics'. Collins, London, UK.

Thompson D'Arcy, W. (1942). 'On Growth and Form', 2nd edition. Cambridge University Press, Cambridge, UK.

Webster, G and Goodwin, B. C. (1981). History and Structure in Biology. *Persp. Biol. Med.* **39–61**.

Webster, G. and Goodwin, B. C. (1982). The Origin of Species: a structuralist approach. *J. Soc. Biol. Struct.* **5**, 15–47.

Weismann, A. (1883). On Heredity. Reprinted (1964). *In* 'A Source Book in Animal Biology'. (T. S. Hall, *Ed.*). Hafner, New York, USA.

Weismann, A. (1885). The continuity of the Germ-Plasm as the Foundation of a Theory of Heredity. Reprinted (1972) *in* 'Readings in Heredity and Development', (J. A. Moore, *Ed.*). Oxford University Press, New York, USA.

Weismann, A. (1904). 'The Evolution Theory'. Edward Arnold, London, UK.

9

A Relational or Field Theory of Reproduction and its Evolutionary Implications

BRIAN C. GOODWIN

Abstract

There can be no adequate evolutionary theory without a causal account of reproduction. The modern synthesis is based upon Weismann's dualistic theory of the organism which postulates a segregated and immortal part of the organism, the germ plasm, carrier of hereditary determinants and generator of all aspects of the adult organism, which itself is mortal. This dualism isolates for special attention that category of inherited particulars, now identified with specific information in the DNA, which correlates with most of the inherited differences between sexually reproducing organisms and hence provides a set of factors (the genes) in terms of which evolutionary change can be described.

The limitations of this viewpoint are two-fold: first, because of the theoretical inadequacy of identifying correlative factors with sufficient causes of (say) organismic morphology, and because of the empirical evidence that gene products do not generate form, there is no way of accounting in causal terms for observed differences of form in organisms by the identification of differences in hereditary factors; and secondly, evolution is not simply about differences between groups of organisms but also about observed regularities, both of form and of generative process, over large taxonomic groups, which neo-Darwinism is unable to explain. This essay examines ways of altering Weismann's dualism so as to overcome these limitations and allow for the development of a causally sufficient account of reproduction, using the field concept to define the relational order which is a basic property of developing organisms. The causal role of hereditary particulars can then be clarified and the meaning of biological universals can be described in molecular and relational terms. From this unitary view of organisms, based upon a generative

BEYOND NEO-DARWINISM
ISBN: 0-12-350080-X

theory of biological organization, comes an understanding of how both development and evolution emerge as aspects of biological process.

INTRODUCTION

Biology is currently going through a period of conceptual turmoil whose long-term consequences are impossible to predict. After a century of almost continuous analytical development, in which the empirical and theoretical consequences of the conceptual scheme established by Darwin and Weismann have been explored with truly extraordinary success, those aspects of organic life which are not readily assimilated within this scheme have returned to haunt the modern synthesis and to question some of its basic assumptions. Foremost among these is the problem of biological form, which includes both the process whereby organisms of specific morphology are generated (morphogenesis), and the logical, structural relationships between organisms of different form (taxonomy). Neo-Darwinism attempts to assimilate both of these within its historically based view of biological process: taxonomy is essentially genealogy, an account of the chronology of species understood in terms of the processes of random variation in genotypes and natural selection of phenotypes; while morphogenesis is the history of individual development as written in the genes, themselves seen as the result of an historical process, morphology then arising by a sort of self-assembly of gene products into higher-order structures.

What resists assimilation in this manner is the stubborn evidence of regularity in, and constraint on, biological form. Whereas historical processes of the type described in neo-Darwinism (random variation in genotypes and selection of the resultant phenotypes by randomly-varying environments) imply that any biological form is possible and survival is the only constraint, the empirical evidence suggests otherwise. As made clear by the pre-Darwinian rational morphologists such as Geoffrey St. Hilaire, Cuvier, Reichert and Owen, the systematic study of comparative morphology indicates that there are basic structural constraints which impose limitations of form over extensive taxonomic groups; examples include the segmental body plan of the arthropods, the invariant features of tetrapod limbs, and the structural homologies that persist despite the great functional changes occurring in the transformation of the reptilian jaw into the mammalian middle ear.

Faced with such phenomena, Darwin suggested that 'the chief part of the organization of every being is simply due to inheritance'. He seems here to be proposing that inheritance provides an explanation of those aspects of morphology which remain unchanged or invariant in lineages, since elsewhere he says 'propinquity of descent — the only known cause of the similarity of organic beings'. This is an extremely limited, not to say deficient, causal account of invariance, since it relies simply upon an undefined conservatism in the historical process of organismic reproduction. In the exact sciences, empirical regularities are the observational basis for the deduction of laws expressing the intrinsic order of systems from which their characteristic organizational properties arise. It is then possible to account rigorously (i.e. to define sufficient conditions for) both invariance and variation, the former being the result of intrinsic order (law-like behaviour), the latter the result of contingencies (e.g. particular initial conditions). Darwin, by focusing his attention on inherited adaptive *differences* between organisms as the primary biological problem, set biology on the path of finding those inherited factors which correlate with observed differences between organisms. These are now identified as the genes, to which are ascribed full causal power in the reproductive process. This confuses necessary with sufficient conditions, for correlative factors, like initial conditions, are only necessary, and for sufficiency we need also descriptions of the invariant relational order in organisms which underlies reproduction and results in the regularities observed in development and evolution.

The absence of any explanatory principles of morphological invariance can be clearly seen in the most influential conceptual scheme for the development of form and pattern in developmental biology, which is the theory of positional information and its interpretation expounded by Wolpert (1971). This theory involves no constraints on form, since all details of structure arise from a gene-based interpretation process which is freely modifiable, in accordance with the neo-Darwinian view that organisms are a result of selection of randomly-variable, infinitely-modifiable forms. Clearly in terms of such a theory, any regularity of form observed over long lineages must be accidental rather than systematic, and so does not present itself as an important problem requiring explanation. Again, any systematic ordering of species into taxonomic schemes involving logical inclusion and exclusion principles must be understood as the historical unfolding of an unconstrained biological potential rather than as evidence of rules or laws operating within organisms. Within such a view, the biological realm inevitably reduces essentially to the study of genes and their products, and problems relating to sources of order arising from organizational principles of the living state above the level of molecules disappear. This is why organisms have disappeared as real entities from contemporary biology: they are simply the vehicles of 'selfish genes', with no autonomous existence or logical priority as structural units embodying and revealing distinctive principles of organization.

While it is always possible that biology is basically different from the exact sciences in being dominated by historical accidents and lacking systematic principles of organization, it seems much more likely that this view has arisen as a result of a conceptual development which stresses one aspect of the subject, the contingent and the particular, at the expense of the other, the necessary and the universal. The conceptual scheme which set the scene for twentieth century biology was Weismann's dualistic description of organisms, which brilliantly saved Darwinism from inconsistency by centering it squarely on a theory of inheritance which firmly excluded Lamarckism. This description makes very specific assumptions about the causes of biological form, since it is essentially a theory of inheritance in terms of growth and development, whose task was to 'trace heredity back to growth' (Weismann, 1883). What I wish to do now is examine this conception in order to identify what appears to be a source of serious confusion in contemporary biology about the causes of morphogenesis, and in particular the role of the genes in this process. This will result in a modification of Weismann's scheme, leading to a theory of the organism, its development and evolution, in which both particulars and universals are given equal emphasis, so that biology takes its place among the exact sciences as a domain of enquiry in which history and law both contribute to the understanding of creative and intelligible process.

WEISMANN'S THEORY OF DEVELOPMENT AND INHERITANCE

The fundamental feature of Weismannism which has profoundly influenced twentieth century biological thought and research is the division of the organism into two quite different parts, the germ plasm and the somatoplasm. The germ plasm was described as a 'highly complex structure' with 'the power of developing into a complex organism' (Weismann, 1885). It also embodies the Darwinian principle of historical continuity of living forms insofar as it 'has remained in perpetual continuity from the first origin of life' (Weismann, 1883). Furthermore, it is the repository of all the specific causes of form observed in the adult organism: each particular part of the organism is caused by a particulate unit ('determinant') in the germ plasm. The somatoplasm, on the other hand, was described as that part of the organism which responded passively to the germ plasm, taking on a form dictated by the specific causal units in the latter. The familiar schematic description of this conception is shown in Fig. 9·1, which illustrates the continuous nature of the germ plasm from generation to generation and the derivative, transient nature of the adult form.

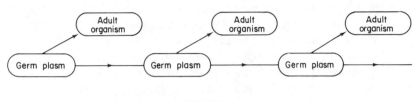

FIG. 9·1

This description clarified a number of points which had been obscure or confused in Darwinism, and it provided a basis for interpreting several observations in cytology which were made at about the time that Weismann was developing his theory. The most important clarification achieved in relation to Darwinism was the unambiguous rejection of Lamarckist-type hereditary mechanisms, for which Weismann could find no evidence; in his scheme there is no path of informational return from adult form to germ plasm between the generations. Thus all inheritance is via the germ plasm. Apart from other limitations which will be considered below, this is actually too strong an assertion as a biological universal, since it is now known that acquired characters in the 'adult' can in fact be inherited in such organisms as the ciliate protozoa (Sonneborn, 1970). However, this aspect of Weismannism is still regarded by neo-Darwinists as one of its most significant features, and it undoubtedly was very important in providing a theory of inheritance of the type required in conjunction with a theory of natural selection. For if the environment directly induced heritable changes of organismic form, then natural selection would be quite unnecessary.

However, there is another aspect of Weismann's scheme which is equally attractive to the contemporary view of organisms and their evolution, and this relates to its implications regarding the specific causes of embryonic development. As a result of developments in cytology towards the end of the nineteenth century, in particular, studies of chromosomes and their behaviour during cell division, it became widely accepted that the cell nucleus contained the material substrate of heredity, so that Weismann naturally located his 'germ plasm' in the nucleus and identified it with chromosomes. Then with the rediscovery of Mendel's laws, the development of genetics, and later the discovery of deoxyribonucleic acid (DNA) as the major vehicle of inherited particulars in higher organisms, the concept of the germ plasm seemed to have achieved full material confirmation. A well defined biochemical substance could be written in for a purely hypothetical concept in Fig. 9·1, continuity between generations then being established by the self-replicating properties of the DNA double helix. The spectacular developments in molecular biology in the 1960s, which led to a detailed understanding of the way in which nucleotide sequences in the DNA give rise to proteins of specific primary structure, provided a further refinement of Weismann's scheme in which DNA was written for 'germ plasm' and protein for 'adult', as in Fig. 9·2. The causal molecular relationships between genotype and phenotype then appeared to be fully established. There remained the problem of accounting for the relationship between proteins and adult morphology, but this was assumed to be solved, in principle at least, by the concept of self-assembly of proteins and other molecular constituents into higher-order structures such as Golgi bodies and membranes, these into cells, and cells into organisms.

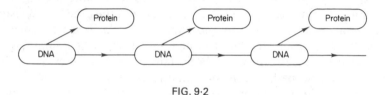

FIG. 9·2

Genetic studies during the first half of the twentieth century had led to a sophistication of Weismann's concept of specific determinants standing in direct causal relationship to specific aspects of form, for it became evident that there is a complex pattern of interactions between the genes and adult morphology. The metaphor now used to describe this relationship is that of a computer programme: the informational content of the DNA of any species (its genotype) is taken to define a 'genetic programme' for embryonic development, so that the arrow from germ plasm to adult in Fig. 9·1 is now understood as a temporal unfolding of specific information from the DNA which controls all the details of protein activity and interaction (Fig. 9·2) resulting in embryogenesis and the development of the adult organism. The question we must now consider is whether this view can be upheld in the face of the evidence.

LIMITATIONS OF THE GENOTYPE–PHENOTYPE DUALITY

As we have seen, the basic feature of Weismann's theory of the organism, which has been preserved in all subsequent modifications, is the dualism which distinguishes between a hereditary part containing the specific causes of organismic form and behaviour, and a non-hereditary part which can respond to hereditary instructions, taking on a structure fully determined by them. Such a view of the reproductive process leads to statements that the genetic instructions are translated into organismic form and behaviour, implying that there is an informational equivalence between the genotype and the phenotype. Similarly, it is said that the phenotype is determined by the genotype, so that 'a theory of development would effectively enable one to compute the adult organism from the genetic information in the egg' (Wolpert and Lewis, 1975). Essentially similar statements were made by Monod (1972) and Jacob (1974), who asserted that all aspects of organismic form are determined by the hereditary particulars encoded in the DNA. In such a conception, the organism effectively disappears from biology as a fundamental entity and is replaced by a collection of sufficient causes in the genome: i.e. the phenotype is reducible to the genotype.

This 'genocentric biology' would be logically valid if the causal chain of command from the DNA extended beyond protein structure to all aspects of organismic morphology. Such an assumption is frequently made, but there is much evidence which makes it a proposition of dubious validity, to say the least. This evidence comes from a great diversity of sources and is discussed in some detail in a publication which analyses the limitations of a view of biological organization based upon the concept of a 'central directing agency' such as Weismann's germ plasm or a genetic programme, conceived as the organizing centre of living entities (Webster and Goodwin, 1982). I shall mention here only a few of these to illustrate the difficulties of the view that the genotype contains the sufficient causes of the phenotype.

A simple counter-example comes from the work of Sonneborn (1970) and others on the ciliate protozoa, in which specific patterns of cortical morphology such as reversed orientation of cilia in a kinety (row) or the presence of a second mouth is inherited by a mechanism that is independent of the genotype. Thus the specific causes of morphology in these organisms certainly do not come solely from the genes, and other biochemical mechanisms, involving nucleation centres in the cortex, have been proposed as proximal causes of cortical morphology. The point that emerges here is that perfectly respectable and comprehensible forces affecting the assembly of proteins can be embodied in cytoplasmic organization; and, in the case of the ciliate protozoa, these can be inherited by virtue of the self-organizing nature of the cell cortex. Of course gene products are involved in this process, for without the synthesis of cortical proteins there could be no reproduction at all. But the causal relationship between cortical proteins and morphology is not direct. Furthermore, species with very different cortical structure can have practically identical proteins, while species with almost identical morphology can differ greatly in their cortical proteins. These and related questions are addressed in an extremely illuminating paper by Frankel (1983).

From these considerations, we see that neither genes nor cytoplasm define sufficient conditions for the specific morphology which is generated in protozoa; both must be considered as sources of informational constraint on biological form, so that reproduction in this large category of organisms does not conform to Weismann's scheme of Fig. 9·1, nor to any of its subsequent modifications. Hence, on these grounds alone, this theory cannot be used as a general description of generation (reproduction) in organisms.

Since the ciliate protozoa are often regarded as very much the bizarre, if not the lunatic, fringe of biology, let us now move to centre stage and consider reproduction and morphogenesis in *Drosophila*. A particularly interesting class of mutations affecting morphology in this insect goes under the name of homoeotic mutations, which have the effect of altering the structure of the appendages that normally appear on certain segments, replacing them partially or wholly by appendages belonging to other segments. For example, the homoeotic mutation *aristapedia* results in the appearance of a leg where there is normally an arista, a part of the antenna (Postlethwaite and Schneiderman, 1971); while in the mutant *bithorax*, a wing appears where a haltere (a small balancing organ) normally arises (Lewis, 1963). Such observations, together with earlier ones on the spontaneous transformation of appendages into one another in insects, have established the concept of homological equivalence among the appendages on different segments in insects. Thus we are led to the idea that ontogenetic (developmental) transformations are identical in kind to phylogenetic transformations of the type underlying the homological equivalence of tetrapod limbs, the one occurring between different parts of one organism and the other between the same parts of different organisms, both revealing the same type of generative constraint on morphogenetic transformations as Bateson (1894) argued so convincingly.

If genes were specific and sufficient causes of morphology, then a fly carrying a mutation would be altered in its form and one with normal or wild-type genes would be morphologically normal. However, it is a commonplace of genetic observation that occasionally an individual fruit fly carrying the mutation aristapedia, say, will have a normal antenna on one side of the body and a mutated morphology, with leg parts appearing where the arista should be, on the other side. This is not explained causally in genetics, but is taken to be an expression of the inevitable variability accompanying such a complex process as morphogenesis. However, such observations warn us that, as in the case of the ciliate protozoa, there is no direct and simple causal relationship between genotype and phenotype of the kind proposed in Weismann's scheme, and that a knowledge of the genetic information in the egg does not allow one to compute the form of the adult organism in simple deterministic terms. Other factors intervene.

There is a further highly significant phenomenon relating to homoeotic mutants which gives us a very important clue about the nature of genetic influences on morphology. It was discovered by Goldschmidt (1935) that if Drosophila embryos at certain stages of their development are exposed for brief periods to non-specific stimuli such as elevated temperature, ether, X-rays, or chemicals such as boric acid, nicotine, vitamins, etc., then the morphology of the adult is altered in specific ways. The interesting aspect of these morphological disturbances is that they fall into the same categories as those produced by mutations, i.e. the non-specific environmental perturbations mimic or copy the genetic perturbations, and so are called phenocopies. Now such an equivalence of causal status between non-specific external stimuli and internal genetic stimuli means that the genetic influences are not, generally speaking, specific causes in the sense of carrying detailed information determining morphology as implied by the genetic programme metaphor. Taking this metaphor seriously, a mutation would be equivalent to a specific change in a programme such as the systematic replacement of one letter or number or operating symbol by another. That such a 'perturbation' would result in a specific disturbance to the computer output is clear enough; but the probability that this specific alteration of output would be mimicked or copied by some general disturbance to the computer, such as transient over-heating or temporarily reduced power, is vanishingly small, unlike the situation observed in phenocopies of homoeotic mutations.

The conclusion we are led to from this type of observation is that organismic morphology is not caused by specific information in the genes in the sense that the output of a computer is caused by a programme. Such a causal relationship *is* valid between the nucleotide sequence of a structural gene and the primary structure of the protein which is generated from it; and if the primary structure uniquely determines tertiary structure, then the three-dimensional configuration of the protein may also be said to be caused by the nucleotide sequence in the genes, given the cytoplasmic environment in which the three-dimensional folding occurs. However, beyond this stage, causal relationships are no longer correctly describable in programmatic terms, as Stent (1982) has clearly argued. Furthermore, the deduction which is to be drawn from the phenocopy data is that gene products affecting morphology are to be understood as stimuli which evoke particular categories of response from a structured, self-organizing process which has a limited repertoire of possible responses (Goldschmidt, 1938).

Environmental stimuli are therefore similarly constrained in the set of morphological responses which they produce, resulting in the phenocopying phenomenon. What, then, is the nature of this structured, self-organizing property of organisms which underlies morphogenesis, giving gene products and environmental stimuli the status of evocators rather than sufficient causes? It derives from what is known as a morphogenetic field.

THE FIELD PROPERTIES OF ORGANISMS

The existence of field properties in developing organisms was demonstrated in a classical experiment by Hans Driesch (1892). This showed that Weismann's causal scheme for the relationship between hereditary factors and adult characteristics was not valid. According to Weismann's original proposals, the specific hereditary determinants in the nucleus of the fertilized egg became partitioned to daughter cells during embryogenesis in such a way that, ultimately, each cell receives only those determinants which specify its final differented state. This is, in embryological parlance, a 'mosaic' theory of development, since it has the consequence that any embryonic cell or group of cells, isolated at some stage in development, can develop only into those parts which are the normal embryological fate of that cell or cell group. An early experiment by Roux on the amphibian egg seemed to confirm this prediction: if one of the cells at the two-cell stage of the embryo is killed by pricking with a hot needle, then the other goes on to produce only half an embryo, either right or left according to which cell is left intact. However, Driesch physically separated the blastomeres (cells) at the two-cell stage of development of a sea-urchin, and he then observed that each of them developed into a completely normal, though half-sized, pluteus larva, the 'free-swimming' stage of the sea urchin life cycle. These larvae then grew to normal size, so that they were indistinguishable from larvae which had undergone undisturbed development. This type of behaviour has since been observed in many different species, and has given rise to the concept of embryonic regulation, the capacity of embryos to recover from spatial perturbation of various kinds by undergoing regulation back to normal morphology.

From these properties of spontaneous spatial self-organization and reorganization has come the concept of the developmental or the morphogenetic field. A field is a spatial domain in which every part has a state determined by the state of neighbouring parts so that the whole has a specific relational structure. Any disturbance to the field, such as removal, reordering, or addition of parts, results in a restoration of the normal relational order so that one whole spatial pattern is reconstituted. This is what is meant by regulative behaviour in embryos. It is a general or universal property of the living state, observed in all types of organisms. In fact, both of the basic phenomena of reproduction (or, to use the old-fashioned term, generation) and regeneration (i.e. healing, making whole) are dependent upon the field properties of living organisms. For the essential feature of these processes is that from a part a whole is generated. This part may be a single cell, the öocyte, as in sexual generation; it may be a group of cells which forms a bud, as in asexual generation in plants or in animals such as hydra; or it may be a single cell which is the whole organism, as in the unicellulars (in which case 'part' means its upper limit, which is the whole, so that one whole organism generates another). Similarly, the part which restores the whole in regeneration may be either many cells cooperating and interacting to restore the missing part, as in axial regeneration in planaria or limb regeneration in insects and urodeles (tailed amphibians such as newts); or it may be a fragment of a cell which restores the whole organism, as in the unicellular ciliate protozoa.

What we see from these examples is that the cell is not the basic biological entity underlying the generative (reproductive) process, and certainly it is a mistake to identify this process with a particular category of cells, the germ cells containing the germ plasm, as Weismann argued; nor is generation to be identified with the properties of the DNA nor the genotype nor the genetic programme as in the contemporary view. Rather, generation is to be understood as a process which arises from the field properties of the living state, with inherited particulars acting to stabilize particular solutions of the field equations so that specific morphologies are generated. Those limitations or regularities of morphological transformation which are observed in homoeosis and in the homologous series of tetrapod limbs are thus to be understood as manifestations of transformational constraints on field solutions; while the inherited variety of form observed in organisms arises from the inheritance of particular influences such as specific gene products or nucleation centres, acting as parameters contributing to the specification of initial and boundary values of the field equations. Exactly what type of field we are concerned with in living organisms is very much a research problem, and more will be said about this in the next section. What we can now observe is that Weismann's dualism isolated for special attention that category of inherited particulars which is now recognized as being carried in the DNA of sexually reproducing organisms. This is the repository of information which correlates with most of the *differences* between organisms, and so it is of particular interest to geneticists whose whole objective is to study what makes one organism hereditarily different to another. However, inheritance and evolution are not simply about differences between organisms and the variety of their adaptive properties, despite the persistent distortion in this direction since Darwin's conceptualization of evolution in these terms, reinforced by the notion of competition as the major driving force of natural selection. The process of evolution manifests both regularity and variation of form. Any adequate theory must therefore account for both aspects of the process, for the unity which gives the whole unfolding an intelligibility, and for the diversity which reveals its morphological potential.

A FIELD THEORY OF THE GENERATIVE PROCESS

Despite the limitations of such schemes, it is useful to construct a diagram which illustrates some modifications which are necessary to correct the limitations of Weismann's conception of the relationship between development and evolution. This is given in Fig. 9·3. From this we can readily extract the formula: particulars + universals = specific form, which any theory of morphogenesis must have, the generative field equations embodying the universals and the parameters containing the particulars. The latter include the two sources of particular influences on organisms, those which are inherited and those which come from the environment. The causal equivalence of these as revealed by the phenocopy phenomenon is immediately apparent from the scheme: both act upon (shown by arrows) the generative field to stabilize or 'select' particular solutions, generating a particular morphology as shown by the arrow from the generative field to the adult organism. The arrows from generation to generation represent the continuity of the living state as embodied in the generative field on the one hand, and the transmission of hereditary particulars on the other. Lamarckian inheritance is not excluded, since the inherited particulars can themselves be aspects of adult morphology, which are then transmitted to the next generation as in the ciliate protozoa.

We can now use this general description of the relationship between development and evolution to describe in more specific terms a field theory of morphogenesis. What will emerge from this is the realization that evolution is itself logically deduced from the properties of the living state, so that the scheme of Fig. 9·3 becomes, not a way of including development in the evolutionary process, but rather a way of understanding how both development and evolution are inevitable aspects of the life process. Biology then becomes grounded in a generative theory of organization.

I shall proceed with the analysis in two steps: first, a formal field description of a relatively simple morphogenetic process, and then a more

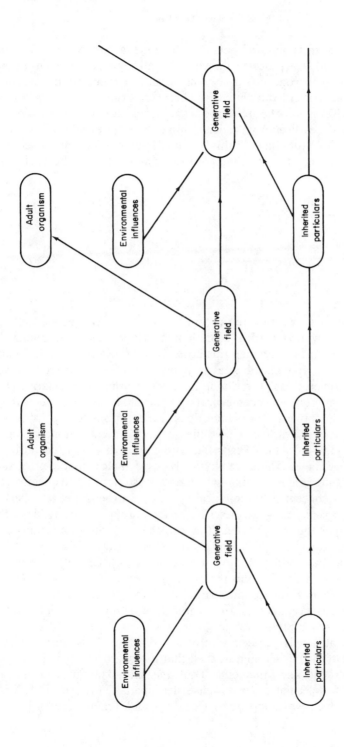

FIG. 9-3

detailed field treatment of related phenomena involving much more molecular detail. The process with which I shall begin also happens to be the beginning of embryogenesis in a great variety of multicellular organisms, ranging from coelenterates up to higher vertebrates. This is a series of regular divisions of the fertilized egg, resulting in a group of cells constituting the early embryo which retain overall spherical form, called the morula. Some of the stages in this process, known as holoblastic cleavage, are shown in Fig. 9·4. Its characteristic is that, during the first several

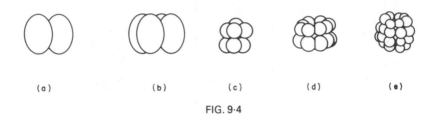

    (a)   (b)   (c)   (d)   (e)

FIG. 9·4

cleavages which may extend to the twelfth, there is a very regular pattern of cell divisions by first vertical and then horizontol cleavage planes. The morphogenetic problem which this poses is the following: if cells behaved independently in holoblastic cleavage, then after first cleavage one would expect blastomeres (cells) to divide in random orientation relative to one another so that no global geometrical order would result. Obviously they do not behave in this way. So where does the constraint come from which causes the observed order? A possible answer is: there is a global field of force of some kind which keeps the whole embryo organized during the early cleavage stages. There are physically respectable candidates for such a force-field and there is nothing in the least mysterious or suspicious about using such a concept in a biological context. Furthermore, it is perfectly acceptable, scientifically, to describe certain properties of fields without knowing either the 'microscopic' (i.e. molecular or atomic) origins or the nature of the forces involved, and there are analytical devices for doing so (e.g. gravitational field and gravitational potential). One of these is to say that, although we do not know the exact properties of the holoblastic field, we do see clear evidence of spatial ordering processes accompanying successive cleavages, such as an organization of microfilaments in a ring around the presumptive cleavage furrow prior to the actual contraction which separates cells physically into two daughter cells, and the earlier orientation of the mitotic spindle such that its equatorial plane contains the ring of contractile microfilaments. Thus one can talk about relative degrees and types of molecular order and disorder, spatially organized. This is the procedure which was adopted by Goodwin and Trainor (1980) in a field description of holoblastic cleavage. Since this has been described elsewhere, I shall confine myself to the briefest exposition in order to bring out certain relevant points.

As described in the section on the field properties of organisms, fields are spatial domains in which the state of one region or volume element is related to that of neighbouring elements, so that the whole domain has a well defined relational structure. Although there are different kinds of fields, and different field equations describing them, one can nevertheless derive a field equation embodying some essential principles relating to order–disorder properties at the molecular level as described above in relation to the orientation and organization of microfilaments, microtubules, and the process of cell division. This equation, in terms of a variable known as an 'order parameter' measuring some aspect of molecular or microscopic order in the field, then has certain solutions, those applying to the spherical geometry of the egg and the early embryo being known as spherical harmonic functions. Using these functions to describe the cleavage process requires a convention relating certain properties of the functions to the cleavage planes, and an obvious one to use is to identify cleavage planes with what are known as nodal lines of the functions, where they take the value zero. Since there is an infinite number of harmonic functions on the sphere, some of which correspond under this convention to cleavage patterns, and others which do not, the question that then arises is how to select the 'correct' functions in terms of a biologically plausible set of constraints or 'selection rules'. This is now the precise form of the question that has been raised throughout this essay in relation to morphogenesis: what are the 'causes' of specific biological form, such as the specific pattern of cell division planes during holoblastic cleavage? Are they to be understood in terms of the properties of specific gene products carrying the information which specifies the pattern as implied by the concept of a genetic programme? We can now give a precise answer to this question.

It turns out that in order to get the sequence of spherical harmonics with nodal lines as shown in Fig. 9·5, corresponding to the pattern of holoblastic cleavage, we need three constraints.

(1) After the pth cleavage, there are 2^p subdivisions of the sphere by the nodal lines, corresponding to the biological constraint that each cell divides into 2 cells so that the total number of cells is a power of 2, at least during the phase of synchronous cleavage (i.e. up to about the twelfth cleavage).

(2) A minimum energy condition, such that the function selected to describe the cleavage pattern always conforms to minimum 'surface energy' (defined in a precise but general way).

(3) A force which selects vertical cleavages whenever there is a choice

resulting from equal surface energies from two solutions (degeneracy), interpreted in terms of the animal–vegetal polarity of the egg.

These constraints then provide a solution to the problem of selecting a particular set of field solutions from a large potential set of possibilities,

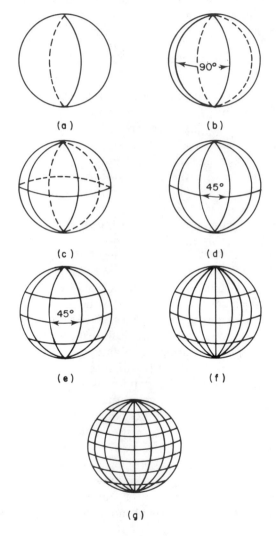

FIG. 9·5

giving a specific morphological pattern. None of them is readily interpreted in terms of the effects of specific gene products produced at particular stages of development as in the genetic programme description. Rather, what we see operating are some very general biological constraints: binary cleavage, energy minimization, and polarity. The first and last are quite basic properties which are to be understood in terms of principles of biological organization at the cellular level, while energy minimization is probably a universal property of morphogenetic processes, conforming to general physical constraints, although it must be emphasized that as yet we cannot measure directly the 'energy' involved, so that this remains an unquantified criterion. Thus the holoblastic cleavage pattern is a spontaneous expression of spatial ordering principles inherent in the living state, described and interpreted in terms of field properties as they apply to this particular condition of organization involving the subdivision of a whole by successive divisions into cellular parts. Both part and whole must be considered in the process, since the parts (cells) embody rules, such as binary division, which impose constraints on the pattern of the whole, while the whole imposes constraints on the parts, resulting in a regularity of cleavage geometry.

There are many modifications of the pattern described in Figs. 9·4 and 9·5, which distinguish different species. A common variation on the basic theme is for the animal blastomeres to be smaller than the vegetal blastomeres; another is for there to be systematic differences in size between the dorsal and the ventral blastomeres, giving bilateral, in contrast to radial, symmetry. These can be accommodated within the general scheme by assuming the influence of secondary fields of force, such as that associated with the animal-vegetal polarity mentioned earlier, or with dorso-ventral asymmetry. The strength of these fields then differs in different species, and these differences may be ascribed to genetic factors.

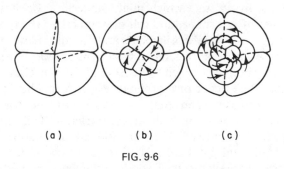

(a)   (b)   (c)

FIG. 9·6

Another modification, which is more profound, is the pattern found in the spiralia, an example of which is shown in Fig. 9·6. Instead of cleavage planes being alternately vertical and horizontal, they alternate between right- and left-handed spirals. Here we find an example of a specific gene product affecting morphogenesis, for in *Limnaea* the wild-type shows a dextral (right-handed) pattern at third cleavage, whereas a mutant cleaves sinistrally at this stage. These patterns are then expressed in the adult in the coiling patterns of the shells, the wild-type being dextral. Here, then, is a case of a gene product, maternally expressed, which 'selects' the handedness of the first spiral cleavage field solution, and subsequently the handedness of the spiral shell structure. However, it would be wrong, according to the above analysis, to say that the gene product generates the pattern. It is the field which generates the possible patterns, which in the case of the spiralia are right- and left-handed spirals, while the gene product selects or stabilizes one of these for expression. Of course, the difference between a spiralian and an organism showing radial cleavage, such as a sea urchin, is presumably also connected with gene products, though we have no direct proof of this. As stated in the previous section, the main source of the heritable *differences* between multicellular organisms is certainly to be traced to genetic information in the DNA. But this is quite a different statement to that asserting that the form of an organism is computable from this genetic information, a statement which implies that all you need to know in solving a field equation is parameter and boundary values. But of course, you also need the equation, and its derivation is not a trivial matter.

MORPHOGENESIS AND VISCO-ELASTIC FIELDS

The field equation used in the treatment of holoblastic cleavage was, as stated, a rather general one which did not attempt to incorporate specific properties of the living state which might underlie processes such as cell division, changes of cell shape, and cell movements of the type involved in morphogenesis. A role of microfilaments and microtubules was implied but not articulated. A much more detailed treatment of morphogenesis in these terms has recently been developed by Oster *et al.* (1980). They start from the observation that the essence of morphogenesis is form-shaping movements of cells and cytoplasm, and that the source of the forces for such mechanical deformation is the cellular cytoskeleton. This is a network of polymers which have already been mentioned in relation to cleavage: microfilaments and microtubules. To these are now added a third type, known as intermediate filaments. The microfilaments are made of actin and myosin and so have contractile properties, while the other polymers act as structural elements giving cytoplasm some rigidity. Together these polymers generate a mobile cytoplasmic scaffolding, the cytoskeleton, which has the behaviour of a visco-elastic gel. The state of the gel can be modified by cross-linking proteins such as actin binding protein, by calcium which affects both the degree of cross-linking and the activity of the contractile microfilaments, and by a number of other factors influencing both the viscosity and the elasticity of the cytogel. What Oster and his colleagues have done is to derive visco-elastic field equations to describe the behaviour of this system and to demonstrate that it has many of the properties required to generate the patterns and forms observed in developing embryos. For example, they have shown that a closed ring of cells, representing a cross-section through the blastula stage of an embryo such as a sea-urchin or a frog (the blastula is the stage after the morula, described above in relation to cleavage), will undergo precisely those movements which characterize gastrulation when the visco-elastic gel is stimulated in a particular way. This stimulation involves a release of calcium in a few cells so that its concentration rises to micromolar levels, initiating the solation of the cytoplasm by reducing the number of cross-bridges in the scaffolding and activating actin contraction. This contraction then stretches neighbouring cells, causing calcium release from bound sites and hence initiating solation and contraction, so that a contraction wave propogates through the tissue due to the stretch–contraction properties of the cytogel (like the knee-jerk reaction). The reason why this contraction wave results in an altered morphology rather than simply being a transient, reversible contraction–relaxing process is that after contraction, the cells stay contracted since the calcium regulation mechanism restores calcium rapidly to very low values (10^{-7}M) so that the gel state is restored by cross-linking and the altered form is 'frozen'. Furthermore, the contractile components of the cytoskeleton are located at the outer surface of the cells so that an asymmetric contraction occurs. The result is an extraordinarily convincing picture of gastrulation, one of the most fundamental of morphogenetic movements.

Oster and his collaborators have now demonstrated how an impressive variety of basic morphogenetic transformations can be accounted for by simple changes in the parameters and boundary conditions of sheets of cells whose behaviour is governed by visco-elastic field equations. They point out that, for each process, it is necessary to define a 'prepattern', i.e. the boundary conditions which are very important in selecting a specific solution of the equations. This prepattern arises from the previous stage of the developmental process, gastrulation having a prepattern defined by the spatial organization of the egg and the cleavage process, for example. With calcium distributions as a major factor in the specification of initial and boundary conditions for the visco-elastic field, it seems very likely that electrical potential differences across membranes, ion-specific channel distributions in cells and current flows of the type demonstrated by Jaffe *et al.* (1974) will be of primary importance in determining the spatial order that initiates and regulates the morphogenetic transformations of the cytogel. These are governed by field equations similar to those used to describe diffusion–reaction processes, but modified by conservation laws relating to currents and charges. There will also be electro-mechanical couplings, so that one might anticipate the development of an electro-visco-elastic model of morphogenesis. However, the different relaxation times of the electrochemical and the mechanical processes will make it possible to treat the two interacting systems in a relatively independent manner. What begins to emerge from this is a truly significant insight into the properties of the living state which underlie both development and evolution.

We can now begin to discern exactly what is meant by biological universals in this context. Morphogenesis occurs in both unicellular and multicellular organisms, so it is a process which is independent of cellular partitions and cell–cell interactions; i.e. cells are not primary units in morphogenesis. The visco-elastic field model makes it clear why this is so: the coordinated behaviour observed in morphogenesis arises from the transmission of the mechanical forces of stress and strain over the cytogel as a *continuum*, independently of membranous partitions. Even if there are cellular partitions, the forces are transmitted not by cell–cell interactions via gap junctions or septate desmosomes or other communication channels across cell membranes, as in diffusion-reaction processes, although these are undoubtedly very important for certain aspects of development in multicellular organisms. The forces governing morphogenesis are transmitted by mechanical interactions throughout the whole system, and these can act over either more or less than one cell diameter since they are governed by stretch–contraction processes whose characteristic length is determined by the properties of the cytogel. Thus the visco-elastic field is a morphogenetic universal, and whether or not organisms are partitioned into cells is a particular which secondarily distinguishes the multicellulars from the unicellulars.

From Fig. 9·3 we see that the generative field, now identified partially as a visco-elastic field, is transmitted from generation to generation, so we may enquire into the nature of this hereditary continuity. Since the field depends upon the properties of certain biopolymers (actin, myosin, tubulin, actin binding proteins, etc.) together with the effects of calcium and other substances which influence the degree of solation and contraction of the cytoskeleton, it is clear that the molecular elements which make up the field are either gene products or are substances whose concentrations depend upon the regulatory activities of gene products. These are biological universals in that the generative (reproductive) process depends upon their presence and activity; and also in that macromolecules such as actin, myosin, and tubulin are among the most highly conserved of all biological polymers. But the identification of such macromolecules in cytoplasm, though necessary in developing an understanding of morphogenesis, is in no sense sufficient to provide an explanation of morphological transformations, whether ontogenetic or phylogenetic, any more than the observation of planetary motion is sufficient to explain elliptical orbits. For what is required is the deduction of the correct relational order which generates the observed phenomena, and this order or organization is not directly observable, though it is real. This logical relational order is what defines the distinctive organizational properties of living organisms, and the proposition of this essay is that the appropriate mathematical description is provided by field equations.

We can now propose that these organizational properties, insofar as they relate to the generation and regeneration of organismic morphology, are to be at least partly understood in terms of the principles of continuum mechanics as expressed in visco-elastic field equations of the type derived by Oster *et al.* (1980). This organization is continuous from generation to generation by virtue of the continuity of the cytoplasm, whose molecular composition is maintained by the activities of genes which are replicated as part of the continuity of the living state. This defines a unity of process, not a duality. That gene products not only contribute to invariant properties of the living state, but also influence which of the possible field solutions is expressed by defining parameters and affecting boundary conditions, means that there is again no sharp duality of role between different categories of genes, but rather a spectrum between universal and particular. The exact structure of this spectrum provides the analytical basis of a rational taxonomy, a systematic classification of morphologies in terms of the relationships between field solutions. For example, as Oster has argued, one can introduce a fundamental classificatory dichotomy between epidermal derivatives which arise from a buckling out of spatially periodic condensations of cells in epidermal sheets, giving rise to feathers and scales; and a buckling in, resulting in structures such as teeth, glands, and hair. Then the distinction between say, feathers and scales, arises from another morphogenetic bifurcation. Thus an understanding of morphogenesis provides the basis for a rational taxonomy, one based upon the logical properties of the generative process rather than a genealogical taxonomy based upon the accidents of history (*see* Webster, this volume). This provides a direct link between a field theory of morphogenesis and the study of the logical relationships of organisms as observed in morphology, the latter being systematically developed in cladistics (Nelson and Platnick, this volume).

The apparent duality of the scheme in Fig. 9·3, which seems to distinguish sharply between universals and particulars, thus finally also dissolves into a unity, a continuum with a spectral decomposition which provides the basis for biological systematics. Development and evolution may then be deducible from the properties of the living state, involving both molecules and fields, which together provide the temporal and the spatial order revealed in ontogenesis and phylogenesis, as well as the diversity expressed within this unity. What form biological field equations will finally take, and what experimental methods will be found to be the most appropriate in exploring long-range biological order, are questions for future research on these central problems. Shifting attention from the dramatically successful study of organismic properties associated with the short-range order (Ångstroms) arising from molecular interactions such as those involved in enzyme catalysis, DNA replication and recombination, self-assembly, metabolic regulation, etc. to the more global (micron–millimetre) domains of spatial order which are fundamental to generative and regenerative processes, will take some time as well as new techniques and theories. But the axiomatics of a generative biology, which could provide the foundation for both the logic of taxonomy (synchronic biology) and of time-dependent process as revealed in development and evolution (diachronic biology), seem now to be emerging from the deeper understanding of biological organization which has arisen from the success as well as the failure of twentieth century biology. The attempt to understand biological process in historical terms, the objective of the modern synthesis based upon the evolutionary paradigm, has failed because of the impossibility of explaining observed biological order and regularity in terms of contingencies and differences, or accidental variation. But the momentum of biological research in this century has been such that many of the ingredients for a rewriting of Weismann's fruitful but limited scheme, uniting development and evolution within a logically-ordered generative process which includes not only hereditary particulars but also biological universals appropriate to different levels of organization in the organism, seems now to be a distinct and exciting possibility. Such a rewriting, if it can be realized, will transform biology from an historical to an exact science.

REFERENCES

Bateson, W. (1894). 'Materials for the Study of Variation'. Cambridge University Press.
Darwin, C. (1859). 'The Origin of Species' 1st edition. Penguin, Harmondsworth.

Driesch, H. (1892). The potency of the first two cleavage cells in the development of echinoderms. *In* 'Foundations of Experimental Embryology' (B. H. Willier and J. M. Oppenheimer, *Eds*) 1964. Prentice-Hall, Englewood Cliffs, New Jersey.

Frankel, J. (1983). What are the developmental underpinnings of evolutionary changes in protozoan morphology? *Brit. Soc. Dev. Biol. Symp.* **6.** (B. C. Goodwin, N. T. Holder, and C. C. Wylie, *Eds*). pp.279–314. Cambridge University Press, Cambridge.

Goldschmidt, R. B. (1935). Gen. und Ausseneigenschaft (Untersuchungen an Drosophila I, II). *Zeits. Indukt. Abst. u. Vererb.* **69**, 38–131.

Goldschmidt, R. B. (1938). 'Physiological Genetics'. McGraw-Hill, New York.

Goodwin, B. C. and Trainor, L. E. H. (1980). A field description of the cleavage process in embryogenesis. *J. Theor. Biol.* **85**, 757–770.

Jacob, F. (1974). 'The Logic of Living Systems'. Allen Lane, London.

Jaffe, L. F., Robinson, K. R. and Nuccitelli, R. (1974). Local cation entry and self-electrophoresis as an intracellular localization mechanism. *Ann. N.Y. Acad. Sci.* **238**, 372–389.

Lewis, E. B. (1963). Genes and developmental pathways. *Am. Zool.* **3**, 33–56.

Monod, J. (1972). 'Chance and Necessity'. Collins, London.

Oster, G., Odell, G. and Alberch, P. (1980). Mechanics, Morphogenesis and Evolution. *Lectures on Mathematics in the Life Sciences* **13**, 165–255.

Postlethwaite, J. H. and Schneiderman, H. A. (1971). Pattern formation and determination in the antenna of the homoeotic mutant *Antennapedia* in *Drosophila melanogaster*. *Devel. Biol.* **25**, 606–640.

Sonneborn, T. M. (1970). Gene action in development. *Proc. R. Soc. London B* **176**, 347–366.

Stent, G. (1982). What is a program? *In* 'Evolution and Development', Dahlem Workshop Report 22, pp.111–113. Springer-Verlag, Berlin.

Webster, G. C. and Goodwin, B. C. (1982). The origin of species: a structuralist approach. *J. Soc. Biol. Struct.* **5**, 15–47.

Weismann, A. (1883). Reprinted in 'A Source Book in Animal Biology' (T. S. Hall, *Ed.*) 1964. Hafner, New York.

Weismann, A. (1885). Reprinted in 'Readings in Heredity and Development' (J. A. Moore, *Ed.*) 1972. Oxford University Press, Oxford.

Wolpert, L. (1971). Positional information and pattern formation. *Cur. Top. Devel. Biol.* **6**, 183–224.

Wolpert, L. and Lewis, J. (1975). Towards a theory of development. *Fed. Proc.* **34**, 14–20.

10

Development and Evolution

PETER T. SAUNDERS

Abstract

While it is no longer unusual for evolutionists to acknowledge that development may have an influence on evolution, few of them actually take it into account, possibly because they consider that our present understanding of developmental biology is insufficient to permit this. The aim of this chapter is to refute this view by example. Two current theories of pattern formation, pre-pattern and positional information, are contrasted, and it is shown that they lead to different predictions about evolution. This demonstrates how the process of development affects the course of evolution, and in a way which is already intelligible. Recent research into the mechanisms which produce the characteristic markings of certain animals indicates that a wide variety of patterns can be understood as variations on a common theme, and it is suggested that this can account for the frequent occurrence of Batesian and Mullerian mimicry. Finally, a brief account is given of progress which has been made towards an understanding of the dynamics of morphogenesis. Two approaches are currently being employed, one involving detailed modelling and the other far more general. A combination of both will be needed if we are to find a solution to the problem of the genesis and evolution of biological form.

In suggesting an explanation for adaptation, the theory of natural selection provides at most a partial explanation for evolution. It is not enough to say that those traits which are favourable will be selected. We have also to explain how they arise, that is, to account for the set of alternatives from which the selection is made. Otherwise we have, in Samuel Butler's (1911) words, 'an *Origin of the Species* with the *"Origin"* cut out'.

Neo-Darwinism, however, contains no theory of the origin of variations. It simply assumes that while they must be small they are to all intents and

BEYOND NEO-DARWINISM
ISBN: 0-12-350080-X

purposes random. A character is deemed to be adequately explained if we can determine, or at least postulate, its selective advantage and evolutionary history.

This approach is based on an extrapolation of Weismann's doctrine of the independence of the germ plasm, according to which the only variations which matter in evolution are random genetic mutations. Now the doctrine is not, in point of fact, unequivocally true (see Pollard, this volume). But even if it were, or if the exceptions were to prove to be of no importance in evolution, the neo-Darwinist position would still not be justifiable. For even if we suppose, for the sake of argument, that the source of all heritable variation is random mutation, it still does not follow that the variations in the phenotype are random. They may be triggered by mutations, but the forms they take are dependent on the properties of the epigenetic system. Consequently the key to an understanding of the origin of variations, and hence to an understanding of evolution itself, must lie in the study of development.

It is significant that more and more neo-Darwinists now appear to feel the need to justify their neglect of developmental biology. For example, Wolpert (1982; see also Wolpert and Stein, 1982) argues that since (in his view) development is pre-programmed in the genes, both development and evolution are understandable in cellular terms and hence in terms of the DNA. This frequently heard argument does not stand up to scrutiny, for to say that the genes are ultimately responsible for development is still not to say that the best way of understanding development is to study genetics. It is surprising how many biologists find reductionism so attractive; if they really believe that all phenomena are best studied at the lowest possible level then they ought to become physicists.

Other writers are willing to concede that development has some influence on evolution, but they insist that it is only a minor one. They maintain, with Mayr (1980), that 'selection is the only direction-giving force in evolution'. Maynard Smith (1969), for example, writes that an understanding of epigenetics is important, but only because it would enable us to estimate how many mutations would be required for a given change to take place, and hence to calculate how much selection pressure would be needed, and over how many generations.

Yet another school of thought is represented by Dobzhansky et al. (1977). They write in the preface of what they claim to be a 'reasonably comprehensive account' of the theory of evolution that developmental biology is a field 'of fundamental importance for evolutionary studies', yet they totally ignore it in the rest of the book. In almost twenty pages of subject index there is not a single entry under either 'development' or 'epigenetics'. They appear to be adopting a point of view which one hears from time to time in conversation: no doubt development has a great deal to tell us about evolution, but at present we simply do not know enough about it to bring it into the discussion. In the meantime the best we can do is to go on as though selection were indeed the only thing that mattered.

These arguments are very beguiling, because they appear both to recognize the importance of the very challenging subject of developmental biology, and yet at the same time provide a justification for continuing with the 'favourite parlour game' (Ruse, 1982, p.134) of the neo-Darwinists: the explanation of anything and everything in terms of largely hypothetical selective advantages. But to say that we do not yet have a complete understanding of development is not to say that we can afford to ignore it. Which blood group one happens to belong to may be determined more or less directly by the genes in the way that Mendelian genetics describes, but most characters arise through long and complex sequences of physical and chemical interactions, some entirely within the embryo, others involving the environment as well. Nor can we evade the issue just by acknowledging that there is not, in fact, a single gene for the trait in question and then proceeding to carry out our calculations as though there were. If all we knew about development were that it is far too complicated to be fully understood even today, with all the experimental and theoretical techniques we have at our command, this would surely be enough to warn us that many neo-Darwinist arguments cannot pass muster.

In any event, the situation in developmental biology is by no means so bleak as the neo-Darwinists would have us believe. Considerable progress has been made. And we do not have to wait for a complete solution before we can incorporate developmental biology into evolutionary theory. Even quite general and preliminary theories about how development proceeds can have important implications for our understanding of evolution. To see how this comes about we begin with a theory whose originator is well known for his insistence that development in general — and his own work in particular — has nothing to say about evolution. (Wolpert, 1982, but see Stock and Bryant, 1981, for a different view).

POSITIONAL INFORMATION

While the hypothesis that cells differentiate according to their position was first put forward around the turn of the century by Driesch (1908), the modern formulation and the name 'positional information' are due to Wolpert (1969). He claims that a developing organism is divided into regions or fields, and that within each of these there is set up some sort of coordinate system which enables every cell to determine its own location. The cell then decides what it is to become on the basis of this information. One relatively straightforward mechanism by which this could occur is that positional information is specified by chemical gradients (one for each dimension of the pattern that is to be laid down) and that the concentrations of the respective substances control the subsequent biochemical reactions within individual cells.

Wolpert illustrates the theory by means of an example known as the 'French flag problem'. Imagine a one-dimensional array of cells, each of which has the potential to become red, white or blue. Our aim is to produce the tricolour pattern of the French flag. Positional information theory suggests a simple way in which this can be done.

Suppose that there is some chemical substance C which controls the colour determining reaction within a cell. We may refer to C as a 'morphogen'. Suppose further that there are thresholds in this reaction such that the cell will become red if the concentration of C is less than one unit, white if it is between one and two units, and blue if it is greater than two units. Then all we have to do is to arrange for a source of C to maintain a concentration of three units at the right hand end and a sink to maintain zero concentration at the left hand end. A linear gradient will be established by diffusion, and it is easily seen that the required pattern will appear (Fig. 10·1(a)).

Note that we have assumed that the participation of the morphogen in the reactions within the cells does not interfere with the gradient. This is perhaps the most distinctive feature of the theory of positional information, the separation between the specification of the information and its interpretation.

Another important aspect is that positional information is taken to be universal, i.e. it is assumed that the mechanism by which position is specified is the same in all fields. Differences between organisms, or between different fields within the same organism, are ascribed to differences in the interpretation of the same positional information.

One consequence of this can be seen in the following thought experiment. Imagine a sheet of cells which are to become an American flag. According to the theory, this requires two gradients. One would be the same as for the French flag, but there would also have to be a second, in some different morphogen C' and more or less orthogonal to the first.

Now suppose some cells are taken from a tissue that was intended to develop into a French flag, and grafted into this sheet, say into the lower right hand corner. They would not exhibit the red and white striped pattern of their new neighbours. Instead, they would 'measure' the local concentration of C and, discovering that they were in the right hand third of the flag, develop into a solid blue patch. They would not react to C', not having been programmed to. Note that it would not matter from which portion of the presumptive French flag the cells had been taken; indeed they need not all have come from the same part.

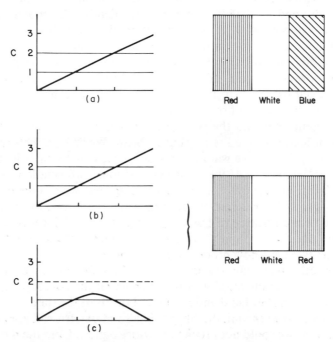

FIG. 10·1 Morphogen concentration C as a function of distance along the tissue, showing the gradients required to produce (a) the French flag using either positional information or pre-pattern, (b) a red-white-red flag using positional information, (c) a red-white-red flag using pre-pattern.

Universality is not strictly essential to the theory of positional information, but without it the theory would be very much weakened. Bryant (1982) argues that one of the accomplishments of Wolpert's work is that it has led developmental biologists to appreciate the underlying similarities of organisms, and this would not have been the case without the universality hypothesis. There is also ample evidence from real grafting experiments to support the claim that positional information is specified in the same way in at least some different fields (*see*, e.g. Wolpert, 1982).

PRE-PATTERN

The theory of positional information postulates a complete separation between specification and interpretation. The coordinate system is the same for all fields, and even if we were able to identify the morphogen and measure its concentration at every point in the field this would tell us nothing about the pattern that would eventually appear.

There is, however, a class of theories in which the morphogen is seen as playing a more active role. Instead of merely providing a reference grid which each cell uses to help it decide its fate, the morphogen contributes directly to the formation of the pattern.

The difference between the two types of theory can be illustrated by a simple modification of the French flag problem. We have, as before, a sheet of cells, only this time we want to produce three equal stripes in only two colours: red, white, red. If we are using positional information, we set up the same linear gradient as for the French flag, but we change the way in which this is to be interpreted. We alter the reactions within the cells in such a way that a concentration of less than one unit still produces red, and one of between one and two units still produces white, but a concentration of greater than two units now produces red, not blue.

An alternative technique would be to use the same thresholds as before, but to change the gradient. If we can arrange that the concentration is less than one unit in the left hand and right hand thirds of the sheet and between one and two units in the middle, then the desired pattern will appear. Note that in this case we could not predict the entire pattern from the morphogen concentration alone, but we would know that it would be symmetric; with positional information we would not even have been able to deduce this (*see* Fig. 10·1(b)(c)).

It is easy enough to see how the monotone gradient required by positional information could be set up, but more explanation is needed for the more complicated gradients associated with pre-pattern theories. The problem was solved by Turing (1952) when he showed that a spatially inhomogeneous morphogen distribution could result from the interaction between chemical reactions and diffusion. Turing was unable to follow his idea through himself because the computers that were available at the time could not cope with the differential equations that arise, but the problem was taken up again by Gierer and Meinhardt (1972), and they and others have demonstrated some of the wide range of patterns that can be generated by quite plausible mechanisms.

PREDICTIONS OF POSITIONAL INFORMATION AND PRE-PATTERN THEORIES

The two theories we have described are both incomplete; in neither case have the morphogens and their reactions been fully elucidated nor the details of the applications to various events in development worked out. Our accounts do not even include all that is known; we have provided only brief outlines, just enough to indicate what the theories are and how they differ. Yet even from what little we have mentioned we can make predictions about what we may expect evolution to have produced. What is more, the two kinds of theory lead to different predictions. This is important because it demonstrates that the predictions arise out of the theories themselves and are not merely consequences of the phenomena the theories are both trying to explain. It also shows that not only does developmental biology have much to contribute to evolutionary theory but that the converse is also true. By observing the forms that have emerged during evolution and the ways in which they have changed, we may draw inferences concerning the process of development.

The most obvious prediction of the pre-pattern theories is that repeated structures should be very common, as they are so easy to produce. Once a reaction-diffusion system has been established it is not at all difficult to arrange even a quite complicated pattern of morphogen concentration. If each peak of concentration is to give rise to a similar feature, then only one threshold is required in the subsequent reactions.

The situation is quite different in the case of positional information. For the same feature to appear more than once in a field, different positional information must somehow be interpreted to yield the same result. There is little simplification to be gained in producing similar features in many different places rather than different ones in each, because different thresholds and subsequent reaction schemes must be used. On this hypothesis it is hard to see why repeated structures should be as common as they are; it is difficult to believe that they should so often be the optimal solution to a problem.

The difference between the two theories is even more striking when we consider the production of changes in numbers. In most pre-pattern theories it is very easy to alter the number or precise distribution of similar features, so much so that a common problem in such theories is that of regulation, i.e. of ensuring that the same number of, say, somites is produced in each individual. All that is required is a small modification in the rate constants or the boundary conditions—just the sort of thing we would expect to result from a single allele substitution. The process which produces the red-white-red flag is somewhat analogous to a standing wave, and it is not at all hard to alter the 'frequency' and hence the number of stripes.

It is considerably more difficult to change to a three-striped flag if the pattern is being specified by positional information. We could, of course, imagine that the reactions within the cells were somehow altered to change the threshold values from 1 and 2 units to 0·6, 1·2, 1·8 and 2·4 units. Alternatively, it might be that the morphogen concentration at the source was increased to 5 units and that at the same time two extra thresholds appeared at 3 and 4 units. Neither of these appears especially plausible. Both demand coordinated changes with no obvious explanation for the coordination. Actually this is a problem which might have been anticipated. Positional information, like neo-Darwinism into which it fits so neatly, is a reductionist theory, and coordination always appears mysterious to reductionists.

One could, to be sure, achieve a three stripe pattern by putting two fields side by side, and even more stripes by even more fields. This is all rather implausible, and, which is even more to the point, it takes us away from positional information and firmly into pre-pattern.

There is, however, at least one sort of change which is more readily produced by a positional information mechanism, and that is one in which the pattern is preserved but the elements that make it up are altered. In terms of the flag metaphor, what we have in mind are changes from the French flag to the Belgian or the Italian. Such changes are probably less likely to occur with a pre-pattern. A mutation which affected the interpretation of the morphogen concentration would also be expected to affect the gradient itself, since specification and interpretation are not separated in the way that positional information supposes them to be.

The implications of these ideas for evolution are—or should be—clear. Suppose, for example, that a certain feature is determined by some kind of pre-pattern mechanism, and suppose that the organism concerned is faced with some environmental challenge. In principle it might cope with this in any one of a number of different ways. If one of these involved nothing more than a change in the number of some component of the trait in question then surely this is the modification we would expect to observe, even if some other conceivable change might have been a better solution to the problem. Hence we would expect changes in number to be common in evolution, as in fact they are. But this is because such changes are relatively easy to bring about, not because they are necessarily optimal (Saunders and Ho, 1981). In other words, the real explanation of the phenomenon is to be found in development rather than in natural selection.

ANIMAL COAT MARKINGS

We have seen how a theory of development can have implications for our understanding of evolution, even if it is framed in quite general terms. More specific theories should make more specific predictions, and to illustrate this we consider one particular example, Murray's (1981) model for the patterns on animals' coats.

The basic idea is that the process is governed by a reaction-diffusion system along the lines originally proposed by Turing (1952). We suppose that there is some morphogen and a single threshold above which the colour will be dark and below which it will be light. The actual pattern that will be laid down in any particular situation can be determined only by integration of the full equations, but Murray shows that it is possible to gain some intuition for what is going on by considering the time development of small perturbations about the steady state concentrations of the relevant substances in a rectangular region of length a and width b. The solutions can be written as linear combinations of eigenfunctions, each of the form

$$\exp(\lambda_{mn}t)\cos(m\pi x/a)\cos(n\pi x/b)$$

The eigenvalues λ_{mn} depend on the dimensions of the domain and also on the constants m and n. If all the λ_{mn} are negative, then the steady state is stable, which means that the perturbations will die out and the end result will be homogeneity. If, on the other hand, at least one of the eigenfunctions is positive, then the largest of them determines which is the dominant spatial pattern (Fig. 10·2). It turns out that if the dominant eigenfunction is one-dimensional, i.e. if either m or n is zero, then the actual pattern formed is very much like that predicted by the linearization. If the dominant eigenfunction is two-dimensional, the pattern will be more complicated and will depend on the boundary conditions and the details of the reaction scheme. But the number of changes of colour in a given direction will still depend on the ratio of the length in that direction to some characteristic parameter of the reaction. The patterns are stable, in the sense that small random variations in the initial conditions yield qualitatively similar spatial patterns.

Figure 10·3 shows the patterns produced in a domain chosen to be a reasonable approximation to the shape of a real animal, especially bearing in mind that the underbelly of a patterned animal tends to be uniformly light-coloured. The scale factor γ here is related to the size of the embryo at the time the pattern is laid down, rather than to that of the adult.

These results have a number of interesting features. If the domain is too

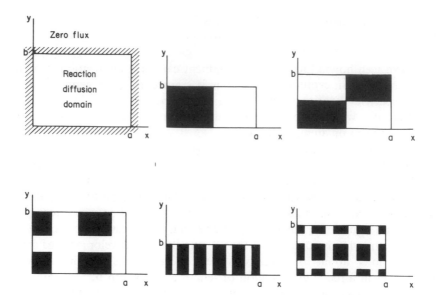

FIG. 10·2 Schematic representation of a pre-pattern model for pattern generation in a rectangle, together with the patterns produced (using the linear approximation) for (m,n) equal to (0,1) and (1,1) (top line) and (2,3), (0,10), (4,8) (bottom line). (Redrawn from Murray, 1981.)

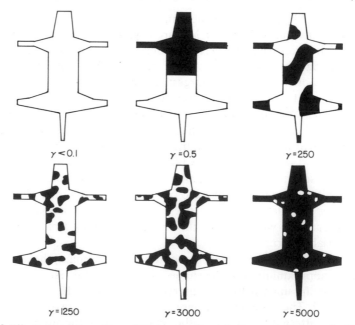

FIG. 10·3 Effect of scale on pattern. The domain dimension is proportional to $\sqrt{\gamma}$. (Redrawn from Murray, 1981.)

large or too small there is no pattern at all, which can account for the homogeneity of coat colour of large animals with long gestation periods and most very small ones with short gestation periods. A long narrow domain can have only a one-dimensional pattern (cf. Fig. 10·2) which may explain why we observe spotted animals with striped tails, but not the converse.

Small variations in the boundary conditions can significantly alter the pattern, although they leave the general nature unchanged; spots are replaced by similar spots and not by stripes or solid colour. This can account for the fact that the markings on patterned animals tend to vary from individual to individual, something that it is rather difficult to do on the basis of positional information. (It is amusing to construct a neo-Darwinist explanation for the variation; perhaps it could be based on the need for an animal to be able to distinguish between others of the same species if kin selection is to be effective.)

A further significance of this work is that showing that these patterns are relatively easy to produce reduces their importance in evolution. It is those features which are difficult to explain that deserve our attention. A good example of this is provided by the zebra, whose most distinctive feature is its familiar stripes. Yet on this model there is nothing unusual about stripes. They can be produced by only a minor variation in the timing of the process which gives rise to the more commonly observed spots on other animals.

Now if we believe that the stripes are a significant trait, we will assign them considerable importance in classification. And since within neo-Darwinism it is assumed that the aim is to achieve a classification which represents phylogeny, this will in turn lead us to suppose that all the zebras have evolved together, as is indeed generally assumed to be the case. If, on the other hand, we see stripes as little more than an accident resulting from a slight change in the timing of some process during development, then we will be less inclined to regard them as a diagnostic feature. It is worthy of note that in a recent careful reappraisal of the genus, Bennett (1980) has come to the conclusion that the different zebras are probably each more closely related to other horses than they are to each other.

MIMICRY

Batesian mimicry is often cited as an example of the wonderful power of natural selection. The Monarch butterfly for example, is foul tasting and so tends not to be taken by birds. The Viceroy lacks this protection, but birds avoid it too, because it so closely resembles the unpalatable Monarch.

The advantage of mimicry is obvious, so where it has evolved we would expect it to persist. But, as is generally the case with neo-Darwinist explanations, this argument does not tell us how the trait evolved in the first place. The pattern of the Monarch's wings is very complex and beautiful, and the Viceroy appears to have gone to a great deal of trouble to copy it. Surely it would have been easier just to have evolved a bad taste. Moreover, it would have given the Viceroy an even greater advantage, because even if the resemblance between mimic and model is very close, the benefit to the mimic is density dependent; it diminishes if the Viceroy becomes relatively more common.

Again it is developmental biology that provides the missing clue. While the patterns of butterflies' wings are indeed very complex in appearance, and while they differ sufficiently from species to species that in almost all cases they can be used for identification, the majority of them can be shown to be variations on a single theme, the so-called nymphalid ground plan. This has been known since the 1920s, but more recently Nijhout (1978) has proposed a mechanism for the patterning. This indicates that for one species to mimic another is far less difficult than it appears and may involve nothing more than one or two minor alterations in rate constants. Contrary to our first impression, therefore, it is probably much easier for the Viceroy to mimic the Monarch than to evolve a bad taste, and we suggest that this is the reason that it is the adaptation with the lower selective value which we observe. (This is an example of the operation of the 'principle of minimum increase in complexity' (Saunders and Ho, 1981).)

GRADUALISM

There has always been disagreement about whether or not sudden jumps or 'saltations' occur in evolution. Much of the recent interest in the question arises from the claim of Eldredge and Gould (1972) that the fossil record can best be explained by the hypothesis that evolution largely consists of short episodes of rapid change interspersed with long periods of stasis. Neo-Darwinists, however, mostly adhere to the traditional view that evolution is a relatively uniform process of slow and continuous change. They hold this view not because of the palaeontological evidence—ever since Darwin (1875, Chapter X) this has generally been acknowledged to be a problem for the gradualist hypothesis—but because their theory demands it. If, as they believe, the individual variations are random, they must perforce be small, for it is exceedingly unlikely that a large random variation could be anything but disastrous for the individual in which it occurred. It is for this reason that the reaction to Goldschmidt's (1940) 'hopeful monsters' is so uniformly dismissive.

Much of the force of this argument disappears when we take development into account. Within neo-Darwinism, with its concentration on separate traits, there is a tendency to see organisms as though they were assembled from pre-shaped components, like cars on a production line. But that is not how organisms are made. As an embryo develops, each organ is constantly adapting both in form and in function to many others. Moreover, during development and even after, when the organism is far less plastic, many cases are known of remarkably large adjustments to damage (*see* e.g. Frazzetta, 1975). The normal working of the epigenetic system depends on its ability to adapt, and this same ability makes large changes possible.

We can also understand how these large changes can occur. Biological systems are maintained and develop by non-linear interactions. The differential equations which describe them typically have solutions with the mathematical property of stability, which in this context is equivalent to what biologists call homeorhesis. If a solution is perturbed by fluctuations below a certain threshold level it will tend to return to its original course, and eventually reach the state it would have achieved had there been no perturbation. This is clearly a necessary property: any developing system is bound to be subjected to perturbations and if it were not resistant to them the process of development would be dangerously unreliable.

If the perturbations exceed the threshold, however, the system will be unable to return to its original course. This may well result in the cessation of development and hence the death of the organism, but it may also cause the system to end up in an alternative developmental pathway which will, like the original one, be stable. (For the existence of alternative steady states *see* Denbigh, Hicks and Page, 1948; Glansdorff and Prigogine, 1971. Alternative pathways are a natural generalization and arise in the same ways.) Thus by considering the properties of the sorts of equations which are generally believed to describe open systems such as organisms, we are led to the view of development portrayed by Waddington (1957) in his 'epigenetic landscape': a pattern of valleys representing the possible developmental pathways, separated by hills which are the thresholds. Of course for Waddington the idea was purely phenomenological and arose out of his study of the behaviour of developing organisms, whereas we can see why we would expect such a landscape to exist.

Homeorhesis is a necessary property of an epigenetic system. But a system which possesses this property will also have the capacity for heterorhesis, i.e. for large, organized change. Such a system will tend not to allow small changes to accumulate but will occasionally permit large ones. Consequently we would expect its evolution to be well described by Eldredge and Gould's 'punctuated equilibria'. (*See* Ho and Saunders, 1979, for a more detailed account of how the process might work.)

Thus the study of development suggests how we can reconcile evolutionary theory with the fossil record. The only neo-Darwinist explanation (apart from a flat denial that real gaps exist) appears to be to postulate mutations in regulatory genes. Since the theory is silent on the question of how these could bring about coordinated and adaptive changes, they are surely no more plausible than the hopeful monsters the neo-Darwinists so readily dismiss.

FORM

An important consequence of Darwin's theory was a diminution of interest in the problem of form. Previously this had been considered to be an important subject in itself, and in particular there had been the influential school of 'rational morphologists' whose concern had been to identify a set of 'archetypes' and use them to produce a natural classification of the organic world. The theory of evolution by natural selection very much changed the emphasis, however, because the structure of an organism came to be thought of as secondary, something to be explained in terms of function.

A notable exception to this trend was D'Arcy Thompson's (1917) *On Growth and Form*. To D'Arcy Thompson it was evident that form could not be accounted for simply by the natural selection of random variations. In physical science, he argued, we account for forms 'by reference to their antecedent phenomena, and in the material system of mechanical forces to which they belong, and to which we interpret them as being due'. And he continued: 'Nor is it otherwise with the material form of living things. Cell and tissue, shell and bone, leaf and flower, are so many portions of matter, and it is in obedience to the laws of physics that their particles have been moved, moulded and conformed.'

In his book he provided many examples to support his assertion. He showed how certain spiral shapes are natural consequences of the modes of growth of the shells which exhibit them. He reminded us that the surfaces of cells are subject to the same forces as are soap bubbles, and take on their shapes accordingly. Our eagerness to postulate some subtle adaptive significance for the complex shape of a medusoid is very much reduced when we discover that a similar shape can be produced by allowing a drop of fusel oil to fall into paraffin. We may marvel at the way in which the trabeculae are aligned along the pressure lines of a bone, but this is not carefully pre-arranged by the genes: a bone is a highly plastic structure and the trabeculae simply fall into line with the forces that are imposed on them, in a perfectly comprehensible fashion. And in the famous chapter on what he called the 'theory of transformations', D'Arcy Thompson showed how quite different, though related, organisms can be deformed one into another, thus showing how wrong it is to look at the evolution of each trait separately—a point also stressed by Dobzhansky (1956). Natural selection does not operate on one character at a time; instead there has been some 'more or less simple and recognizable system of forces' in control.

D'Arcy Thompson's ideas appear to have had little direct influence on twentieth century biologists. As Bonner (1961) writes, they 'were heretic in 1917 and it must be admitted that, for partly different reasons, they remain so to-day.' Yet they have not been forgotten; the abridged version of *On Growth and Form* is still in print. Somehow, throughout the long period in which the main thrust of biological research has been in other directions, D'Arcy Thompson's elegantly written treatise has served to remind biologists that there was an important aspect of their subject to which they would eventually have to return. It will never be easy to elucidate the contributions that the laws of physics and chemistry make to biological form, but sooner or later these questions will have to be faced; they cannot be put off indefinitely.

And indeed, interest in these problems is now increasing. Considerable progress has, for example, been made towards an understanding of the mechanism by which tissue folding occurs in processes such as gastrulation (Gierer, 1977; Odell *et al.*, 1981, Oster *et al.*, 1981. See the article by Goodwin (this volume) for an account of the latter model and its possible implications for evolution). Saunders and Trinci (1979) have applied some of D'Arcy Thompson's own work directly to explain the shapes of certain microorganisms.

Now such models are obviously of considerable interest, and we may expect that they will contribute a great deal towards our understanding of development and so of evolution. If Odell and his co-workers are correct in their view of tissue folding, they should be able to discover that certain shapes and changes of shape are more likely than others to occur, quite apart from any considerations of selective advantage. This sort of work does have its limitations, however. For one thing, all the examples that have so far been produced are essentially two-dimensional, and it is not at all clear that it is practicable to attempt to simulate a three dimensional situation on even a very large computer. The predictions are also naturally specific to the model and only applicable to processes which it adequately describes. Moreover, for the study of evolution what we really need is a theory which could tell us the entire range of possibilities. No mechanistic model can do this, because if a certain form does not arise this may mean nothing more than that we have not been clever enough to find a way of producing it. Only a general theory can tell us what is forbidden.

It is, of course, one thing to see the need for—or at any rate the desirability of—such a theory, but quite another actually to find one. Just such an attempt has been begun by Thom (1972) in developing catastrophe theory. This is now better known through its applications to quite different problems, but its original aim is clear from the title of Thom's book: *Stabilité Structurelle et Morphogénèse: Essai d'une Théorie Générale des Modèles*.

In one sense, Thom has returned to the centuries old problem of archetypes. There is, however, an important difference. To the rational morphologists, archetypes were entities, along the lines of Platonic ideals, which were somehow given. The task of the scientist was to identify the archetypes and apply them to produce a natural classification of organisms. Thom, in contrast, is attempting not just to identify but to explain, to proceed deductively rather than inductively. The claim is that the very way in which forms appear in nature—in biology and in physics alike—makes it inevitable that certain forms and transitions of form will be preferred. The remainder of this section will be devoted to what is necessarily a very much over-simplified account of Thom's ideas; the reader who wants a fuller understanding will have to read Thom's own book, though it might be easier to start with the introduction given by Saunders (1980).

We begin by making the assumption that the shapes of organisms are not determined by complex sequences of genetic switches in individual cells, each cell acting independently of its neighbours. Instead, we suppose that form arises through processes which involve a whole region of tissue. This is the point of view both of D'Arcy Thompson and also of the other workers we have mentioned in this section.

If we assume that neighbouring cells influence each other and, for the most part, develop in similar ways, we may infer that the processes can generally be modelled by differential equations. Now it is well known that even if the equations involve only smooth, well behaved functions, and the boundary conditions are also smooth, the solutions can still develop discontinuities and so produce sharp boundaries separating distinct regions. One familiar example is the shock wave associated with supersonic flight; this divides the region in which the aircraft can be heard from that in which it cannot. Another example is the bright cusp-shaped curve that can be seen when light is reflected from the edge of a cup on to the surface of the coffee.

This immediately suggests a relatively simple and robust way of generating form. We have only to arrange that the differential equations that describe the process are the sort that can give rise to a sharp boundary and that the initial conditions are suitably chosen. We then start the process going, and the boundary—which is to say the form—will appear. We do not have to do very much to produce a caustic, after all: we just hold the cup to the light and the rest happens automatically.

Of course so far we have nothing especially new. It has been known for well over a century that cusped boundaries are often associated with the wave equation. Hence if we could suggest some physico-chemical system that led to the wave equation we would be able to offer an explanation for the appearance of some cusp-shaped feature of an organism. This is all well and good, but it isn't the general theory we are looking for. There are many other differential equations we might consider instead, and they might lead to all sorts of different shapes.

It is here that catastrophe theory makes its contribution. Essentially, what it tells us is the following: No matter how many differential equations we use to model our system, the number of different shapes that can be generated in the way we have outlined is remarkably small. The complete set can be seen in the geometry associated with the seven 'elementary catastrophes'.

This is a very powerful result, although we must not exaggerate it. It does not apply to quite all dynamical systems, so exceptions are possible. It does, however, apply to a very large class, including almost all of those we encounter in biological modelling, and this range is extended if we make the sorts of assumptions (like the 'quasi-steady state approximation') which are often considered justifiable in other contexts. Hence while especially complicated interactions may provide counter examples, the forms on Thom's list have a real claim to be considered as the fundamental building blocks of morphogenesis.

Even so, catastrophe theory is not the solution to the problem of form, only a step towards it. For one thing, the elementary catastrophes are not the whole story; there are also 'generalized catastrophes', which have not yet been analyzed in any detail. And even to the extent that the seven elementary catastrophes do pick out the possible forms they are not to be thought of as archetypes in the sense that was intended by the rational morphologists. As Thom (1972, p.8) puts it: '. . . the model attempts only to classify local accidents of morphogenesis, which we will call *elementary catastrophes*, whereas the global macroscopic appearance, the form in the usual sense of the word, is the result of the accumulation of many of these local accidents'. The theory thus predicts that 'independent of the substrate of forms and the nature of the forces that create them' forms will be built up from a small set of basic units. It is left to the scientist, however, to elucidate how the larger scale process takes place, using the list of elementary catastrophes as clues. And indeed, in reading Thom's own work, one is bound to be struck by the extent to which he himself relies on analysis beyond that which his own theory provides.

Catastrophe theory is a part of the branch of mathematics known as topology, and is therefore qualitative. It tells us, for example, that in a situation in which only two dimensions are significant we are to expect nothing more complicated than cusps, but it says nothing about the precise shapes or orientations. This is, however, a strength of the theory rather than a weakness, because in biology the problem of form is much more topological then geometric. Individuals of one species generally differ in the details of their shapes but resemble each other closely in the numbers of various organs and limbs and so forth, and in the ways in which these are connected to each other. There is also strong evidence that it is topology that is fundamental in development. In the famous illustrations in *On Growth and Form*, D'Arcy Thompson showed how certain organisms or parts of organisms which appear to be quite different are in fact topologically nearly identical. And Wolpert (1982) and his co-workers have found that in the growth of a chick limb the basic pattern—the topology— is specified first and the precise dimensions—the geometry—only afterwards, and in an apparently separate process. Surely if the problem we are trying to solve is topological, we may expect to apply topological techniques in solving it.

Thom's work represents an attempt to develop a theory to answer certain specific questions about organisms—in this case having to do with the shapes they are likely or unlikely to possess—rather than to construct complete models. Another example is the 'field theory' described by Goodwin (this volume). Now there are still many to whom this approach is anathema. 'Whereof we cannot write down differential equations, thereof we must be silent' appears to be their credo. This is a remarkably blinkered view, given the difficulties involved in studying biology. Living organisms are, to say the least, highly complex. We can model some aspects of them quite successfully by conventional methods, but it is not to be expected that we will ever be able to construct detailed models which are accurate enough to represent the organisms adequately, general enough that we may be confident that we have covered all the possibilities, large enough to include all the factors that can affect the phenomena, and tractable enough that we can solve them. The models put forward by Odell *et al.* deal with only one particular type of folding and in an especially simple case, yet they require vast amounts of computer time. To derive general statements about morphology in this way must be completely out of the question. It is not a matter of choosing the one true path. Useful and interesting results have already been obtained by conventional modelling and we may anticipate with confidence that more are to come. But we should not allow the successes of this method, nor the fact that we may feel more comfortable with it because it is familiar to us from physics, to blind us to the need for theories which offer the opportunity to answer those questions with which reductionist theories are inherently unable to cope.

CONCLUSION

It is unlikely that there will ever be a general theory of development, any more than there will be a general theory of evolution. Instead, there will be different theories and, indeed, types of theory, each dealing with some different situation or problem. For example, while it is doubtful that positional information has the universal applicability that Wolpert (1982) suggests it has, there is evidence that it does apply to at least some processes in development. So it will probably continue to co-exist with pre-pattern theories, and others as well.

Now this has an important consequence for biology. If development and evolution could each be described simply, then the link between the two could be expected to be simple as well. We might therefore be content to leave it to a few specialists to concern themselves with the problem of connecting them, and we could be confident that they would discover some relatively straightforward modification of evolutionary theory which would adequately take into account the role of development. Perhaps this would take the form of a set of constraints, as some present day writers appear to believe.

But because neither subject is simple, it cannot be a matter of two essentially independent disciplines with a few well defined connections between them. Instead, the two are going to have to move closer together. The evolutionary implications of developmental biology will only be fully recognized if developmental biologists are encouraged to discuss the possible evolutionary significance of their work, and not to think of evolution as the preserve of population geneticists. Conversely, evolutionary biologists will have to become accustomed to considering the epigenetic aspects of the problems they are studying. The examples we have seen show that this is not beyond the range even of our current knowledge. It is, moreover, essential, if the study of evolution is to cease becoming a 'parlour game' and go back to being a science.

REFERENCES

Bennett, D. K. (1980). Stripes do not a zebra make. Part I: A cladistic analysis of equus. *Syst. Zool.* **29**, 272–287.

Bonner, J. T. (1961). Introduction to abridged edition of D'Arcy Thompson (1917) 'Growth and Form'.

Bryant, S. V. (1982). Introduction to the symposium: Principles and problems of pattern formation in animals. *Am. Zool.* **22**, 3–5.

Butler, S. (1911). 'Evolution Old and New'. A. C. Fifield, London.

Darwin, C. (1875). 'The Origin of Species' 6th edition. John Murray, London.

Denbigh, K. G., Hicks, M. and Page, F. M. (1948). The kinetics of open reaction systems. *Trans. Faraday Soc.* **44**, 479–493.

Dobzhansky, Th. (1956). What is an adaptive trait? *Am. Nat.* **90**, 337–347.

Dobzhansky, Th., Ayala, F. J., Stebbins, G. L. and Valentine, J. W. (1977). 'Evolution'. San Francisco, Freeman.

Driesch, H. (1908). Science and Philosophy of the Organism. A. and C. Black, London.

Eldredge, N. and Gould, S. J. (1972). Punctuated Equilibria: An alternative to phyletic gradualism. *In* 'Models in Paleobiology' (T. J. M. Schopf, *Ed.*), pp.82–115. Freeman Cooper, San Francisco.

Frazzetta, T. H. (1975). 'Complex Adaptations in Evolving Populations'. Sinauer, Sunderland, Massachusetts.

Gierer, A. (1977). Physical aspects of tissue evagination and biological forms. *Quart. Rev. Biophys.* **10**, 529–593.

Gierer, A. and Meinhardt, H. (1972). A theory of biological pattern formation. *Kybernetik* **12**, 30–39.

Glansdorff, P. and Prigogine, I. (1971). 'Thermodynamic Theory of Structure, Stability and Fluctuations'. Wiley, New York.

Ho, M. W. and Saunders, P. T. (1979). Beyond neo-Darwinism—An epigenetic approach to evolution. *J. Theor. Biol.* **78**, 573–591.

Maynard Smith, J. (1969). The status of neo-Darwinism. *In* 'Towards a Theoretical Biology' '2. Sketches' (C. H. Waddington, *Ed.*), pp.82–89. Edinburgh University Press, Edinburgh.

Mayr, E. (1980). Some thoughts on the history of the evolutionary synthesis. *In* 'The Evolutionary Synthesis' (E. Mayr and W. B. Provine, *Eds*), pp.1–48. Harvard University Press, Cambridge, Massachusetts.

Murray, J. D. (1981). A pre-pattern mechanism for animal coat markings. *J. Theor. Biol.* **88**, 161–199.

Nijhout, H. F. (1978). Wing pattern formation in Lepidoptera: a model. *J. Exp. Zool.* **206**, 119–136.

Odell, G. M., Oster, G. F., Alberch, P. and Burnside, B. (1981). The mechanical basis of morphogenesis. 1. Epithelial folding and invagination. *Devel. Biol.* **85**, 446–462.

Oster, G. F., Odell, G. M. and Alberch, P. (1981). Mechanics, morphogenesis and evolution. *In* 'Lectures on Mathematics in the Life Sciences', (G. F. Oster, *Ed.*), Vol. 13, pp.165–255. American Mathematical Society, Providence.

Ruse, M. (1982). 'Darwinism Defended'. Addison-Wesley, Reading, Massachusetts.

Saunders, P. T. (1980). 'An Introduction to Catastrophe Theory'. Cambridge University Press, Cambridge.

Saunders, P. T. and Ho, M. W. (1981). On the increase in complexity in evolution. II. The relativity of complexity and the principle of minimum increase. *J. Theor. Biol.* **90**, 515–530.

Saunders, P. T. and Trinci, A. P. J. (1979). Determination of tip shape in fungal hyphae. *J. Gen. Microbiol.* **110**, 469–473.

Stock, G. B. and Bryant, S. V. (1981). Studies of digit regeneration and their implications for theories of development and evolution of vertebrate limbs. *J. Exp. Zool.* **216**, 423–433.

Thom, R. (1972). 'Stabilité Structurelle et Morphogénèse'. Benjamin, Reading, Massachusetts. (English translation by D. H. Fowler (1975): 'Structural Stability and Morphogenesis'. Benjamin, Reading, Massachusetts.)

Thompson, D'A. W. (1917). 'On Growth and Form'. Cambridge University Press, Cambridge.

Turing, A. M. (1952). The chemical basis of morphogenesis. *Phil. Trans. R. Soc. London B* **237**, 37–72.

Wolpert, L. (1969). Positional information and the spatial pattern of cellular differentiation. *J. Theor. Biol.* **25**, 1–47.

Wolpert, L. (1982). Pattern formation and change. *In* 'Evolution and Development' (J. T. Bonner, *Ed.*), pp.169–188.

Wolpert, L. and Stein, W. D. (1982). *In* 'Learning, Development and Culture' (H. C. Plotkin, *Ed.*), pp.331–342. Wiley, London.

Section V

FORM AND FUNCTION

11

Environment and Heredity in Development and Evolution

MAE-WAN HO

Abstract

The integration of development and evolution occurs through the organisms' experience of the environment. The registering of experience and its eventual assimilation are the basis of adaptive evolution.

The material link between development and evolution is the hereditary apparatus which realistically includes both maternal/cytoplasmic effects and nuclear genes. The oocyte cytoplasm is at once a carrier of heredity independently of nuclear genes and the necessary communication channel between the environment and nuclear genes in the coordination of developmental and evolutionary processes.

Pattern formation in the *Drosophila* thorax serves as a concrete illustration of the general argument. Some experimental work bearing on canalization and genetic assimilation is reviewed and discussed in relation to (a) the developmental genetics of pattern formation, (b) the problem of the epigenetic origin of spatial organization and (c) the epigenetic mechanisms of differential gene expression and their relevance to genetic assimilation.

INTRODUCTION

The purpose of a theory of evolution is to account for the origin and transformation of organisms. Thus, to define evolution as the change in allele frequencies in populations from the outset — as neo-Darwinists do — is to defeat the purpose of the theory before we begin.

BEYOND NEO-DARWINISM
ISBN: 0-12-350080-X

Contrary to what is commonly supposed, the genotype–phenotype dichotomy, or Weismann's barrier, which conceptually separates evolution from the development of organisms (and hence from environmental influences), does not thereby absolve us from a theory of phenotypes. Natural selection, after all, acts on phenotypes. This leads us back to development, and to the role, not only of genes, but of the environment in the shaping of phenotypes. It is impossible to evade the problem which continues to confront us: how *organisms* evolve.

The physiological nature of development was emphasized by Goldschmidt alone among neo-Darwinists (especially in his *Physiological Genetics*, 1938). The rates of its reactions are almost unavoidably subject to influence from both environmental and genetic factors. Therefore, even a theory of evolution based solely on natural selection must reckon with developmental physiology, and with the interactions between genetic and environmental factors in morphogenesis.

Physiological considerations of development however, fit at best uneasily within the narrow confines of neo-Darwinism. This has prompted us to outline an alternative approach (Ho and Saunders, 1979; 1982a; 1982b). Our project is a logical extension of physiological genetics, and is continued in the present chapter. Here, I shall elaborate on an epigenetic framework which integrates development and evolution through the organisms' experience of the environment. The registering of experience and its eventual assimilation are the basis of adaptive evolution. An important feature which distinguishes the framework from neo-Darwinism is our concept of heredity. It includes, besides nuclear genes, maternal and cytoplasmic effects which provide the material link between development and evolution. The egg cytoplasm is both the carrier of heredity independently of the nuclear genes and the necessary interface between nuclear genes and the environment in the coordination of developmental and evolutionary processes.

In order to give focus to the discussion, much of it (especially in the second half of the Chapter) will be centred around pattern formation in the *Drosophila* thorax, which serves as a concrete illustration of the general argument. In part, this reflects our own research interests. But the main justification is that pattern formation is recognized as the major problem in development, and the *Drosophila* thorax has been subject to the most intensive investigations within the classical genetics framework. Both the successes and limitations of the latter will therefore stand in the sharpest relief.

DEVELOPMENTAL PHYSIOLOGY, THE 'NORM OF REACTION' AND THE 'EPIGENETIC LANDSCAPE'

The concept of the 'norm of reaction' was first introduced by Woltereck (1919) from his studies on cyclomorphosis in cladocerans. These freshwater animals such as *Daphnia* undergo a cycle of transformations involving a number of morphologically different generations throughout the year. Woltereck showed that the same range of variations could be produced in the laboratory by controlled feeding. Hence the term *Reactionsnorm* was coined to describe the whole range of phenotypes (reactivity) of a single genotype under different environmental conditions. The existence of the norm of reaction has the immediate consequence that an organism is not uniquely defined unless its environment is specified. The environment exerts necessary formative influences on development. This simple truth has been obscured by the Weismannist[1] fallacy that organisms are uniquely determined by their genotype, and that environmental influences are merely 'disturbances' to be overcome on the way to realizing the ideal phenotype (*see* Ho and Saunders, 1982a).

Inherent in the concept of the norm of reaction is the notion of a reacting system. It is a physiological system governed by principles which ensure both stability and the capacity for organized change (Ho and Saunders, 1979). As evolution is in essence a series of organized changes in development, it follows that development contains the potential for evolution. So it was that Goldschmidt (1940) envisaged large organized changes in development could result from 'systemic mutations'—these 'hopeful monsters' becoming in one step the precursors of a new evolutionary lineage.

As time passed, Goldschmidt's systemic mutations—without the physiology—became accomodated (albeit uneasily) within the orthodoxy. Waddington (1957) attempted to re-instate developmental physiology by inverting Goldschmidt's argument, drawing attention to environmental effects on development which parallel the genetic, and reopening once again the enquiry concerning the precise role of the environment in evolution.

Like Goldschmidt, Waddington saw development as a totality of biochemical and cellular interactions. This dynamic system—the famous but imprecise 'epigenetic landscape'—is characterized by a topography: a set of equilibrium pathways ('valleys') separated by thresholds ('hills'), such that development is normally *canalized* or buffered against disturbance. Thus many genetic mutations or environmental fluctuations will not alter the end result of development appreciably. However, with large disturbances—genetic or environmental—the topography can change so that alternative developmental pathways can be traversed, leading to different end-results.

The epigenetic system is not the additive outcome of individual gene action but a characteristic, at least, of the species[2] (Ho and Saunders, 1979). Proof of this assertion is provided by the phenomenon of phenocopies, a term coined by Goldschmidt (1935) to describe the mimics of morphogenetic mutants produced in normal wild-type strains of organisms by environmental agents. Phenocopies occur in all groups of organisms. Within a given species, the same kinds of phenocopies are found in all varieties or strains, indicating that different genotypes *per se* are irrelevant to the existence of underlying systemic constraints which define a dynamic *structure* of development. The latter ultimately determines the kinds of variations available to evolution. If we wish to understand evolution, the epigenetic system above all should be the object of our enquiry.

In order to explain phenocopies within the accepted orthodoxy, some neo-Darwinists, including Goldschmidt (1938) and Waddington (1957), postulated the existence of 'modifier' genes which as the name implies, modify the expression or penetrance of the phenocopy response, typically in a continuous 'polygenic' manner. The response itself is supposed to be controlled primarily by one or more 'major' genes. The hypothesis of modifier genes appears to be supported by the observation that different strains exhibit greater or lesser sensitivity to the agents inducing phenocopies (Goldschmidt and Piternick, 1957). Our own investigations to be described later indicate that this is an oversimplified explanation.

PARALLEL BETWEEN ARTIFICIAL AND NATURAL VARIATIONS

The phenomenon of phenocopies itself has a longer history dating back to the last third of the nineteenth century (*see* Shapiro, 1976). Extensive studies were carried out, for example, on wing pattern in butterflies in order to uncover the physiological basis of seasonal polyphenism—the existence in the same population of strikingly different seasonal generations. Temperature was found to be one of the controlling environmental factors and in many species of multivoltine Lycaenidae, it was possible to reproduce seasonal forms by appropriate temperature treatments at specific developmental stages. In some experiments (Standfuss, 1896; Fischer, 1901) phenocopies were obtained which resembled different geographic races or species in which the wing pattern was apparently genetically determined.

It turns out that parallels between artificially induced modifications and naturally occurring variations are extremely common. A very large body of observations (first reviewed by Rensch, 1929) show that structural divergences known or presumed to be hereditary are often correlated with environmental differences, so much so that a number of ecological rules[3] have been formulated on this basis. More importantly, however, these same structural modifications can be artificially induced in related organisms by simulating the appropriate environmental conditions in the laboratory.

A well known study is that of Beebe (1907) who found that doves of the genus *Scardafella* increase in dark pigmentation in nature from Mexico to Brazil. The point of least pigmentation coincides with the area of lowest humidity and increases in either direction as the humidity increases. By exposing the lightly pigmented form to very humid conditions, he demonstrated that pigment was acquired gradually through a series of moults till finally a stage was reached which was darker than any known in nature. Another case is the honey bee *Apis mellifera* in Europe which exhibits a north–south cline in a number of morphometric traits. Alpatov (1925, cited in Robson and Richards, 1936) showed that cold temperatures in the laboratory induced the same effects as those found in nature. Recently, experimental transplants of red-winged blackbird eggs between nests in northern and southern Florida, and from Colorado to Minnesota, showed that a significant proportion of the geographical differences in nestling development is 'non-genetic'. The author (James, 1983) concludes that if 'natural selection is maintaining the cline of character variation . . . the genetic and non-genetic components of phenotypic variation must covary'.

Many examples are known in which parallels between natural and artificial variations are attributed to special or local environmental conditions.

Woltereck induced helm modifications in *Daphnia* which paralleled those existing in natural races under similar environmental conditions. Races that lived in poor environments displayed those characters that can be invoked, in part, in any race cultured in a poor nutrient. Conversely, the characteristic features of races in rich environments were produced in almost any race maintained on a rich nutrient. Nonetheless, those respective features of the natural races appear more or less 'fixed' as they can persist side by side for a long time under identical culture conditions in the laboratory.

In summary, the naturally existing variations—many of which are regarded as clear adaptations by Darwinists and neo-Darwinists alike—are adaptive to the very environment capable of eliciting the parallel artificial modification. This strongly suggests that the environment plays a central role in the origin and evolution of the adaptation itself. In retrospect, this should occasion little surprise. An adaptation is but a particular functional relationship between the organism and its environment. Functions in turn arise naturally out of organism–environment interactions (Ho and Saunders, 1982b). So it is entirely plausible that adaptations should originate from interactions between organism and environment. The missing link is how the somatic changes resulting from those interactions could become hereditary in the sense that they seem to anticipate the environmental conditions for which they are an adaptation.

THE EXPERIENCE AND ASSIMILATION OF
NEW ENVIRONMENTS

The link between environment and heredity in the origin of adaptation is explained by Waddington (1957) as follows. A population of organisms experience a new environment and respond developmentally in a novel fashion, that is, a phenocopy appears. As the population is heterogeneous in modifier alleles, individuals will respond to varying degrees. If the response is adaptive, there will be selection for the modifier alleles which increase the initial intensity of the response and then (or at the same time) regulate it so that the same degree of response will occur within a range of intensity of the environmental stimulus. In other words, the response is canalized. Later, genetic assimilation takes place: the response now occurs in the absence of the environmental stimulus.

Canalization is explained as the result of selection for modifier alleles. As for genetic assimilation, Waddington was unclear as to what sort of mechanisms were involved. It could be the end-point of selection for modifier alleles which lowers the threshold for the response so much that the latter occurs spontaneously. Alternatively, genetic assimilation could be due to a chance mutation in a single (major) gene which fixes the response genetically, as it were.

There has been a recent revival of interest in genetic assimilation as a model for the evolution of new species (*see* for example, Shapiro, 1976; Matsuda, 1982), based to varying extent on Waddington's strictly neo-Darwinist hypothesis. Matsuda (1982) proposes the genetic assimilation of environmentally induced heterochronic modifications in talitrid amphipods and salamanders by the fixation of hormonal states normally subject to environmental control. He rejects the existence of modifier genes in favour of single mutations for genetic assimilation, thus reviving Baldwin's (1896) theory of organic selection. The latter states that after somatic modifications were produced by the environment, *coincident* genetic mutations also arose which gave the same phenotypes. If those phenotypes were adaptive, they would be favoured by natural selection. The major flaw in the argument is that there is really no connection between the environmentally induced modification and genetic assimilation — unless the former causes coincident mutations to arise more frequently in some way. Furthermore, as both the somatic modification and the coincident genetic mutation result in the same phenotype, how could natural selection distinguish between the two?

Waddington's hypothesis has the virtue that it does connect the initial environmentally induced modification to the final genetically assimilated adaptation. Moreover, it is amenable to experimental testing whereas organic selection is not, for it is impossible to predict in advance whether coincident mutations will occur in any one experiment.

Among the first experiments on genetic assimilation, Waddington (1956) succeeded in producing assimilated lines of the phenocopy *bithorax* in *Drosophila* by means of generations of ether-treatment and selection. These lines, when analyzed, seemed to involve both single major gene mutations as well as polygenic modifiers. The same applies to subsequent work on other phenocopies (Bateman, 1959a,b).

Whatever the detailed mechanisms involved in canalization and genetic assimilation, the modifier genes model presupposes that the starting population contains variation in modifier genes, and that there is selection for the phenocopy, either naturally or artificially. The specific prediction which could be made is that canalization would not occur in a genetically uniform population or where there is no selection for the phenocopy. Thus, Bateman (1959a,b) reported no progress in the frequency (penetrance) of certain phenocopies of mutants affecting the *Drosophila* wing in inbred lines after a small number of generations of heat treatment and selection. But neither was there any progress in the massbred line until the high mortality of the phenocopies was reduced by shortening the period of heat treatment. No parallel checks on the mortality of phenocopies were done in the inbred lines. In one inbred line selected for a wing venation phenocopy for nine generations, progress was recorded in the last 3 generations. But that experiment was terminated with the remark, 'The fact that a parallel trend was also shown by the unselected line indicated, however, that the apparent response was in fact merely secular' (Bateman, 1959b, p.448). This happened to be the only instance in which an unselected control was maintained. The latter, as we shall see, is crucial to the interpretation of these experiments.

MATERNAL AND CYTOPLASMIC EFFECTS —
AN ALTERNATIVE ROUTE OF INHERITANCE

One complicating factor which has been consistently overlooked in all investigations on phenocopies is the effect of the environmental stimulus on the hereditary constitution of the organisms. In particular, the possibility cannot be ruled out that cumulative cytoplasmic or maternal influences could lead to an increase in the phenocopy response in successive generations. This would indeed explain Bateman's (1959b) findings cited above, as well as some of her results in the genetic analysis of assimilated lines which clearly implicated maternal/cytoplasmic components.

Since the egg cytoplasm and cortical organization are conditioned by both maternal genes and environment (Cohen, 1979), it seems reasonable to conclude that the special environments experienced by the mother in successive generations may be passed on cumulatively in the egg. Only the modern counterpart of the medieval spermists—the *'omni ex* DNA' school[4] of geneticists and developmental biologists—would deny this possibility *a priori*.

Environmentally induced modifications of the cytoplasm which are transmitted across generations are by no means unknown. Many of these were studied extensively in *Paramecium* by Jollos (1921), Sonneborn (1970) and Beale (1957), in *Amoeba* by Danielli (1958) and in *Aspergillus* by Jinks (1957; 1958) and others. In all the above systems, considerable interactions between cytoplasm and nuclear genes were found, which often resulted in large phenotypic effects. (As we shall see later, cytoplasmic-nuclear interactions are central to the process of development.) Among higher organisms, Harrison (1928) induced heritable changes in the pigmentation of the cabbage white butterfly pupae by means of orange light, and Fujii (1978) reported the transmission of serum calcium disorders to the offspring of parathyroidectomized female rats up to the F_4 generation. Recently, Damjanov and Solter (1982) identified cytoplasmically transmitted factors which modify the development and malignancy of teratomas in mice. Considerations such as these led us to suspect that cumulative cytoplasmic effects may be responsible for the canalization of novel phenotypic responses to environmental challenge (Ho and Saunders, 1979). This mechanism, operating independently of, or in addition to natural selection, could have important consequences on the rate of phenotypic evolution.

CYTOPLASMIC EFFECTS AND THE CANALIZATION OF THE *BITHORAX* PHENOCOPY

In order to test our hypothesis, we re-investigated the *bithorax* phenocopy studied by Waddington (1956) (Fig. 11·1). By exposing *Drosophila* embryos to ether, homeotic transformations of the *meta*thoracic to *meso*thoracic structures are induced which mimic mutants of the *bithorax* genes. We chose this phenocopy because it is intimately connected with pattern formation—the major problem in development. Elucidation of the mechanisms involved in its canalization and genetic assimilation would therefore be a significant contribution to our understanding of the relationship between evolution and development.

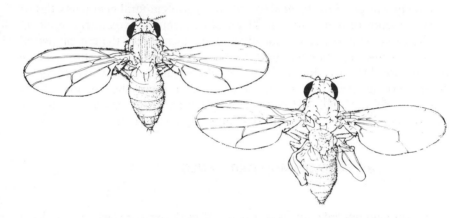

FIG. 11·1 The bithorax phenocopy. (A normal fly is depicted on the left, and an extreme phenocopy on the right.)

We followed Waddington's (1956) procedure as closely as possible but without selection in a massbred and an inbred line of the same genetic background. The results have been published in detail (Ho *et al.*, 1983a) so I shall only review the salient points here.

According to the modifier genes model, there should have been no increase in the phenocopy response in successive generations of ether treatment, as there was no selection for the phenocopy. Actually, a steady increase in the frequency of the phenocopy was found in both inbred and massbred lines. The question arose as to whether 'indirect' natural selection for the phenocopy was taking place. Our own observations as well as those of Waddington (1956) indicate that the bithorax flies were if anything naturally selected against. Statistical analyses of our data further demonstrated that generations of ether treatment increased phenocopy frequencies in a direct and cumulative way, independently of the effects on viability.

Direct evidence for cytoplasmic effects was obtained in that after the first six generations of ether treatment, embryos from a cross between treated females and control males showed the same increased tendency to phenocopy as embryos of the long-term treated line; whereas the reciprocal cross gave embryos which were no more responsive than controls.

A striking feature of our findings is the great similarity in the results obtained in the two different lines in all aspects. This strongly implies a basic identity in the epigenetic processes underlying the total response to long-term ether treatment in both lines.

A clue to some of the epigenetic processes affected by long–term ether treatment was provided by analyses of the spatiotemporal characteristics of the phenocopy response (see Fig. 11·2). Long-term ether treatment results in an increased tendency to phenocopy and an extension of the critical period for phenocopy induction into both earlier and later embryonic stages. These effects are in turn due to the cumulative influence of ether on a 'prepatterning' event in the early embryo which establishes the main body segments of the adult (Ho et al., 1983a,b,c). Once again, the results obtained in the massbred and inbred lines were very similar.

THE ROLE OF MODIFIER GENES

Our results do not rule out the existence of modifier genes which may have contributed to the increase in phenocopy response in successive generations. Rather, we draw attention to other mechanisms such as cytoplasmic inheritance—for which positive evidence was obtained—that may have profound effects on the rate of phenotypic evolution, independently of modifier genes.

The reason we cannot unequivocally rule out modifier genes is simply that our inbred line was tested by isoenzyme electrophoresis and found to contain residual genetic variation (Ho et al., 1983a). Therefore, it may also have contained variation in modifier genes. As we have indicated, however, the most notable feature of our results is the consistency in the response of the two different lines in all aspects. Even though the inbred line was not isogenic the amount of genetic variation had been much reduced by the inbreeding regime. To explain the results in terms of selection for modifier alleles, one must assume that almost precisely the same alleles were still

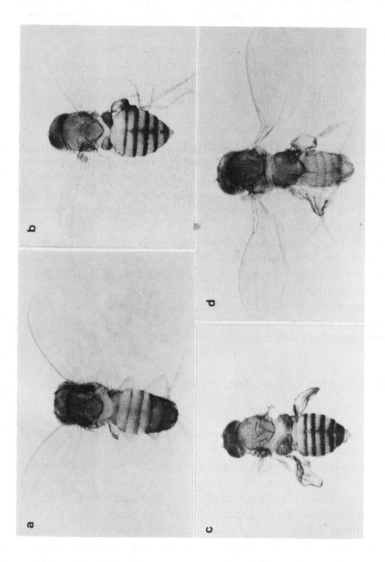

FIG. 11-2 Dorsal view of some of the bithorax transformations induced by ether treatment. A continuous range of tranformations is typically observed. The detailed spatial extent of the transformed spots (with respect to various landmarks on the thorax) differs according to the precise timing of ether treatment. (From Ho *et al.*, 1983b.)

present in the inbred as in the mass bred line. The only reasonable alternative is to recognize the existence of systemic, organismic properties common to both lines, which do not depend on specific alleles in specific genes. These properties are in part dependent on cytoplasmic constitution which may in turn be subject to environmental modification. We conclude that in our experiment, at least, the effect of any modifier genes would have been small compared to that of cumulative cytoplasmic influence.

Modifier genes have been invoked in order to explain the phenomenon of genetic assimilation within an accepted framework. Although modifier genes have been identified for certain mutations, their role in phenocopy responses is far from clear. The major factors influencing the penetrance and expression of the bithorax phenocopy, for example, turn out to be the timing of ether-treatment in relation to a more or less invariant critical period for phenocopy induction, and the tendency for the females in some lines to deposit a high proportion of fertilized eggs which have been retained in the body and are therefore no longer sensitive to ether (*see* Ho *et al.*, 1983b). None of the factors specifically modify the amount of penetrance or the intensity of expression.

Modifier genes (especially the 'polygenic' ones) are frequently assigned by genetic analysis to 'all chromosomes'. If they are ubiquitous, then they obviously cannot serve as an explanation for specific responses. As a category of explanation they have little or no predictive power: any experimental result will be consistent with the action of *some* putative modifier genes. This too conveniently obscures the underlying developmental physiology that should be the object of our enquiry.

GENES AND CYTOPLASM IN PATTERN FORMATION

In this and subsequent sections, I try to relate heredity and development at the level of mechanisms affecting pattern formation in particular. This in turn enables us to see how evolution and development may be connected through heredity. I stress that the argument is a general one and the concentration on pattern formation in *Drosophila* is simply in order to illustrate how in a specific case the generalities can be given substance.

So much is known concerning the genes affecting development that one might easily be led into thinking that development can be understood when all the genes affecting all the characters of an organism have been identified and mapped to parts of chromosomes. Yet, as many biologists including Goldschmidt (1938) realize, the enumeration of genes without the elucidation of the underlying physiology contributes little to our understanding of development. There is another reason why development is not reducible to genes, however, even if it were possible to catalogue all the genes *and* their physiological functions. This will become evident as we concentrate the mind as before on pattern formation in the *Drosophila* thorax, of which a great deal is known in genetics terms.

The bithorax gene complex (BX-C) (Lewis, 1982) is a cluster of tightly linked loci cytogenetically defined by a series of homeotic mutants. The latter lead to specific transformations of compartments or segments in the thorax and abdomen. For example, mutations in the *bithorax (bx)* and *postbithorax (pbx)* loci respectively transform the anterior and posterior compartments of the *meta*thorax to homologues of the *meso*thorax. (Both these mutations are mimicked in the bithorax phenocopies described earlier.) Many BX-C mutants exhibit 'polarity' effects—causing the transformation of particular segment(s) to a more posterior or more anterior segment. The loci themselves are interestingly arranged roughly in the order of the body segments each affects. This suggests a model of gene function involving the activation of increasingly more genes in each segment as one proceeds caudally (Lewis, 1982). Thus in the mesothorax no genes are active as it represents a ground state; whilst in the last abdominal segment all the genes are active.

The expression of the BX-C complex is in turn affected by at least ten other loci which are not linked with the cluster (*see* Garcia-Bellido and Capdevila, 1978; Lewis, 1982; Duncan, 1982; Struhl, 1982). Mutations in some of those loci cause homeotic transformations of all or nearly all body segments, and furthermore, interact with mutations in the BX-C complex in a multiplicative way. For example, mutants of the *extra sex comb (esc)* locus cause all body segments including the head to transform to the last abdominal segment, typically A8; whereas deletion of the entire BX-C results in the transformation of the segments posterior to the mesothorax— including the A8—to mesothorax. In the double mutant, all the segments are transformed to the mesothorax—as the last abdominal segment is mesothoracic in the single BX-C deletion mutant. This implies that a hierarchy of gene function is involved in segmental determination (*see* Struhl, 1981). In analogy with the well known prokaryotic operon model, the BX-C complex appears to contain the operator and structural genes, whereas the other loci seem to code for various inducers or repressors (Lewis, 1982; Struhl, 1981; Ingham, 1981).

The genetics of segmental determination in *Drosophila* is fascinating, and continues to be most extensively and thoroughly investigated. As more and more loci affecting segment specification are discovered, however, the simple analogy with prokaryotic gene function inevitably breaks down. One result of this is an 'explanatory' system which tends to identify increasingly small bits of anatomy with increasingly minute bits of DNA—an atomistic extension of the 'one gene one character' concept of the early Mendelians (*see* for example Duncan, 1982). In the simple analogues of the operon model, spatial organization is largely explained in terms of differential gene activation in different segments due to hypothetical gradients of repressors or inducers[5]. One fundamental question which is never really addressed is how the spatial distribution of hypothetical repressors/inducers becomes established in the first place, since the genes coding for these molecules must have been present in every nucleus of the embryo (cf. Løvtrup, this volume). The epigenetic origin of spatial organization is a major problem in developmental biology that is not reducible to differential gene activation. The solution, I believe, lies in the maternally imposed organization of the mature oocyte.

This idea is not new, and could be traced back to a theory espoused by Boveri, Loeb and Morgan (*see* Davidson, 1968; Cohen, 1979) which proposed that the egg cytoplasm is responsible for the form of the embryo in the rough, and that the role of mendelian factors is merely to determine the details of the individual development. In different forms, the morphogenetic role of cytoplasmic and cortical organization has been emphasized by many other eminent embryologists within the first half of the present century (Wilson, 1925; Spemann, 1938; Dalcq, 1938; Horstadius and Wolsky, 1936; to name but a few).

Cytoplasmic and maternal determinants of development are indeed well-known (Raven, 1961; Sander, 1976; Wolsky and Wolsky, 1976; Wessels, 1977; Cohen, 1979). In all organisms except mammals, embryogenesis right up to late gastrulation—the 'phyletic' stage—is completely under the control of the egg cytoplasm (Davidson, 1968). During that time there may be transcription of the zygotic genes but certainly no expression of their products. From our point of view, the most important aspect is that crucial determinants of polarity and symmetry are laid down in the egg cytoplasm or cortex during oogenesis (Wessels, 1977; Raven, 1961). The cortical and cytoplasmic organization of the mature egg contains not only all the instructions necessary for making the phyletic body plan, but also a precise time-schedule for doing so (Wessells, 1977).

It would be wrong, however, to imagine that the body plan is preformed in the cytoplasm or cortex. This is not necessary; nor is it desirable in most cases. Indeed, the prodigious regulatory capabilities of the majority of eggs and early embryos show that the picture of detailed preformation is decidedly false. Mosaicism (or apparent detailed preformation) becomes established always after a determination event that sets up a kind of 'prepattern' (Stern, 1953) of the major body regions (*see* Ho *et al.*, 1983b,c). The time at which the event occurs varies. In so-called 'determinate' eggs, such as those of the ascidian *Styela,* and the annelid, *Tubifex,* this occurs probably during oogenesis or immediately after fertilization. In other eggs capable of varying degrees of regulation, it occurs later during embryogenesis. It will be instructive to consider prepatterning once again, in *Drosophila.*

As mentioned earlier, we have analyzed the spatiotemporal characteristics of the bithorax phenocopies induced by ether in some detail. The results suggest to us that ether disrupts a prepatterning event which occurs at the surface of the embryo during the precellular blastoderm stages. Other evidence for the existence of such an event include the following. Ligation of *Drosophila* embryos before, but not after, blastoderm formation results in gross disturbances to segmentation (Schubiger, 1976) implying that segments are determined at blastoderm formation. Clonal analyses also indicate that segments are distinct at or soon after the blastoderm stage, at which time, the anterior and posterior compartments of the thoracic segments are also determined (Steiner, 1976; Wieschaus and Gehring, 1976; Garcia-Bellido *et al.*, 1976). That the prepattern affects primarily the cytoplasm rather than the underlying nuclei is corroborated by the finding that blastoderm nuclei are totipotent, and can become incorporated into any kind of differentiated cells, whereas blastoderm cells are determined in their developmental fate (Illmensee, 1978). This result also implies that nuclear genes become involved— activated or otherwise—in pattern formation secondarily, through cytoplasmic-nuclear interactions. The latter in effect transfer a global prepattern locally to individual nuclei or groups of nuclei which only then could express their genes differentially.

The case is sufficiently strong that a prepatterning event is required for segmental determination. Segments are therefore not specified by a preformed maternal organization. What is laid down during oogenesis, so far as spatial organization is concerned, are the principal axes of the future organism: the antero-posterior and dorso-ventral polarities, both of which can be drastically disturbed by mutations in maternal effect genes (Wieschaus, 1980).

As Davidson (1968) points out, cytoplasmic localization of developmental 'determinants' is universal, though the precise nature of the determinants remain to be elucidated. The necessity for cytoplasmic localization in metazoan development may also be viewed in the light of the particular problem involved in 'translating' linear instructions—presumably coded in the genes—into a three-dimensional spatial domain which is the organism. It may be that what is required is no more than an initial symmetry breaking in the form of a polarity, such as the animal-vegetal polarity of the amphibian egg, or the antero-posterior polarity of the *Drosophila* egg. This, together with a train of cytoplasmic reactions, activated by sperm-entry (*see* Jaffe, 1979) then generate a prepattern of the main body regions in the early embryo. The overall dynamics involved in prepattern generation are unknown (although we were able to make inferences concerning the reactions at a local level affecting metathorax prepatterning (Ho *et al.*, 1983b,c)). They could be the sort envisaged in Goodwin's field theory description (this volume). Whatever the precise nature of prepatterning, it is clearly pre-requisite for pattern formation.

CANALIZATION AND GENETIC ASSIMILATION — SPECULATION ON MECHANISMS

Our investigations on the bithorax phenocopy showed that ether disrupts the early prepatterning event in the *Drosophila* embryo. The effect of ether is inherited cumulatively through the cytoplasm of the oocyte, and is manifest as an increase in the phenocopy response due to alterations in the timing and rates of reactions involved in prepatterning. At this moment, one can only speculate concerning the sort of cytoplasmic 'localization' responsible for the effect of ether. The seemingly continuous nature of the latter leads one to suspect that cumulative changes in the concentration of some membrane protein(s) or lipids—which take part in the prepatterning event—may be involved. Thom (1968) describes a molecular mechanism of facilitation or 'memory' which may be applicable here. In essence, a metabolic state S results in the synthesis of a molecular structure M. If M persists beyond the state S, it will subsequently facilitate a return to the state S. Such a mechanism could be the basis of the facilitation of the phenocopy response in successive generations of ether treatment.

Cytoplasmic inheritance thus appears to be responsible for canalization; the mechanism of genetic assimilation, however, is still unclear. I am inclined to doubt that random or fortuitously coincident mutations could be responsible. I am not aware that ether treatment increases mutation rates in general, so that for instance, mutations to *white eyes* or *dumpy wings*, as well as to *bithorax* are above the spontaneous background level. If mutations in 'major' genes accompany genetic assimilation, they could well result from 'instructive' processes in the sense that the environmental stimulus increases the likelihood for the right mutations to occur. The instructive process does not have to be as unsubtle as 'reversed translation' (unlike systems described in the following chapter, no specific protein or RNA sequences are involved in the environmental signal). When we bear in mind that gene expression depends on nuclear–cytoplasmic interactions, then it is not difficult to envisage a feedback to the genome from the cytoplasm due to alterations in concentrations of proteins and metabolites which favour alternative gene expression states by the same mechanism of molecular memory stated above. The latter, if sufficiently intense, could then be stably inherited (i.e. become genetically assimilated).

One clear example is antigen expression in *Paramecium aurelia* (Beale, 1957). These organisms exist in a number of stably inherited gene-expression states each characterized by the presence of a different antigen. The states are determined by cytoplasmic factors which in turn depend on environmental conditions. A given cytoplasmic state favours the expression of alleles at one particular locus. Certain cytoplasmic states in turn are controlled by the very genes whose expression are favoured by the same cytoplasmic states[6]. This system underlines the reciprocal interactions between nucleus and cytoplasm in the determination of phenotype.

Recent investigations using recombinant DNA technology have brought to light a plethora of cellular mechanisms affecting gene expression (Brown, 1981; Pollard, this volume). Many involve genomic changes such as gene loss, gene amplification and DNA rearrangement. Some are directed by the environment and serve useful functions while others occur apparently at random to no specific purpose. Genomic change is not necessary for stable alterations in gene expression, however. Binding of different repressors to a single operator gene can result in the mutually exclusive and stably inherited expression of alternative structural genes (Johnson *et al.*, 1981).

Genetic assimilation may therefore be achieved either by directed genomic change or the fixation of a cytoplasmic state favouring altered gene expression. In classical genetic analyses, these would be identified as 'major gene mutants' and maternal/cytoplasmic components respectively. Re-examination of earlier results (Waddington, 1956; Bateman, 1959a,b) indicate frequent instances of both. Yet another mechanism for genetic assimilation may be envisaged when we are dealing with sexual organisms which are as a rule genetically very diverse. This may be described as a kind of organic selection via cytoplasmic-nuclear incompatibility reactions. It is possible that as the environmental stimulus continues to modify the cytoplasm, the latter becomes incompatible with some of the nuclear genes or genotypes. Organic selection will then operate through the elimination of genotypes which give lethal or harmful combinations with the cytoplasm. A final state will be reached when all the incompatible genes are eliminated and nuclear-cytoplasmic compatibility is re-established. The assimilated phenotype will often end up as a compromised version of the original (e.g. the enlarged-haltere line obtained by Waddington (1956) from the bithorax phenocopy). Nuclear-cytoplasmic incompatibilities are well known in interspecific crosses (Moore, 1955). These result in developmental arrest often at gastrulation—the stage at which nuclear gene products are first used. By implication, intraspecific crosses succeed because the genes and cytoplasm are compatible.

Genetically, assimilation by organic selection will probably appear to be the result of polygenic factors mapped to all chromosomes, though maternal/cytoplasmic components and selection of *pre-existing* rare variants may also be implicated.

The alternative hypotheses presented above for genetic assimilation could be tested. If an instructive event is involved, assimilation will take place as readily in an isogenic line as in a massbred line, whereas the reverse is likely to be true if organic selection via cytoplasmic-nuclear incompatibility is involved. The application of recombinant DNA analyses in conjunction with more conventional genetic methods will no doubt provide a definitive picture of genetic assimilation at the molecular level.

CONCLUSION

When a truly physiological approach is adopted, one invariably arrives at an extended framework encompassing not only nuclear genes, but cytoplasm and environment (cf. Goodwin, this volume). Only such a system is capable of epigenesis and evolution.

Epigenesis depends on a series of nuclear-cytoplasmic and organism-environmental interactions. Cytoplasmic localization, crucial to embryonic development, is itself the result of an earlier ovarian epigenetic process. Davidson (1968) points out that the only preformation involved in development is the nuclear genome itself. So it is that in the neo-Darwinian scheme, in which development is seen as an unfolding of a preformed genetic programme, there can be no evolution except by fortuitous random mutation and natural selection. Within the framework presented in this chapter, development and evolution are formally and materially connected. Organisms develop in accordance partly with the assimilated experiences of their forebears and partly with their own experiences. Development evolves through the internalization of new environments. The material link between organism and environment, and development and evolution alike is the hereditary apparatus which realistically includes both cytoplasm and nuclear genes. The cytoplasm registers the somatic imprint of experienced environments which can be transmitted to the next generation independently of the nuclear genes. At the same time, it acts as a true communication channel between the environment and the nuclear genome in the coordination of developmental and evolutionary processes.

There has been a recent explosion in molecular genetics research following initial breakthroughs in recombinant DNA technology. The genomic content of every organism is for the first time susceptible to being read base by base from beginning to end. Yet the first glimmerings have already yielded major surprises. Forever exorcized from our collective consciousness is any remaining illusion of development as a genetic programme involving the readout of the DNA 'master' tape by the cellular 'slave' machinery. On the contrary, it is the cellular machinery which imposes control over the genes. The central role of protein–protein and protein–nucleic acid interactions in the regulation of gene expression is reinforced many times over by the detailed knowledge which has recently come to light in both eukaryotic and prokaryotic systems. The classical view of an ultraconservative genome—the unmoved mover of development—is completely turned around. Not only is there no master tape to be read out automatically, but the 'tape' itself can get variously chopped, rearranged, transposed and amplified in different cells at different times.

The emergent picture is that of an extremely fluid genome amidst a potential chaos of mechanisms all threatening to subvert the existing order (Dawid, 1982). This only deepens the age-old enigma of the stability and repeatability of development itself. I believe the solution lies at least partly within the epigenetic framework.

Stability and repeatability reside in the dynamics of the epigenetic system in two senses. First, it is the automatic result of physicochemical reactions of which the system is composed—and the physicochemical environment in which the system is in turn embedded (cf. Saunders, this volume). Second, it is due to assimilated experiences held jointly in the nucleus and cytoplasm. These introduce regular biases into developmental reactions which may otherwise behave in a non-committal or unpredictable way (cf. Goodwin, this volume). Assimilated experiences therefore anticipate the environments to be experienced.

But the hereditary apparatus is not and cannot be completely prescient (see Plotkin and Odling-Smee, 1982). New or unforeseen environments will crop up during epigenesis of the present generation. The developing organism must react and adjust to existing circumstances. The multiplicity of mechanisms involved in gene expression poses the insurmountable problem of how the astronomical amount of 'information' required to control the mechanisms could be accommodated in the genome. The answer which organisms have discovered for themselves since the beginning of time is to rely largely on clues from the environment. So it is that the organisms adapt and the environment serves.

ACKNOWLEDGEMENT

I am very grateful to Professor Alexander Wolsky for comments on an earlier draft of this manuscript.

Notes

(1) As Darwin was not a Darwinist and Marx not a Marxist, so Weismann too was not a Weismannist. He admitted the possibility of the inheritance of variations produced as the result of interactions involving physiological changes, as distinct from mere injuries (*see* Matsuda, 1982, for an interesting discussion on this point).

(2) Homeotic transformations — the substitution of normal structures by their serial homologues — are probably common to the entire arthropod phylum, as they have been observed in regenerating appendages in crustacea as well as in insects. St. George Mivart (1871) described varieties of pigeons which possess long feathers and partial fusion of the third and fourth digits in the legs. Both of those conditions are characteristic of the wings. This suggests the intriguing possibility that homeosis may reflect a fundamental organization of all developments.

(3) The best known are as follows: *Bergman's rule* states that in nearly related warm-blooded animals, the larger live in the north and the smaller in the south. *Allen's rule* states that the extremities (i.e. feet, ears, tail) of mammals tend to be shorter in colder climate in closely allied races. *Gloger's rule* states that southern races tend to be black, brown or grey and especially rust red whereas northern races tend to be pale.

(4) I am much indebted to Professor Lewis Wolpert for suggesting the term.

(5) Static gradients of hypothetical morphogens are a key feature of many current models of pattern formation. Since gradients are universal, the burden of explanation is placed entirely on different hypothetical threshold values of the same morphogen activating different genes, thus resulting in spatial pattern. The archetype is Wolpert's positional information model (*see* Saunders, this volume). The alternative view places greater emphasis on dynamic processes which establish a relatively more specific 'prepattern' of spatial organization (Stern, 1953).

(6) Recent investigations show that the alternative antigenic states involve transcription of different messenger RNA species (Preer *et al.*, 1981).

REFERENCES

Baldwin, J. M. (1896). A new factor in evolution. *Am. Nat.* **30**, 441–541.

Bateman, K. G. (1959a). The genetic assimilation of the dumpy phenocopy. *J. Genet.* **56**, 341–351.

Bateman, K. G. (1959b). The genetic assimilation of four venation phenocopies. *J. Genet.* **56**, 443–474.

Beale, G. H. (1957). Antigen system of *Paramecium aurelia. Int. Rev. Cytol.* (G. H. Bourne and J. F. Danielli, *Eds*), Vol. VI, pp.1–23. Academic Press, New York.

Beebe, C. W. (1907). Geographic variation in birds. *Zoologica* **1**, 1–41.

Brown, D. D. (1981). Gene expression in eukaryotes. *Science* **211**, 667–674.

Cohen, J. (1979). Maternal constraints on development. *In* 'Maternal Effects in Development' (D. R. Newth and M. Balls, *eds*), pp.1–28. Cambridge University Press, Cambridge.

Dalcq, A. M. (1938). 'Form and Causality in Early Development'. Cambridge University Press, Cambridge.

Damjanov, I. and Solter, D. (1982). Maternally transmitted factors modify development and malignancy of teratomas in mice. *Nature (Lond)* **296**, 95–96.

Danielli, J. F. (1958). Studies of inheritance in amoebae by the technique of nuclear transfer. *Proc. R. Soc. London B.* **148**, 321–331.

Davidson, E. H. (1968). 'Gene Activity in Early Development'. Academic Press, New York.

Dawid, I. (1982). Genomic change and morphological evolution. Group report. *In* 'Evolution and Development' (J. T. Bonner, *Ed.*), pp.19–39. Springer-Verlag, Berlin.

Duncan, I. McK. (1982). Polycomblike: a gene that appears to be required for the normal expression of the bithorax and Antennapedia gene complexes of *Drosophila melanogaster*. *Genetics* **102**, 49–70.

Fischer, E. (1901). Experimentelle Untersuchengen über die Vererbung erworbener Eigenschafter. *Allg. Zeit. f. Ent.*

Fujii, T. (1978). Inherited disorders in the regulation of serum calcium in rats raised from parathyroidectomised mothers. *Nature* **273**, 236–238.

Garcia-Bellido, A. and Capdevila, M. P. (1978). The initiation and maintenance of gene activity in a developmental pathway of Drosophila. *Symp. Soc. Dev. Biol.* pp.3–21.

Garcia-Bellido, A., Ripoll, P. and Morata, G. (1976). Developmental compartmentation of the dorsal mesothoracic disc of *Drosophila*. *Dev. Biol.* **48**, 132–147.

Goldschmidt, R. (1935). Geographische Variation und Artbildung. *Nature* **23**, 169–176.

Goldschmidt, R. (1938). 'Physiological Genetics'. McGraw Hill, New York.

Goldschmidt, R. (1940). 'The Material Basis of Evolution'. Yale University Press, New Haven.

Goldschmidt, R. B. and Piternick, L. K. (1957). The genetic background of chemically induced phenocopies in *Drosophila*. *J. Exp. Zool.* **135**, 127–202.

Harrison, J. W. II. (1928). Induced changes in the pigmentation of the pupae of the butterfly *Pieris napi* L., and their inheritance. *Proc. R. Soc. (London) B* **102**, 347–353.

Ho, M. W. and Saunders, P. T. (1979). Beyond neo-Darwinism: an epigenetic approach to evolution. *J. Theor. Biol.* **78**, 573–591.

Ho, M. W. and Saunders, P. T. (1982a). Adaptation and natural selection: mechanism and teleology: *In* 'Towards a Liberatory Biology' (S. Rose, *Ed.*), pp.87–104. Allison and Busby, London.

Ho, M. W. and Saunders, P. T. (1982b). The epigenetic approach to the evolution of organisms — with notes on its relevance to social and cultural evolution. *In* 'Learning, Development, and Culture' (H. C. Plotkin, *Ed.*), pp.343–361. Wiley, London.

Ho, M. W., Tucker, C., Keeley, D. and Saunders, P. T. (1983a). Effects of successive generations of ether treatment on penetrance and expression of the bithorax phenocopy in *Drosophila melanogaster*. *J. Exp. Zool.* **225**, 357–368.

Ho, M. W., Bolton, E. and Saunders, P. T. (1983b). The bithorax phenocopy and pattern formation. I. Spatiotemporal characteristics of the phenocopy response. *Exp. Cell Biol.* **51**, 282–290.

Ho, M. W., Saunders, P. T. and Bolton, E. (1983c). The bithorax phenocopy and pattern formation. II. A model of prepattern formation. *Exp. Cell Biol.* **51**, 291–299.

Hörstadius, S. and Wolsky, A. (1936). Studien über die Determination der Bilateral Symmetrie des jungen Seeigelkeimes. *Roux Arch.* **135**, 69–113.

Illmensee, K. (1978). *Drosophila* chimeras and the problem of determination. *In* 'Genetic Mosaics and Cell Differentiation' (W. J. Gehring, *Ed.*), pp.51–69. Springer-Verlag, Berlin.

Ingham, P. W. (1981). Trithorax: a new homeotic mutation of *Drosophila melanogaster* II. The role of *trx*⁺ after embryogenesis. *Wilhelm Roux Arch.* **190**, 365–369.

James, F. C. (1983). Environmental component of morphological differentiation in birds. *Science* **221**, 184–186.

Jaffe, L. F. (1979). Control of development by ionic currents. *In* 'Membrane Transduction Mechanisms' (R. A. Cone and J. E. Dowling, *Eds*), pp.199–231. Raven Press, New York.

Jinks, J. L. (1957). Selection for cytoplasmic differences. *Proc. R. Soc. London B* **146**, 527–540.

Jinks, J. L. (1958). Cytoplasmic differentiation in fungi. *Proc. R. Soc. London B* **148**, 314–321.

Johnson, A. D., Poteete, A. R., Lauer, G., Sauer, R. T., Ackers, G. K. and Ptashne, M. (1981). λ repressor and cro—components of an efficient molecular switch. *Nature (London)* **294**, 217–223.

Jollos, V. (1921). Experimentelle Protistenstudien. I. Untersuchungen über Veriabilität und Vererbung bei Infusorien. *Arch. Protistenk.* **43**, 1–222.

Lewis, E. B. (1982). Developmental genetics of the bithorax complex in *Drosophila*. *In* 'Developmental Biology Using Purified Genes' ICN-UCLA Symposia on Molecular and Cellular Biology vol. XXIII (D. D. Brown and C. F. Fox, *Eds*), pp.1–20. Academic Press, New York.

Matsuda, R. (1982). The evolutionary process in talitrid amphipods and salamanders in changing environments, with a discussion of 'gene assimilation' and some other evolutionary concepts. *Canad. J. Zool.* **60**, 733–749.

Mivart, St. George (1871). 'On the Genesis of Species'. Macmillan, London.

Moore, J. A. (1955). Abnormal combination of nuclear and cytoplasmic systems in frogs and toads. *Adv. Genet.* **7**, 132–182.

Plotkin, H. C. and Odling-Smee, F. J. (1982). Learning in the context of a hierarchy of knowledge gaining processes. *In* 'Learning, Development, and Culture' (H. C. Plotkin, *Ed.*), pp.443–471. Wiley, London.

Preer, J. R., Jr., Preer, L. B. and Rudman, B. M. (1981). RNAs for the immobilization antigens of *Paramecium*. *Proc. Nat. Acad. Sci. U.S.A.* **78**, 6776–6778.

Raven, C. P. (1961). 'Oogenesis: The Storage of Developmental Information'. Pergamon Press, Oxford.

Rensch, B. (1929). 'Das Princip geographicscher Rassendreise und das Problem der Artbildung'. Bornträger, Berlin.

Robson, G. C. and Richards, O. W. (1936). 'The Variation of Animals in Nature'. Longmans, Green, London.

Sander, K. (1976). Specification of the basic body pattern in insect embryogenesis. *Adv. Insect Physiol.* **12**, 125–238.

Schubiger, G. (1976). Adult differentiation from partial *Drosophila* embryos after egg ligation during stages of nuclear multiplication and cellular blastoderm. *Devel. Biol.* **50**, 467–488.

Shapiro, A. M. (1976). Seasonal polymorphism. *In* 'Evolutionary Biology' (T. Dobzhansky, M. R. Hecht, and W. C. Steere, *Eds*), pp.250–332. Appleton-Century-Crofts, New York.

Sonneborn, T. M. (1970). Gene action in development. *Proc. R. Soc. (London) B.* **176**, 347–366.

Spemann, H. (1938). 'Embryonic Development and Induction'. Yale University Press, New Haven.

Standfuss, M. (1896). 'Handbuch der pal'darctischen Gross-Schmetlerlinge für Forscher und Sammler G. Fischer'. Jena.

Steiner, E. (1976). Compartments in the developing leg imaginal discs of *Drosophila melanogaster. Wilhelm Roux Arch.* **180**, 9–30.

Stern, C. (1953). Genes and developmental patterns. Proc. Int. Congr. Genet. 9th (Suppl) *Caryologia* **6**, 355–459.

Struhl, G. (1981). A gene product required for correct initiation of segmental determination in Drosphila. *Nature* **293**, 36–41.

Struhl, G. (1982). Genes controlling segmental specification in the *Drosophila* thorax. *Proc. Nat. Acad. Sci. U.S.A.* **79**, 7380–7384.

Thom, R. (1968). Comments by René Thom. *In* 'Towards a Theoretical Biology' Vol. 1 (C. H. Waddington, *Ed.*), pp.32–41. Edinburgh University Press, Edinburgh.

Waddington, C. H. (1956). Genetic assimilation of the bithorax phenotype. *Evolution* **10**, 1–13.

Waddington, C. H. (1957). 'The Strategy of the Genes'. George Allen and Unwin, London.

Wessells, N. K. (1977). 'Tissue Interactions and Development'. W. A. Benjamin, Menlo Park, California.

Wieschaus, E. (1980). A combined genetic and mosaic approach to the study of oogenesis in *Drosophila. In* 'Development and Neurobiology of *Drosophila*' (O. Siddiqui, P. Babu, C. Hall, and J. Hall, *Eds*), pp.85–94. Plenum Press, New York.

Wieschaus, E. and Gehring, W. (1976). Clonal analysis of primordial disc cells in the early embryo of *Drosophila melanogaster. Devel. Biol.* **50**, 249–263.

Wilson, E. B. (1925). 'The Cell in Development and Heredity'. Macmillan, New York.

Wolsky, M. de I. and Wolsky, A. (1976). 'The Mechanism of Evolution: A New Look at Old Ideas'. S. Karger, Basel.

Woltereck, R. (1919). 'Variation und Arbildung. Analytische und experimentelle Untersuchungen an pelagischen Daphniden und anderen Cladoceran. I. Teil. Morphologische, entwicklungsgeschichtliche und physiologische Variations — Analyse'. A. Francke, Bern.

12

Is Weismann's Barrier Absolute?

JEFFREY W. POLLARD

Abstract

The rise of the neo-Darwinian theory of evolution was facilitated by Weismann's contention that the germline was inviolate from the soma. In contradiction to Weismann's doctrine, Steele (1979) has recently proposed a plausible genetic mechanism whereby acquired characters may be inherited. This 'somatic selection' hypothesis states that mutant somatic information, if selected sufficiently for expression within the soma, will be transmitted to the germline in the form of RNA captured by endogenous retroviral vectors. Once in the germline, the RNA will be transcribed to DNA by reverse transcriptase and become integrated into the germline DNA.

This article re-examines Weismann's doctrine not only in the light of the somatic selection hypothesis but also in the broader scope of contemporary molecular genetics. Data are presented which show Weismann's barrier to be permeable to genetic sequences in much the way envisaged by Steele and also that under certain circumstances specific complex acquired characters may be inherited. Although the relationship between these genetic elements and the physiological characters transmitted is not established, models are discussed that could harmonize the data and at the very least suggest future definitive experiments.

INTRODUCTION

Application of Weismann's theory of the 'continuity of the germplasm' (1893) and its corollary that there is a barrier functionally separating the soma from the germline means that somatically altered genetic information is specifically excluded from consideration in evolutionary mechanisms.

BEYOND NEO-DARWINISM
ISBN: 0-12-350080-X

Thus, historically, Weismann's barrier provided the conceptual foundation for Mendelian genetics and its offshoot, the Synthetic or neo-Darwinian theory of evolution. The latter states that random variations in the germline genes produce gradual phenotypic changes upon which natural selection acts and that this is both *necessary* and *sufficient* to account for evolutionary change (Huxley, 1943; Dobzhansky *et al.*, 1977). In contrast, however, many authors have felt that in order to explain adaptation, there must be a feedback relationship between the environment and evolutionary processes, over and above the simple selection of more fit individuals (Waddington, 1961; Ho and Saunders, 1979; Steele, 1979, and this volume). It is therefore appropriate as we approach the centenary of the publication of Weismann's book 'The Germplasm: A Theory of Heredity' that the inter-relationship between the environment and the germline should be reconsidered. Such attempts, until recently, would have been dismissed, not simply because of the lack of convincing experimental evidence but also because of the absence of rational hypotheses to explain how a transfer of genetic material could be effected. Indeed, Mayr (1959) used the distance that genes would have to travel between peripheral end-organs and the germline as one of the major arguments against such a transfer occurring. However, with the increasing realization of the dynamic nature of eukaryotic genomes, the high frequency of occurrence of moveable genetic elements (Temin and Engels, 1984), plus the proposal of at least one rational hypothesis (Steele, 1979) which can explain the transfer of somatic genetic information to the germline, Weismann's barrier once again comes under scrutiny.

THE SOMATIC SELECTION HYPOTHESIS

One of the greatest recent incentives to the re-examination of the question of the inheritance of acquired characteristics was the somatic selection hypothesis of Steele (1979) Basically the theory is 'somatic-Darwinism' and is founded on the combination of two hypotheses: Burnet's (1959) clonal selection theory of immunity, and Temin's (1971; 1976) proto- and pro-virus hypothesis. It is known that mutant somatic information can confer on a stem cell an altered phenotype such that under the appropriate conditions this cell will be preferentially selected until it may dominate the whole somatic compartment. Under conventional Darwinian hypotheses this mutant somatic information is of no consequence, except in as much as it confers fitness on the individual. However, Steele (1979) invoked Temin's proto-virus hypothesis—which suggests that RNA tumour viruses are pathological offspring of naturally-occurring endogenous retroviruses involved in inter-cellular communication—and proposed that endogenous retroviruses targeted to the germline may act as vectors for the mutant somatic information by capturing RNAs from the somatic cells and transducing them to the germline. Once in the germline, the passenger RNA may be copied by the virally encoded reverse transcriptase (RNA-dependent DNA-polymerase) into DNA which is able to align with the parent gene and to replace it (or segments of it) by recombination.

Biologically, this hypothesis is best illustrated in the immune system. The lymphoid compartment contains progenitor lymphocytes displaying a vast array of antigenic specificities in part encoded by the germline genes and in part arising by somatic mutation (Cohn et al., 1981). When an antigen interacts with a precursor cell that particular clone is stimulated to proliferate, such that in cases of chronic antigen exposure it can dominate the whole lymphoid system (Burnet, 1959). Thus, once the clone is selected by virtue of its unique antigenic specificity, it may constitute a substantial percentage of the somatic compartment. Under these circumstances, Steele (1979) suggested that this information would be captured from the lymphocyte cytoplasm as mRNA, by retroviruses during their maturation and packaging and transduced to the germline. Once in the germline the retroviruses would infect either the germ cells or be carried in the seminal fluid and infect the fertilized embryo, with the resultant introduction of the somatic information into the germline by reverse transcription of the mRNA molecules into DNA. The progeny, therefore, would develop with a biased immune repertoire encoded in their germline genes.

Steele (1979) did not consider his theory restricted solely to the immune system, however, but to be a general theory of information transfer. Conceptually it is more difficult to envisage in circumstances where clonal domination does not attain, e.g. in the brain, where cell division is almost entirely restricted, or in the liver which shows limited proliferative capacity. However, during development a cell bearing a particular variation may be preferentially selected, e.g. in the gut, cells resistant to an environmental toxin may be selected, or as is the case in the immune system, a specific B-cell may select a complementary T-cell. Steele (1979) also suggested that the concept of 'idiotypy' (i.e. a somatically variable sensor protein) may not only be restricted to the immune system, but may be a common feature of adaptive systems. There is little data bearing on this point, but it is interesting to note that the liver cytochrome P-450 detoxifying system shows many similarities to the immunoglobulin system (Nerbert, 1979). These include existence of multiple, unique P-450 forms that appear to anticipate synthetic chemicals not known in nature and the presence of a specific memory. Elucidation of the corresponding genetic parallels, however, will have to await the molecular cloning of the cytochrome P-450 genes. Non-dividing cells are also presumably able to respond adaptively to environmental conditions by gene amplification in a fashion similar to that demonstrated by drug resistant cells in culture. For example, cells selected for resistance to increasing concentrations of the folic acid analogue, methotrexate, an inhibitor of the enzyme dihydrofolate reductase (DHFR), have increasingly higher levels of DHFR until it can reach five per cent of the total cell protein. This increase in DHFR is reflected by increased cytoplasmic mRNA levels produced as a consequence of a several hundred-fold increase in DHFR gene number (Schimke *et al.*, 1978). This amplification may also involve mutant DHFR genes (Lewis *et al.*, 1982). A similar case of amplification was observed for the gene encoding the multifunctional CAD protein, which contains the first three enzymatic activities of UMP synthesis, in cells resistant to N-(phosphonoacetyl)-L-aspartate an inhibitor of aspartate transcarbamylase (Wahl *et al.*, 1979). Thus it may be that gene amplification first demonstrated for ribosomal genes (Long and Dawid, 1980)—which also show environmental adaptation (Ritossa, 1976)—is a general feature of eukaryotic adaptive cellular responses. Interestingly, gene duplication also seems to be the mechanism for the evolution of multigene families (Jeffries, 1982) and remarkably in some circumstances an altered gene within the multigene family can result in conversion of the other genes in the complex (Slighthon *et al.*, 1980; Ohta, 1980).

One may, therefore, hypothesize that in non-dividing cells a mutant gene conferring resistance, for example, to an environmental toxin, may be amplified to a sufficient level that the mRNA is captured by a retrovirus and transferred inter-cellularly, both within the organ and back to the germline, followed by integration into the genome by the mechanism discussed above.

In the next two sections I propose to consider experimental evidence not only relevant to Steele's somatic selection hypothesis, but also to the more general questions of whether there are direct environmental influences on the germline and of the permeability of Weismann's barrier to somatically derived genetic elements.

EXPERIMENTAL EVIDENCE SUPPORTING THE ENVIRONMENTAL ALTERATION OF GENOTYPES

It is generally accepted that maternal influences may be inherited. These influences include amongst many documented cases specific immune responses, behavioural traits, alterations in endocrine regulation and susceptibility to gastic ulcers (Denenberg and Rosenberg, 1967; Solomon, 1977; Fujii, 1978; Skolnick *et al.*, 1980). These examples are rarely considered genetic, despite the fact that they often breed through several generations (Fujii, 1978) and despite the statements by many authors that they do not have satisfactory explanations for the phenomena (Denenberg and Rosenberg, 1967; Skolnick *et al.*, 1980). Consequently I will restrict myself to the less easily dismissed examples of apparent inheritance through the male line produced as a result of experimental manipulation.

The immune system

In a series of experiments designed to test Steele's somatic selection hypothesis, Gorczynski and Steele (1980, 1981) found that acquired neonatal tolerance to foreign major histocompatibility (MHC) antigens could be inherited through the male line. The design of the experiments exploited the classical studies of acquired tolerance to foreign antigens of Medawar and co-workers (Billingham *et al.*, 1956). Thus, neo-natal mice (strain A) were made tolerant to foreign histocompatibility antigens (strain B) by injecting 5×10^7 cells, each of bone marrow and spleen cells from an F_1 animal (A × B). Thereafter, mice were given biweekly injections of donor lymphoid tissue throughout the study to ensure chimerism and a very deep state of tolerance necessary in Steele's view (1979) to maximize the possibility of capture of somatic information and consequent transfer to the germline. At eight weeks of age, male mice were bred to untreated females of the same strain and the progeny tested for tolerance using a spleen cell mediated lympholysis (CML) assay. In this assay spleen cells (from strain A and its progeny) are incubated *in vitro* for five days with inactivated (X-irradiated) 'stimulator' strain B spleen cells. At this point the culture is harvested and tested for cytotoxic T-cell activity in a 4-hour Cr^{51} release test against Cr^{51} labelled strain B target cells. A high response indicates normal cytotoxic activity, whilst a low response is indicative of tolerance to strain B cells. This assay is generally accepted to parallel the acceptance or rejection of allografts (Gorczynski and Macrae, 1979) used in earlier studies on tolerance, but is much quicker to perform, allowing the screening of large numbers of progeny.

In the first study (Gorczynski and Steele, 1980), tolerance was induced in male CBA (H-2, K^kD^k) mice by neonatal injections of (CBA × A/J)F_1 lymphoid cells (H-2, $K^kD^{k/d}$). At eight weeks the animals were bred to normal non-tolerant CBA females and several waves of progeny produced. These progeny were further bred, either by brother–sister mating, or by mating to normal females, and all the progeny tested by a spleen mediated

CML assay. In the first generation of progeny from tolerant fathers, a significant proportion of offspring (50–60 per cent) failed to produce detectable anti-A/J cytotoxic responses and this characteristic bred through to the second generation when 20–40 per cent failed to show significant responses against A/J antigens (Fig. 12·1). Responsiveness to third party

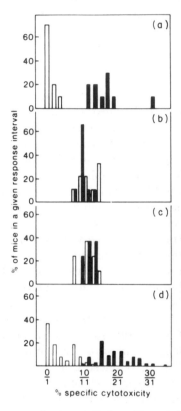

FIG. 12·1 Cytotoxic responses of spleen cells from 10 normal F_1(a) 9 CBA imported from Cumberland View Farms, (b) 8 CBA bred at the Ontario Cancer Institute and 63 first generation tolerant progeny. Open columns represent the response to A/J antigens (left of intervals) and solid columns responses to C57BL/6J antigens (right of intervals). The response of the tolerant progeny to A/J is significantly different from the control CBA, but not from the F_1 response. (Data from Gorczynski and Steele (1980) with permission.)

antigens, C57BL/6J and B10.A(2R), however, was unaffected, showing that the inheritance of the tolerant phenotypes was a specific image of the fathers. In the second group of experiments (Gorczynski and Steele, 1981), congenic mice were made tolerant to two independent MHC loci. Thus tolerance to two different H-2 haplotypes was induced in neonatal B10.SgN (H-2, $K^b D^b$) mice by intraperitoneal injections of lymphoid cells from

(B10.SgN × B.10.D$_2$)F$_1$ (H-2, K$^{b/d}$D$^{b/d}$) and (B10.SgN × B.10.BR)F$_1$ (H-2, K$^{b/k}$K$^{b/k}$). Breeding from tolerant fathers demonstrated transmission of the tolerance state for each haplotype independently but often simultaneously at a high frequency to both first and second generations (Fig. 12·2). Interestingly, once tolerance to a single haplotype is transmitted, reversion

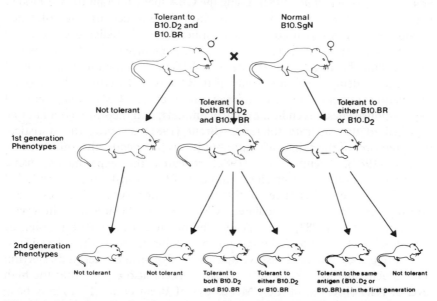

FIG. 12·2 Schematic diagram of the phenotypes of the progeny of B10.SgN mice made simultaneously tolerant to B10.D$_2$ and B10.BR antigens and bred to normal females. Note that mice displaying tolerance (hyporesponsiveness) to one antigen do not revert to being tolerant to the other antigen. (After Gorczynski and Steele, 1981.)

to tolerance to the other haplotype is never observed. In fact, the inheritance of each haplotype displays independent assortment and although the data on this point are not complete, the behaviour of each character is indicative of a single Mendelian dominant allele.

The above experiments investigated T-cell responses, but in a further study, B-cell responsiveness was tested (Steele and Gorczynski, 1981). Neonatal CBA males were rendered tolerant to sheep red blood cells (SRBC) and progeny tested for their ability to mount a primary anti-SRBC antibody (B-cell) response. It was observed again that a high proportion of both first and second generation progeny were unresponsive to SRBC in a similar fashion to their fathers. T-cell responsiveness, using a spleen CML assay, was also investigated. But contrary to the previous studies, the major observation here was a perturbation of the T-cell response, rather than a specific hyporesponsiveness. This may reflect the problem of establishing full tolerance to SRBC in this study, and can be rationalized by invoking network theory models of immunoregulation (Jerne, 1974).

Several attempts using broadly similar protocols, have been made to repeat the experiments of Gorczynski and Steele (1980) with both positive (Mullbacher *et al.*, 1983) and negative results (Brent *et al.*, 1981, 1982; Smith, 1981; Hasek *et al.*, 1982). Using the CBA made tolerant to A/J model, Brent *et al.* (1981) were unable to show inheritance of the tolerance phenotype. This study used both a peripheral blood mediated CML assay and the more stringent criteria of allograft acceptance to test for tolerance. In reworking the data, however, Steele (1981) claimed that the study showed a significant delay in rejection time of (CBA × A/J) F_1 skin in the progeny of tolerant fathers. He also pointed out that the CML assay showed hyporesponsiveness, even in the control animals, and therefore could not be expected to demonstrate tolerance. Brent (1981) disputed the statistical interpretation of the skin graft data, but it was independently confirmed by Josevic (1982). Brent *et al.* (1982) repeated the experiments, using experimental protocols even closer to those of Gorczynski–Steele and failed to find positive inheritance of the tolerant phenotype using both skin grafting and a spleen mediated CML assay. Meanwhile, however, Gorczynski *et al.* (1983) had repeated the previous experimental design of CBA tolerant to A/J and showed not only inheritance of the tolerance to A/J antigens by the CML assay, but also a significantly delayed rejection time, albeit small, of skin grafts. These workers also argued that the high level of lysis observed in the CML assays of Brent *et al.* (1982) may have obscured the tolerant phenotype and also pointed out the dangers of interpretation if only a single target: effector cell ratio is taken, since not all the assays are directly proportional to dilution. In this group of experiments, Gorczynski *et al.* (1983) also tested progeny derived from breeding normal untreated males with normal females that had borne at least three litters to tolerant fathers. In this case the progeny, surprisingly, still displayed specific tolerance to A/J antigens. This raises the possibility that there is a maternal derived origin for the transmission of characters observed through tolerant males. This could be produced by, for example, immunization of the mothers by the sperm or seminal fluid of the tolerant males, to infection of oocytes by viral vectors carried in the sperm or to an immune response to the previous carried tolerant fetuses. In the former case it is difficult to explain the apparently Mendelian character of the transmission seen in the B10 tolerance experiments. But clearly more experiments, perhaps involving the transplantation of fertilized ova between females, need to be performed to resolve this issue.

Independent data to support the Gorczynski–Steele experiments have been obtained in a limited study by Mullbacher *et al.* (1983), who showed that the non-specific hyporesponsiveness of cytotoxic T-cells to alpha-virus and alloantigens induced by tolerizing neo-natal mice to gamma-inactivated alpha-virus, was transmitted through the male line to some first generation mice, not exposed to the tolerizing antigens.

There are precedents in the literature to the work of Gorczynski and Steele (Guttman *et al.*, 1963, 1964; Kanazawa and Imai, 1974). In the study by Guttman and co-workers (1963; 1964) C3H males were made tolerant to A strain histocompatibility antigens and mated to $(A \times C3H)F_1$ females. The backcross progeny, together with normal backcross offspring were tested for the ability to support growth of an $(A \times C3H)F_1$ passaged tumour and for the ability to accept $(A \times C3H)F_1$ skin. Amongst the normal progeny, 8 per cent allowed growth of the tumour and 3 per cent accepted the skin, whilst in the tolerant progeny 90 per cent accepted the tumour and 100 per cent the skin. In secondary testing of the tolerant progeny with $(C57B1/1 \times C3H)F_1$ skins, all the animals rejected the skin, indicating the transmission of tolerance was specific. This study was rather limited, however, and both Steinmuller (1967), in an uncontrolled study, and Brent *et al.* (1982) failed to reproduce the data.

In other experiments in the immune system, putative evidence for the inheritance of acquired characters has also been obtained. Steele (1979) interpreted the data on the inheritance of rabbit idiotypy (Kelus and Gell, 1968; Oudin and Michell, 1969; Eichmann and Kindt, 1971; Winfield *et al.*, 1973), and concluded that in situations when the antigen was administered prior to mating there was evidence for the positive transmission of idiotypes (somatically variable antibody domains). Similarly, in rabbits the inheritance of eye-defects induced by treating adult animals with fowl-anti-rabbit eye antiserum, was found through the male line for at least nine generations. These defects became progressively worse in each generation, and presumably represented an on-going auto-immune disease (Guyer and Smith, 1920; Guyer, 1928). Parallel experiments were performed by damaging the eyes of adult rabbits and again the auto-immune disease (shown by the presence of anti-lens antiserum) was transmitted through the male line (Guyer, 1928).

Systems other than the immune system

In this section demonstrating the inheritance of acquired characters, the responses to the environment documented can be arbitrarily divided into specific effects and apparently random events. In the former group, the best documented are those involving the so-called 'cybernin' molecules (Guilleman, 1978), i.e. those regulatory molecules involved in complex inter-active roles with one another in order to effect regulation of the whole system.

In a long-term study, the sub-diabetic state (abnormal glucose tolerance) induced by the drug alloxan, was found in the untreated progeny and could be inherited for at least three generations with equal efficiencies through both the male and female lines (Goldner and Spergel, 1972 for review). The induced phenotype seemed to involve a diminution in number and altered morphology of pancreatic beta, but not alpha, cells resulting ultimately in lower plasma levels of insulin (similar to the alloxan treated fathers), and consequently a progressively deteriorating glucose tolerance rate. Similar results were also obtained following surgically induced diabetes. In a second group of studies, thyroid function was impaired by treatment of neonatal rats with large dosages of thyroxine, or by thyroidectomy. This impaired thyroid function resulted in lower levels of both free and protein bound thyroxine, abnormal development and pituitary function. In both cases the offspring of treated males showed disturbances in development despite the fact that they had never been exposed to their fathers or undergone any treatment (Bakke et al., 1975). These effects, although somewhat different in the two experimental treatments, included increased pituitary and thyroid weights and a reduced responsiveness to thyroid stimulating hormone release factor (TRH). With the exception of the TRH response the effects seemed to be a mirror image of the father's. Bakke et al. (1975) interpreted this data as an alteration in the thyrostat set point of the offspring, which, since it may be inherited through at least three generations, presumably involves a genetic alteration transmitted from the fathers.

Broadly similar breeding protocols were followed in experiments in which male mice were treated with opiates. This treatment resulted in the progeny having an accommodation to an initial test dose of morphine (Friedler, 1974). The parental treatment was considered to have altered the regulation of the hypothalamic–hypophyseal axis in the progeny. Alterations of the neuroendocrine regulation of diapause were achieved by treatment of the butterfly, Pieris brassicae, with LSD and this effect bred through the male line for at least two generations (Vuillaume and Berkaloff, 1974). Offspring were also more resistant to the toxic effects of LSD, which the authors interpreted as an enhanced detoxifying mechanism in the progeny.

At first sight, many of the above examples are of the instructionalist type (i.e. damage from outside), and therefore specifically excluded from Steele's selectionist hypothesis (1979) and by the tail-cutting experiments of Weismann (1893) or by mutilation according to custom in humans. However, all the experiments appear to involve complex regulatory phenomena, true also of the studies of neo-natal tolerance to MHC antigens, and therefore control sequences rather than structural genes may have been transmitted to the progeny.

Apart from these examples showing a directional component in the transmission of phenotypes, there are a number of apparently non-directional, environmentally induced genetic events. The best documented of these are the induction of genotrophs in flax following growth of plants under different environmental conditions (Cullis, 1983). Two extremes of genotypes can be observed, the large type characterized by tall plants with hairless septa of the seed and small plants with hairy septa (Durrant, 1971). Biochemically they can be distinguished by nuclear DNA amounts, ribosomal RNA cistron number and isozyme patterns of peroxidase and alkaline phosphatase (Cullis, 1979 a, b). Once established, the stable genotrophs behave as genetically distinct lines and in fact the new isozyme patterns of peroxidase segregate as expected for a Mendelian locus with dominant and recessive alleles and with the large genotroph being a homozygous dominant (Cullis, 1979a). Cullis (1983) suggested the amplification of gene sequences could be produced by environmentally induced disproportionate replication of certain regions of the genome. The resultant phenotypes appear not to be specific environmental adaptations (although since only a small number of phenotypic characters have been studied it is not possible to completely define adaptive under these experimental conditions), but rather to represent an environmental increase in the rate of genetic change. Nevertheless, despite this lack of directionality, since specific gene sequences have been shown to be affected, this system reflects is the best documented example of an environmentally altered genotype. It may not, however, represent a genetic penetrance of Weismann's barrier in the strictest sense, since it has yet to be shown if the genetic effects (transposable elements?) originate in the soma, or are induced directly within the germline, as suggested by Cullis (1983). It is, however, in contravention to Weismann's doctrine since this excluded the possibility of *any* environmental effects on the genome.

Taken together, the above data strongly suggest that there can be environmental modifications of genotype passed through both the male and female line and that in some cases this may constitute a transfer of genetic material through Weismann's barrier. The next section will discuss the types of genetic sequences that may be involved in such a transfer.

SACRED AND PROFANE GENES

Historically much of the intellectual rejection of Lamarckian ideas was due to the impossibility of envisaging a means of transfer of genetic information from the soma to the germline, coupled with the concept of a fixed, rigid, particulate genome. However, it is now realized that the genome is more fluid than previously anticipated and that there are an ever growing number of profane genes. In this section, I will somewhat selectively range over the more recent observations on the fluid genome and in the context of Weismann's barrier, comment on their significance.

Moveable genetic elements have recently been established to be present in mammalian cells (Temin and Engels, 1984; Shapiro and Cordell, 1982), as well as bacteria, yeast, insects and plants (Calos and Miller, 1980). They therefore appear to be ubiquitous. It is attractive to argue that these moveable genetic elements are involved in gene regulation and development (Temin, 1980; Davidson and Posakony, 1982) since, for example, in *Drosophila* they constitute a considerable fraction of the genome (Young, 1979) and produce a large amount of RNA (Cameron *et al.*, 1979). In fact, the Alu group of dispersed repetitive sequences that appear to be moveable elements, are represented in the human genome 500 000 times (Schmid and Jelinek, 1982). In rats similar elements are not only repeated many times, but subclasses have been identified inserted into the beta-tubulin pseudogene (Lemischka and Sharp, 1982) and a growth hormone gene where the sequence is an intron (Barta *et al.*, 1981). Classes of these elements also show developmental specificity in their expression and it has been suggested that they may identify gene sequences to be expressed in particular tissues (Sutcliffe *et al.*, 1982). Other examples of their role in developmental regulation may be in mating strain switching in yeast (Nasmyth and Tatchell, 1980) and in the generation of antigenic diversity in trypanosomes (Pays *et al.*, 1981). Functional roles in the development of higher eukaryotes have yet to be unequivocally demonstrated, however (Calos and Miller, 1980), and it may be that these elements merely replicate and transpose parasitically (Orgel *et al.*, 1980).

Several authors have previously proposed that there is circulating informational DNA involved in cellular communication (Stroun *et al.*, 1977 for review). Whether these sequences are similar to moveable genetic elements, experimental artefacts, or other forms of genetic interchange will have to wait for molecular cloning. However there is little doubt, although it has yet to be proved, that moveable genetic elements exchange between cells. Whether this is a natural consequence of life-style of these sequences or a consequence of cellular function is not known. Transposition may not necessarily involve DNA, but may be as RNA packaged within a viral coat (retrovirus?). (Flavell and Ish-Horowitz, 1983). In fact transposition of the family of Alu-dispersed repetitive elements (Schmid and Jelinek,1982) appears to involve an RNA intermediate (Van Arsdell *et al.*, 1982; Hammarström *et al.*, 1982. And according to Lemischka and Sharp (1982) 'since there are 500 000 dispersed repetitive elements in human DNA the insertion of small repetitive elements in the germline of mammals must not be uncommon'. In this context it is worth noting that a different class of moveable genetic element, the P-transposon in *Drosophila*, readily integrate into the germline and in fact are more active here than in somatic cells (Temin and Engels, 1984). In one case also an experimentally manipulated Xanthine dehydrogenase gene was introduced into the P-transposon and microinjected into an oocyte where it became integrated and corrected the rosy locus (ryl) defect in both the adult and its offspring (Rubin and Spradling, 1981). More startling is the speculation that genes may cross species barriers by transposition, perhaps captured in a viral vector (Lewin, 1982). Examples of this are the presence in legumes of leghaemoglobin, whose gene so closely resembles the structure of vertebrate globin that the probability of chance similarity is vanishingly small but instead points to a common evolutionary origin. But since in the plant kingdom only legumes have this gene, it must have been established by gene transfer from vertebrates (Hyldig-Nielson *et al.*, 1982; Jeffries, 1982). Similar interpretations have been made for the similarities of histone H3 and H4 and their spacer sequences between two sea urchin species (Busslinger *et al.*, 1982). If this type of transposition is occurring between species, it conclusively demonstrates that Weismann's barrier is permeable to functional genetic elements.

Temin (1970; 1976) postulated that retroviruses were evolutionary offshoots of normal cellular moveable genetic elements involved in intercellular communication. Sequence analysis has supported this idea, since there are striking sequence homologies between retroviruses and transposable elements (Temin, 1980; 1983; Temin and Engels, 1984; Shimotohno *et al.*, 1980; Majors *et al.*, 1980; Flavell and Ish-Horowitz, 1981). It is also known that retroviruses can transduce host genes between and within cells (Baxt and Meinkoth, 1970). Indeed, the recombination between the virus and a host oncogene is thought to be the cause of oncogenicity in the RNA tumour viruses (Weinberg, 1981; Bishop, 1981). Since all mammalian cells contain retroviruses and these are vertically transmitted, it is clearly feasible that this gene transduction is the normal role of endogenous retroviruses, perhaps, as suggested by Temin (1976), as a means of inter-cellular communication.

Steele's hypothesis (1979), however, focused on the capture of passenger RNAs endogenous retroviruses rather than covalent linkage of somatic cell genes, for three reasons. First, it provides a mechanism whereby a representative selection of genes being actively expressed may be sampled. This ensures, as opposed to the covalent linkage model (unless there is some specialized ability to recombine with actively transcribing DNA), that specific expressed genetic information (as intact mRNA molecules) will be packaged at high efficiency within a majority of the endogenous viral particles derived from that cell. Secondly, viral genomes would only be able to accommodate a limited number of covalently bound cellular genes and these would be situated at the site of viral integration, rather than at the normal cellular locus, and be flanked by viral long term repeats and consequently under different regulation. Thirdly, it is probable that the characteristic of an endogenous natural vector would be to facilitate the transfer of genetic information and not to disrupt host genes.

Relevant to the speculation that endogenous retroviruses are vectors for host information, is the observation that retroviral particles carry, apart from their own genomes, a variety of non-covalently bound RNA molecules, including a range of small RNAs and mRNAs (Sawyer and Dahlberg, 1973; Ikawa *et al.*, 1974). Similarly retroviral-like particles have been demonstrated in *Drosophila melanogaster* which contain RNA homologous to the transposable element copia (Shiba and Saigo, 1983). Thus, by capture, transduction to the germline and reverse transcription, somatically expressed information may be introduced into the germline. Recently, evidence consistent with such a process has been obtained by the observation that some pseudogenes (genes functionally dead owing to the presence of nonsense sequences within their coding regions) bear evidence of having been generated by reverse transcription of mRNA molecules (Hollis *et al.*, 1982; Nishoka *et al.*, 1980; Lemischka and Sharp, 1982), tRNAs (Reilly *et al.*, 1982), as well as some nuclear RNAs (Van Arsdell *et al.*, 1982; Hammarström *et al.*, 1982). A point that is rarely emphasized, however, is that since these pseudogenes are found in *all* somatic DNA they must have been introduced into the germline. Since adult globin and immunoglobulin mRNAs are not thought to be expressed in oocytes or sperm, these may have been introduced from the somatic compartment by a retroviral vector. It cannot be excluded, however, that adult genes may be expressed at a low level in germ cells (or putative germ cells of the embryo) and in fact, Temin and Engels (1984) have stated that there must be considerable reverse transcriptase activity within germ cells. Nevertheless, it may be argued that although by definition pseudogenes are functionally dead they represent the error level of a frequently occurring normal transfer process from soma to germline. These processed pseudogenes also presumably offer sites for recombination during meiosis and during evolution for a drift into new, functional genes with different coding properties.

If it is accepted that one mode of transfer of somatic information is as RNA, then it is pertinent to question the nature of the information carried. It appears that many of the systems that show inheritance of acquired characters are, rather than specific single gene changes (as was discussed for the transfer of rabbit idiotypes (Steele, 1979)), complex regulatory systems, e.g. the B and T cell interactions in the establishment of immune tolerance or the 'cybernin' systems. One class of molecule that may be a candidate for regulatory information transfer is the small nuclear RNA molecules. These show larger complexity changes during development than heterogenous nuclear RNAs in general and they have been implicated in the regulation of gene batteries (Davidson and Britten, 1979; Davidson and Posakony, 1982). Amongst this group are the molecules involved in RNA splicing, the transcripts of Alu-repetitive elements and a large range of RNAs of unknown function. Introns may also belong to this group, prior to their being degraded, and their concentration presumably reflects the level of mRNA export. It is possible that such elements may be captured by retroviruses and transduced to the germline. In some cases introns are similar to the Alu-like repetitive elements thought to integrate into the germline in this fashion and sub-classes of which show tissue specific regulation and have been implicated as identifier sequences in gene regulation (Sutcliffe *et al.*, 1982; Davidson and Posakony, 1982). Interestingly, introns have structural similarities to viroids which are known to pathologically affect regulation in plants without the intercession of a viral encoded RNA or protein (Diener, 1981). Retroviral particles contain a large number of unrelated small RNAs (Sawyer and Dahlberg, 1973; Faras *et al.*, 1973) and molecular cloning should establish the relationship, if any, between these, introns and other regulatory nuclear RNAs.

PERSPECTIVES

In the above sections some of the evidence for the permeability of Weismann's barrier has been discussed. The data suggests that genetic elements may pass from the soma to the germline and also that the environment may activate genetic alterations within the germline. This data includes the ubiquitous nature of transposable elements, the presence of pseudogenes, the examples of horizontal gene transfer, and the induction of genotrophs in flax.

These examples have emerged in the early days of recombinant DNA technology and probably represent the tip of the iceberg. Many of the examples, however, are of genetic elements with unknown function, or dead genes, and as yet, with the exception of horizontal gene transfer, specific, functional genetic information has not been demonstrated to be transmitted from the soma to the germline. Data has also been presented that demonstrated the specific transfer of phenotypic information from the soma to the germline. The weakness in the interpretation of these latter experiments lies in the unknown and complex nature of the information transmitted, which lies outside the realm, at present, of molecular genetic analysis.

The challenge therefore still remains to rigorously test the hypothesis that *specific* genetic characters may be selected in the soma and transmitted to the germline. Such experiments are technically feasible, at least for simple characters. Thus exogenous genes are taken up and expressed with surprising ease by somatic cells, oocytes and developing embryos (Jaenisch *et al.*, 1981; Wagner *et al.*, 1981b; Constantini and Lacy, 1981). In the latter cases, the introduction of foreign genes results in their being incorporated into progenitor germ cells and consequently into the germline of the adult, and in some cases are expressed in the progeny (Jaenisch *et al.*, 1981; Constantini and Lacy, 1981). Cline and co-workers (1979) have claimed to have introduced a foreign gene encoding drug resistance into haematopoietic stem cells and to have reintroduced these cells into foster animals. By virtue of their acquired drug resistance these cells were preferentially selected by appropriate drug treatment to the animals. Since the donor gene was different from the host gene and has been molecularly cloned, a feasible experiment would be to breed from the male animals and probe for the foreign gene sequence. In the case of Cline's mice, because of the particular treatment they received, it is unlikely that they would be fertile, nevertheless less destructive experiments along these lines could be devised and, if positive, would provide definitive evidence for the inheritance of acquired characters.

This article has focused on the movement of genetic elements from the soma to the germline. An alternative genetic explanation has been proposed to explain the data on page 294–304 (Campbell, 1982; Campbell and Zimmerman, 1982). Campbell's hypothesis is based on a postulated feedback between cell surface receptor proteins and their genes. Thus when a modulator ('cybernin-like' molecules such as drugs, hormones or specific cell factors) interacts with its receptor during a specific developmental period, a signal is relayed to the nucleus which activates DNA modifying enzymes to alter the structure of the receptor gene. This alteration results in both a quantitative and qualitative expression of the receptor on the cell surface, and hence alters the cell's sensitivity to future signals; a process referred to as automodulation. If such receptors were displayed on germ cells then environmental alterations of cybernin molecules could produce specific inherited effects. In this theory, therefore, there is not a transfer of genetic material, but a direct environmental effect on the germline DNA. Indeed, a similar but less specific mechanism has been postulated for the environmental modulation of genotype in flax (Cullis, 1984). If automodulation is to be a general theory, however, a large number of different receptors would need to be expressed on both sperm and oocytes, perhaps more than could be accommodated on the cell surface or by the transcriptional activity of the cell. But, nevertheless, there is evidence at least for the expression of neuroendocrine receptors on germ cells (Kusano *et al.*, 1977).

Many criticisms have been levelled at the concept of inheritance of acquired characters on a theoretical level. Several of these have been dealt with above, e.g. the vector problem. But one criticism still remains paramount; that is the problem of genetic chaos or more prosaically, the stability of linkage relationships or alternatively the maintenance of heterozygosity. It is hard to comment on this, since by essence of selection of characters analyzed by conventional techniques the observed stability of linkage relationship may have been unintentionally selected and thus may not be such a general feature, as thought. Similarly, if the events documented above are rare, they would represent 'statistical noise' in genetic experiments which deal with populations, but nevertheless they may be of absolute importance in evolution which deals with individuals. Mendelian genetic experiments, also by virtue of good experimental practice, are always performed under constant conditions, but these hardly represent real evolutionary scenario. In many circumstances, in fact, when environmental conditions are changed, or when laboratory stocks are outbred, Mendelian rules tend to break down. This is often attributed to activation of transposable elements, e.g. in the MHC locus in the mouse (Bailey, 1966) and during hybrid dysgenesis in *Drosophila* (Rubin *et al.*, 1982). It is also safe to say that there is very little genetic understanding, if any, of complex phenotypes. Thus, by simple, and until now necessary bias of classical genetic experiments, a large proportion of genetic variability may be waiting to be analyzed.

The available data on the dynamic genome already poses a challenge to conventional genetic analysis (Temin and Engels, 1984). This includes not only the penetration of Weismann's barrier, but also the role of transposition and gene amplification, etc. within the germline. For example, Lemischka and Sharp (1982) calculated that pseudogenes for beta-tubulin arose at $1/3 \times 10^6$ years. If pseudogenes evolve into new genes at a significant rate (Proudfoot, 1981), this process must have substantial impact on evolution. Similarly in one well documented case in *E.coli*, and its prophage P1, over 90 per cent of the mutations were not due to classical base substitutions, but to insertion element mediated events (Arber *et al.*, 1981). Rather similar data has been obtained for certain loci in *Drosophila* (Temin and Engels, 1984, for review). Thus molecular genetics will not only document new phenomena, but demand of population genetics a quantitative evaluation of their occurrence and behaviour within populations, before a final judgment can be made of their significance in evolution.

If, as a result of these considerations, the genetic penetration of Weismann's barrier is found to be a frequent process, it poses a number of questions and offers some new perspectives. The first of these is the specificity of the process. Are specific genes or control elements for specific states transmitted, or is the process simply a means of amplifying genetic variability? The available data suggests that both events are occurring. If specific transfer does occur, then a feedback between the environment (use) and the germline exists, giving a directional component to evolution. Thus, by playing on the enormous potentiality for variation within an organism, somatic characters will be selected that provide adaptation in an environment, perhaps to many individuals simultaneously within a small population in a local area; thus providing a concentration of a particular phenotype. Since the data also indicates the transfer of complex traits involving several cell types, then it provides a means whereby parallel evolution can occur through the co-selection of one somatic variant by another. This need for unrelated mutations to occur synchronously to explain phenotypes that are not adaptive unless fully functional (e.g. the so-called organs of extreme perfection) poses one of the major problems for neo-Darwinism, since the chance of several mutations for a specific trait occurring in a single germ cell is extremely remote. Thus, the inheritance of acquired characters provides a rationale, not only for the occurrence of co-selected traits, but also for the genetic assimilation of variant developmental patterns. Finally, the germline apparently is immortal compared to somatic cells (see Medvedev, 1981 for discussion), which implies that it cannot accumulate the same mutational load as the soma. This could be explained by a continuous genetic proof-reading against somatic genes, the great majority of which would not contain mutations. Therefore, the inheritance of acquired characters may provide an answer to the continuity of the germplasm first proposed by Weismann.

ACKNOWLEDGEMENTS

I would like to acknowledge the many stimulating discussions held over several years with Dr. Ted Steele. My thanks also to Drs. Reg Gorczynski and J. Campbell for sending me manuscripts prior to their publication and to Dr. Brian Gardiner for critical comments on this manuscript.

REFERENCES

Arber, W., Humberlin, M., Casper, P., Reif, H. J., Iida, S. and Meyer, J. (1981). Spontaneous mutations in the Escherichia coli prophage P1 and IS-mediated processes. *Cold Spring Harbor Symp. Quant. Biol.* **45**, 38–40.

Bailey, D. W. (1966). Heritable histocompatibility changes: Lysogeny in mice? *Transplanation* **4**, 482–483.

Bakke, J. L., Lawrence, N. L., Bennett, J. and Robinson, S. (1975). Endocrine syndrome produced by neonatal hyperthyroidism, hypothyroidism, or altered nutrition and effects seen in untreated progeny. *In* 'Perinatal Thyroid Physiology and Disease' (D. A. Fisher and G. H. Burrow, *Eds*), pp.79–116. Raven Press, New York.

Barta, A., Richards, R. I., Baxter, J. D. and Shine, J. (1981). Primary structure and evolution of rat growth hormone gene. *Proc. Nat. Acad. Sci. U.S.A.* **78**, 4867–4871.

Baxt, W. G. and Meinkoth, J. L. (1978). Transfer of duck cell DNA sequences to the nucleus of 3T3 cells by Rous sarcoma virus. *Proc. Nat. Acad. Sci. U.S.A.* **75**, 4252–4256.

Billingham, R. E., Brent, L. and Medawar, P. B. (1956). Quantitative studies on tissue transplantation immunity. III: Actively acquired tolerance. *Phil. Trans. R. Soc. (London) B.* **239**, 357–414.

Bishop, J. M. (1981). Enemies within: the genesis of retrovirus oncogenes. *Cell* **23**, 5–6.

Brent, L. (1981). Lamarck and immunity: The tables unturned. *New Scientist* **90**, 493.

Brent, L., Chandler, P., Fiertz, W., Medawar, P. B., Rayfield, L. S. and Simpson, E. (1982). Further studies on supposed Lamarckian inheritance of immunological tolerance. *Nature* **295**, 242–244.

Brent, L., Rayfield, L. S., Chandler, P., Fiertz, W., Medawar, P. B. and Simpson, E. (1981). Supposed Lamarckian inheritance of immunological tolerance. *Nature* **290**, 508–512.

Burnet, F. M. (1959). 'The Clonal Selection Theory of Acquired Immunity'. Cambridge University Press, London and New York.

Busslinger, M., Rusconi, S. and Birnsteil, M. L. (1982). An unusual evolutionary behaviour of a sea urchin histone gene cluster. *EMBO J.* **1**, 27–33.

Calos, M. P. and Miller, J. H. (1980). Transposable elements. *Cell* **20**, 579–595.

Cameron, J. R., Loh, E. Y. and Davis, R. W. (1979). Evidence for transposition of dispersed repetitive DNA families in yeast. *Cell* **16**, 739–751.

Campbell, J. H. (1982). Autoevolution. *In* 'Perspectives on Evolution' (R. Milkman, *Ed.*), pp.190–201. Sinauer Press, New York.

Campbell, J. H. and Zimmermann, E. G. (1982). Automodulation of Genes: A proposed mechanism for persisting effects of drugs and hormones in mammals. *Neurobehavioral Toxicol. Teratol.* **4**, 435–439.

Cline, M. J., Stang, H., Mercola, K., Morse, L., Ruprecht, R., Browne, J. and Salser, W. (1980). Gene transfer in intact animals. *Nature* **284**, 422–425.

Cohn, M., Longman, R. and Gleckeler, W. (1980). Diversity 1980. *In* 'Immunology 1980' (M. Fougereau and J. Dausset, *Eds*), pp.153–220. Academic Press, London and New York.

Constantini, F. and Lacy, E. (1981). Introduction of a rabbit β-globin gene into the mouse germline. *Nature* **294**, 92–94.

Cullis, C. A. (1979a). Segregation of the isozyme of flax genotrophs. *Heredity* **42**, 237–246.

Cullis, C. A. (1979b). Quantitative variation of ribosomal RNA genes in flax genotrophs. *Biochem. Genet.* **17**, 391–401.

Cullis, C. A. (1984). Environmentally induced DNA changes. *In* 'Evolutionary Theory: Paths into the Future' (J. W. Pollard, *Ed.*). John Wiley and Sons, London and New York. 203–216.

Davidson, E. H. and Britten, R. J. (1979). Regulation of gene expression: Possible role of repetitive sequences. *Science* **204**, 1052–1059.

Davidson, E. H. and Posakony, J. W. (1982). Repetitive sequence transcripts in development. *Nature* **297**, 633–635.

Denenberg, V. H. and Rosenberg, K. M. (1967). Nongenetic transmission of information. *Nature* **216**, 549–550.

Diener, T. O. (1981). Are viroids escaped introns? *Proc. Nat. Acad. Sci. U.S.A.* **78**, 5014–5015.

Dobzhansky, Th., Ayala, F. J., Stebbins, G. L. and Valentine, J. W. (1977). 'Evolution'. W. H. Freeman, San Francisco.

Durrant, A. (1971). Induction and growth of flax genotrophs. *Heredity* **27**, 277–298.

Eichmann, K. and Kindt, T. J. (1971). The inheritance of individual antigenic specificities of rabbit antibodies to Streptococcal carbohydrates. *J. Exp. Med.* **134**, 532–552.

Faras, A. J., Garapin, A. C., Levinson, W. E., Bishop, J. M. and Goodman, H. M. (1973). Characterization of the low-molecular weight RNAs associated with the 70S RNA of Rous sarcoma virus. *J. Virol.* **12**, 334–342.

Flavell, A. J. and Ish-Horowitz, D. E. (1981). Extrachromosomal circular copies of the eukaryotic transposable elements *copia* in cultured Drosophila cells. *Nature* **292**, 591–595.

Flavell, A. J. and Ish-Horowitz, D. E. (1983). The origin of extrachromosomal circular *copia* elements. *Cell* **34**, 415–419.

Friedler, G. (1974). Long-term effects of opiates. *In* 'Perinatal Pharmacology: Problems and Priorities' (J. Dancis and J. C. Hwang, *Eds*), pp.207–216. Raven Press, New York.

Fujii, T. (1978). Inherited disorders in the regulation of serum calcium in rats raised from parathyroidectomised mothers. *Nature* **272**, 236–238.

Goldner, M. G. and Spergel, G. (1972). On the transmission of alloxan diabetes and other diabetogenic influences. *Ad. Metabolic Diseases* **6**, 57–72.

Gorczynski, R. M. and MacRae, S. (1978). Suppression of cytotoxic response to histocompatible cells. 1. Evidence for two types of T-lymphocyte-derived suppressors acting at different stages in the induction of a cytotoxic response. *J. Immunol.* **122**, 737–776.

Gorczynski, R. M. and Steele, E. J. (1980). Inheritance of acquired immunological tolerance to foreign histocompatibility antigens in mice. *Proc. Nat. Acad. Sci. U.S.A.* **77**, 2871–2875.

Gorczynski, R. M. and Steele, E. J. (1981). Simultaneous yet independent inheritance of somatically acquired tolerance to two distinct H-2 antigenic haplotype determinants in mice. *Nature* **289**, 678–681.

Gorczynski, R. M., Kennedy, M., MacRae, S. and Ciampi, A. (1983). A possible maternal effect in the abnormal hyporesponsiveness to specific alloantigens in offspring born to neonatally tolerant fathers. *J. Immunol.* **131**, 1115–1120.

Guillemin, R. (1978). Peptides in the brain: The new endocrinology of the neuron. *Science* **202**, 390–402.

Guttman, R. D. and Aust, J. B. (1963). A germplasm transmitted alteration of histocompatibility in the progeny of homograft tolerant mice. *Nature* **197**, 1220–1221.

Guttmann, R. D., Vosika, G. J. and Aust, J. B. (1964). Acceptance of allogenic tumor and skin grafts in backcross progeny of a homograft tolerant male. *J. Nat. Cancer Inst.* **33**, 1–5.

Guyer, M. F. (1928). 'Being Well Born: An Introduction to Heredity and Eugenics'. Constable and Company, London.

Guyer, M. F. and Smith, E. A. (1920). Studies on cytolysins: II. Transmission of induced eye-defects. *J. Exp. Zool.* **30**, 171–216.

Hammarström, K., Westin, G. and Pettersson, K. (1982). A pseudogene for human U4 RNA with a remarkable structure. *EMBO J.* **1**, 737–739.

Hasek, M., Holan, V. and Kousalova, M. (1982). On the problems of the genetic effects of immunological tolerance. *In* 'Evolution and Environment' (V. J. A. Novak and J. Mlikovsky, *Eds*), pp.399–404. Praha: CSAV.

Ho, M. W. and Saunders, P. T. (1979). Beyond neo-Darwinism—An epigenetic approach to evolution. *J. Theor. Biol.* **78**, 573–591.

Hollis, G. F., Hieter, P. A., McBride, D., Swan, D. and Leder, P. (1982). Processed genes: A dispersed human immunoglobulin gene bearing evidence of RNA-type processing. *Nature* **296**, 321–325.

Huxley, J. (1942). 'Evolution: The Modern Synthesis'. George Allen and Unwin, London.

Hyldig-Nielson, J. J., Jensen, E. O., Paludan, K., Wiborg, O., Garrett, R., Jorgensen, P. and Marcker, K. A. (1982). The primary structures of two leghaemoglobin genes from soybean. *Nucleic Acid Res.* **10**, 689–701.

Ikawa, Y., Ross, J. and Leder, P. (1974). An association between globin messenger RNA and 60S RNA derived from Friend leukaemia virus. *Proc. Nat. Acad. Sci. U.S.A.* **71**, 1154–1158.

Jaenisch, R., Jahner, D., Nobis, P., Simon, I., Lohler, J., Harkers, K. and Grotkopp, D. (1981). Chromosomal position and activation of retroviral genomes inserted into the germline of mice. *Cell* **24**, 519–529.

Jeffries, A. J. (1982). Evolution of globin genes. *In* 'Genome Evolution', (G. A. Dover and R. B. Flavell, *Eds*), pp.157–175. Academic Press, London.

Jerne, N. K. (1974). Towards a network theory of the immune system. *Ann. Immunol. (Inst. Pasteur)* **125c**, 373–389.

Josovic, J. (1982). Lamarckian inheritance—The statistical arguments. MSc. Dissertation, City of London Polytechnic.

Kanazawa, K. and Imai, A. (1974). Parasexual–sexual hybridization. Heritable transformation of germ cells in chimeric mice. *Japan J. Exp. Med.* **44**, 227–234.

Kelus, A. S. and Gell, P. G. H. (1968). Immunological analysis of rabbit anti-antibody systems. *J. Exp. Med.* **127**, 215–234.

Kusano, K., Miledi, R. and Stinnakne, J. (1977). Acetylcholine receptors in the oocyte membrane. *Nature* **270**, 739–741.

Lemischka, J. and Sharp, P. A. (1982). The sequences of an expressed rat α-tuberlin gene and a pseudogene with an inserted repetitive element. *Nature* **300**, 330–335.

Lewin, R. (1982). Can genes jump between eukaryotic species? *Science* **271**, 42–43.

Lewis, J. A., Davide, J. P. and Melera, P. W. (1982). Selective amplification of polymorphic dihydrofolate reductase gene loci in Chinese hamster lung cells. *Proc. Nat. Acad. Sci. U.S.A.* **79**, 6961–6965.

Long, E. O., and Dawid, I. B. (1980). Repeated genes in eukaryotes. *Annu. Rev. Biochem.* **49**, 727–764.

Majors, J. E., Swanstron, R., Delorbe, W. J., Payne, G. S., Hughes, S. H., Ortiz, S., Quintrel, N., Bishop, J. M. and Varmus, H. E. (1980). DNA intermediates in the replication of retroviruses are structurally (and perhaps functionally) related to transposable elements. *Cold Spring Harbor Symp. Quant. Biol.* **45**, 731–738.

Mayr, E. (1959). Where are we? *Cold Spring Harbor Symp. Quant. Biol.* **24**, 1–14.

Medvedev, Z. A. (1981). On the immortality of the germline: Genetic and Biochemical Mechanisms. A review. *Mech. Ageing and Devel.* **17**, 331–359.

Mullbacher, A., Ashman, R. B. and Blanden, R. V. (1982). Induction of T-cell hyporesponsiveness to Bebaru in mice, and abnormalities in the immune responses of progeny of hyporesponsive males. *Aust. J. Biol. Med. Sci.* **61**, 187–191.

Nasmyth, K. M. and Tatchell, K. (1980). The structure of transposable yeast mating type loci. *Cell* **19**, 753–764.

Nerbert, D. W. (1979). Multiple forms of inducible drug-metabolising enzymes: A reasonable mechanism by which any organism can cope with adversity. *Mol. Cell. Biochem.* **27**, 27–46.

Nishoka, Y., Leder, A. and Leder, P. (1980). Unusual α-globin-like gene that has cleanly lost both globin intervening sequences. *Proc. Nat. Acad. Sci. U.S.A.* **77**, 2806–2809.

Ohta, T. (1980). Evolution and variation in multigene families. *Biomaths* **37**, 1–131.

Orgel, L. E., Crick, F. H. C. and Sapienza, C. (1980). Selfish DNA. *Nature* **288**, 645–646.

Oudin, J. and Michell, M. (1969). Idiotypy of rabbit antibodies. 1. comparison of idiotypy of antibodies against *Salmonella typhi* with that of antibodies against other bacteria in the same rabbits, or of antibodies against *Salmonella typhi* in various rabbits. *J. Exp. Med.* **130**, 595–617.

Pays, E., Van Meirvenne, N., LeRay, D. and Stewart, M. (1981). Gene duplication and transposition linked to antigenic variation in *Trypanosoma brucei. Proc. Nat. Acad. Sci. U.S.A.* **78**, 2673–2677.

Proudfoot, N. (1981). Pseudogenes. *Nature* **286**, 840–841.

Reilly, J. G., Ogden, R. and Rossi, J. J. (1982). Isolation of a mouse pseudo tRNA gene encoding CCA — a possible example of reverse flow of genetic information. *Nature* **300**, 287–289.

Ritossa, F. (1976). The bobbed locus. *In* 'The Genetics and Biology of Drosophila' (M. Ashburner and E. Novitski, *Eds*), Vol. 1b, pp.801–846. Academic Press, London.

Rubin, G. M. and Spradling, A. C. (1982). Genetic transformation of drosophila with transposable element vectors. *Science* **218**, 348–353.

Rubin, G. M., Kidwell, M. G. and Bingham, P. M. (1982). The molecular basis of P-M hybrid dysgenesis: The nature of induced mutations. *Cell* **29**, 987–994.

Sawyer, R. C. and Dahlberg, J. E. (1973). Small RNAs of Rous sarcoma virus: Characterization by two dimensional polyacrylamide gel electrophoresis and fingerprint analysis. *J. Virol.* **12**, 1226–1237.

Schimke, R. T., Kaufmann, R. J., Alt, F. W. and Kellems, R. F. (1978). Gene amplification and drug resistance in cultured murine cells. *Science* **202**, 1051–1055.

Schmid, C. W. and Jelinek, W. R. (1982). The Alu family of dispersed repetitive sequences. *Science* **216**, 1065–1070.

Shapiro, J. A. and Cordell, B. (1982). Eukaryotic mobile and repeated genetic elements. *Biol. Cell* **43**, 31–54.

Shiba, T. and Saigo, D. (1983). Retrovirus-like particles containing RNA homologous to the transposable element copia in *Drosophila melanogaster*. *Nature* **302**, 119–124.

Shimotohno, K., Mizutani, S. and Temin, H. M. (1980). Sequence of retrovirus provirus resembles that of bacterial transposable elements. *Nature* **285**, 550–554.

Skolnick, N. J., Ackermann, S. H., Hofer, M. A. and Weiner, H. (1980). Vertical transmission of acquired ulcer susceptibility in the rat. *Science* **208**, 1161–1163.

Slightom, J. L., Blechl, A. E. and Smithies, O. (1980). Human fetal γ- and α-globin genes: Complete nucleotide sequences suggest that DNA can be exchanged between these duplicated genes. *Cell* **21**, 627–638.

Smith, R. N. (1981). Inability of tolerant males to sire tolerant progeny. *Nature* **292**, 767–768.

Solomon, J. B. (1971). 'Foetal and Neonatal Immunology'. North-Holland, Amsterdam.

Steele, E. J. (1979). 'Somatic Selection and Adaptive Evolution: on the Inheritance of Acquired Characters', 1st edition. Williams-Wallace Productions International, Toronto.

Steele, E. J. (1981a). 'Somatic Selection and Adaptive Evolution: on the Inheritance of Acquired Characters', 2nd edition. University of Chicago Press, Chicago.

Steele, E. J. (1981b). Lamarck and immunity: a conflict resolved. *New Scientist* **89**, 360–361.

Steele, E. J. and Gorczynski, R. M. (1981). Inheritance of acquired somatic modifications of the immune system. *In* 'Cellular and Molecular Mechanisms of Immunologic Tolerance', (T. Hraba and M. Hasek, *Eds*), pp.381–397. Marcel Dekker, New York.

Steinmuller, D. (1967). Behaviour of skin allografts in back cross progeny in tolerant males. *J. Nat. Cancer Inst.* **39**, 1247–1251.

Stroun, M., Anker, P., Maurice, P. and Gahan, P. B. (1977). Circulating nucleic acids in higher organisms. *Int. Rev. Cytol.* **51**, 1–48.

Sutcliffe, J. G., Milner, R. J., Bloom, F. E. and Lerner, R. A. (1982). Common 82-nucleotide sequence unique to brain RNA. *Proc. Nat. Acad. Sci. U.S.A.* **79**, 4942–4946.

Temin, H. M. (1971). The protovirus hypothesis: Speculations on the significance of RNA-directed DNA synthesis for normal development and for carcinogenesis. *J. Nat. Cancer Inst.* **46**, III–VII.

Temin, H. M. (1976). The DNA provirus hypothesis. *Science* **192**, 1075–1080.

Temin, H. M. (1980). Origin of retroviruses from cellular moveable genetic elements. *Cell* **21**, 599–600.

Temin, H. M. and Engels, W. (1984). Movable genetic elements and evolution. *In* 'Evolutionary Theory: Paths into the Future'. (J. W. Pollard, *Ed.*). Wiley, London. pp.173–201.

Van Arsdell, S. W., Dennison, R. A., Bernstein, L. B., Weiner, A. M., Manser, T. and Gesteland, R. F. (1981). Direct repeats flank three small nuclear RNA pseudo-genes in the human genome. *Cell* **26**, 11–17.

Vuillaume, M. and Berkaloff, A. (1974). LSD treatment of *Pieris brassicae* and consequences on the progeny. *Nature* **251**, 314–315.

Waddington, C. H. (1961). 'The Nature of Life'. George Allen and Unwin, London.

Wagner, E. F., Stewart, T. A. and Mintz, B. (1981a). The human β-globin gene and a functional viral thymidine kinase gene in developing mice. *Proc. Nat. Acad. Sci. U.S.A.* **78**, 5016–5020.

Wagner, T. E., Hoppe, P. C., Jollick, J. D., Schell, D. R., Hodinka, R. L. and Gault, J. B. (1981b). Microinjection of a rabbit β-globin gene into zygotes and its subsequent expression in adult mice and their offspring. *Proc. Nat. Acad. Sci. U.S.A.* **78**, 6376–6380.

Wahl, G. M., Padgett, R. A. and Stark, G. R. (1979). Gene amplification causes overproduction of the first three enzymes of UMP synthesis in N-(phosphonacetyl)-L-aspartate-resistant hamster cells. *J. Biol. Chem.* **254**, 8679–8689.

Weinberg, R. A. (1980). Origins and roles of endogenous retroviruses. *Cell* **22**, 643–644.

Weismann, A. (1893). 'The Germ-Plasm: A Theory of Heredity'. Scott Publishing Co. Ltd, London.

Winfield, J. B., Pincus, J. H. and Mage, R. G. (1972). Persistence and characterization of idiotypes of pedigreed rabbits producing antibodies of restricted heterogeneity to pneumococcal polysaccharides. *J. Immunol.* **108**, 1278–1287.

Young, M. W. (1979). Middle repetitive DNA: A fluid compartment of the Drosophila genome. *Proc. Nat. Acad. Sci. U.S.A.* **76**, 6376–6378.

Note Added in Proof

It has been recognized for many years that in animals capable of regenerating gametes from somatic tissue following injury that these gametes must contain somatic mutations and so, in the strictest sense, contravene Weismann's doctrine. Buss (1983. *Proc. Nat. Acad. Sci. USA* **80**, 1387–1391) however, has recently pointed out that the formation of gametes from somatic tissue is the normal mode of development in protists, fungi, plants and at least nineteen phyla of animals. Thus, in these groups, the concept of sequestration of gametes during embryogenesis and their protection from environmental conditions proposed by Weismann, is unfounded and in these examples somatic genetic variations may well be selected for, and become inherited.

Section VI

NATURE AND MIND

13

Artificial Intelligence and Biological Reductionism

MARGARET A. BODEN

Abstract

Anti-reductionism is not anti-scientific if it offers positive suggestions about empirically-based concepts to explain phenomena not explicable in currently accepted terms. Artificial intelligence (AI) is anti-reductionist in a number of respects. Computational concepts can provide explanatory power over and above that of the more basic theories in the life sciences, while being entirely compatible with them. Human and animal intelligence, and psychological phenomena in general, can usefully be thought of in these terms. Behaviourist psychology and neurophysiology cannot express the phenomena concerned, because their vocabulary has no room for the concepts of *representation* or *intentionality*. But AI is concerned with symbol-manipulating systems, and these concepts are theoretically central to it. AI can be useful in physiology, by clarifying what are the computational tasks which the nervous system is performing, and it may even illuminate some aspects of evolutionary and morphogenetic biology. A computational approach to life and mind is entirely compatible with notions of human freedom. By its emphasis on the subject's representation of the world, it counters the mechanization of the world-picture brought about by the natural sciences.

Reductionists usually see anti-reductionists as enemies of science, as a warring intellectual army whose aim is to oust science and substitute mystery. For the reductionist assumes that all problems in a given domain are soluble in terms of available types of concepts, preferred as empirically well grounded and ontologically basic—an assumption the anti-reductionist

BEYOND NEO-DARWINISM
ISBN: 0-12-350080-X

denies. In truth, however, anti-reductionists often wish to oust mystery and substitute science. For the mysterious, though unknowable by familiar methods of enquiry, is potentially intelligible in other ways (hence the promise implicit in titles like *'The Mystery of the Red Room'*). Only the mystical is unknowable *tout court*. Science and mysticism thus make strange bedfellows, but science and mystery can enter into fruitful liaison.

Far from being attempts to oppose or undermine science, anti-reductionist claims may be (though admittedly they are not always) intellectual challenges aimed at strengthening it. Anti-reductionist critiques are partly negative exercises, intended to show the incapacity of current approaches to express or explain the problematic phenomena. But ideally, the anti-reductionist will also offer positive suggestions about what alternative types of concept might be better suited to this task. If these concepts are clear, empirically fruitful at least on their own level, and grounded in or demonstrably compatible with the more basic scientific concepts favoured by the reductionist, they may hope to be welcomed into the scientific community instead of being shunned as intellectual outlaws.

With respect to living things, the reductionist temperament shows itself in such assumptions (for instance) as that neo-Darwinism can answer all evolutionary questions, molecular biology all questions about individual morphogenesis, stimulus-response psychology all questions about behaviour, and neurophysiology all questions about the mind. These assumptions are sometimes made explicit. More often, they remain unspoken — but they are no less powerful for that.

Such forms of biological reductionism have been countered by various anti-reductionist critiques, some of which are represented in this volume. The negative poles of these critiques have much in common, but their positive aspects are more diverse. They do not deny the truth of neo-Darwinism, but see it as conceptually inadequate for addressing certain important biological problems. Positively, they offer other concepts they regard as better suited to these problems. These differ from case to case, but in general they are not *alternatives* to the reductionist's conceptual base so much as theoretical *complements* to it.

I shall concentrate on the positive exercise, suggesting ways in which *computational* concepts (drawn from computer science and artificial intelligence, or 'AI') might offer explanatory power that is not provided by the more basic theories in the life-sciences — theories with which these concepts are nonetheless entirely compatible. They can be useful to biologists, to physiologists, and to ethologists, but their most obvious application is to psychology: the science not of body, but of mind.

The reason for this is that AI uses the methodology of computer programming to study the content and function of various sorts of knowledge (Boden, 1977). It focuses on the structure and organization of internal models, or symbolic representations, and on how these can be transformed in complex and flexible ways so that the representational system can cope with changing and largely unknown situations. It is this flexibility, and the large degree of autonomy that it involves (so that the system can achieve its ends, within limits, irrespective of external conditions), that constitute *intelligence.*

Accordingly, AI programs are quite unlike the more familiar types of computer program (dealing with wages, tax-returns and the like), which are rigid, unimaginative, and inflexible—in a word, stupid. AI programs specify computations enabling computers to do such things as: recognizing objects seen in widely varying positions or lighting conditions; planning complex tasks involving unpredictable conditions; conversing (by teletype) in natural language; understanding spoken speech; making sensible guesses where specific knowledge is not available . . . and the like. It should be evident from these examples that 'computation' here does not mean 'counting', but *any* symbolic process of inference, comparison, or association. The symbolism may be numerical (for counting is one example of computation), or it may be of some other form (such as verbal, visual, or logical).

AI is sometimes thought of as a branch of cybernetics, and if one accepts Wiener's (1948) broad definition of cybernetics as 'the science of control and communication in the animal and the machine', then so of course it is. Cybernetics itself was an anti-reductionist exercise. Negatively, it claimed that the concepts of physics and chemistry alone could not express the nature and functioning of homeostatic control in biological organisms or man-made mechanisms. Positively, it showed how these matters could be expressed, and rigorously discussed, in terms of new—information-processing—concepts which were logically independent of their embodiment in a given body or artifice.

That a steam-engine needs a part functioning as a governor is a truth of cybernetics. That the governor is embodied in two metal balls linked to a rotating shaft is, rather, a truth of nineteenth century engineering—for *any* mechanism with equivalent functional properties would do. The physics of centrifugal forces explains how it is possible for the steam-governor to do its job, much as biochemistry explains how it is possible for blood-sugar or body-temperature to be regulated. But physics and chemistry concern the possibilities for embodiment of cybernetic principles, rather than their essential nature. That must be expressed at a different level, in information-processing terms defining signals and feedback.

Wiener himself suggested that not only biological functions, but psychological regulations too, could be cybernetically explained. In an influential discussion of behaviour, purpose, and teleology written with biologist colleagues, Wiener used the key terms so that teleological or purposive behaviour was synonymous with 'behaviour controlled by negative feedback', by which he meant behaviour that is 'controlled by the margin of error at which the [behaving] object stands at a given time with reference to a relatively specific goal . . . The signals from the goal are used to restrict outputs which would otherwise go beyond the goal' (Rosenblueth, Wiener and Bigelow, 1943, p.19). A paradigm case of 'purposive' behaviour on this account would be, he said, that of a machine designed to track a moving luminous goal, or target.

But truly purposive behaviour involves more than this. It is guided by internal representations of the 'goal', of plans of action for reaching the goal, and of various criteria (including moral values and personal preferences) for evaluating these plans, wherein the overall goal may not correspond at all closely to *any* external state of affairs or target (Boden, 1972).

For example, people can engage in the highly complex behaviour of going on a unicorn-hunt, and several centuries ago they commonly did so. But, clearly, we could not explain their activities in terms of feedback signals from the fleeing unicorns — for there never were any unicorns. Rather, our explanations must be in terms of the hunters' ideas and inferences about unicorns, including their beliefs that unicorns are likely to be found in forests, their heads resting in the laps of virgins, and that their horns have magical properties. We need to show how these ideas and inferences might be represented, organized, and accessed, in such a way as to generate the behaviour — and experience — of the people concerned.

These are computational matters, in the sense defined above. They cannot be expressed in the terminology of traditional cybernetics, which focuses on feedback and adaptive networks, and which defines information-processing in quantitative rather than qualitative terms. It could not express the difference between a mermaid and a unicorn, for instance, nor the very different planning activities involved in hunting them (does one head for the seaside or for a forest?). To understand intelligence we need computational concepts, specifically designed to define complex transformations within qualitatively distinct symbolic representations such as these.

Such concepts are supplied by AI, for it is concerned with symbol-manipulating systems wherein the degree — and the nature — of match-mismatch between representation and reality (or even possibility) may vary enormously. So AI differs importantly from classical cybernetics, and is helpfully described as a 'cybernetic' enterprise only in the broadest sense.

AI then, sees minds (whether animal or human) as symbolic systems containing many internal representations of aspects of the world (and possible worlds), and a variety of rules for building, changing, comparing, and inferring from them. Psychological questions, on this view, concern the structure and content of mental representations, and the ways in which they can be generated, augmented, and transformed. Thinking, experience, and motivation — and psychological differences between individual people — are all grounded in computational processes.

One main strength of the AI-approach to psychology is its emphasis on rigor. AI offers precisely definable concepts, because a program has to be expressed clearly (as a set of instructions defining specific symbol-manipulations) if the computer is to accept it. In aiming for rigor — indeed, in setting a new standard for rigor in psychological explanation — AI differs from other anti-reductionist approaches to the life sciences. The positive sides of these critiques commonly invoke concepts which, while they may have some phenomenological plausibility and intellectual resonance, are not expressed with clarity and are therefore difficult to apply in a scientific research-programme (Lakatos, 1970). One example is Piaget's concept of 'equilibration' mentioned below.

Another strength of the computational approach to psychology is its highlighting of *process* as well as structure. This derives from the fact that a program has to tell the computer not only what result to produce but also *how* to produce it. Non-computational psychologists often take psychological change for granted, assuming that it can be sufficiently specified by stating the initial and final mental states involved. However, the process of mental transformation is itself theoretically problematic. In a programming context, a failure to suggest any way in which the change might be effected will show up as a glaring gap in the program, a gap over which the uninstructed computer is unable to leap. Some computational account of how to make the leap must be supplied if the program is to function.

AI is relevant not only to psychology but to neurophysiology also, and this in at least two ways. First, it helps us achieve a clear account of what computational tasks the brain and peripheral nervous system must be performing. This can help the neurophysiologist in formulating questions and hypotheses about which bodily mechanisms are performing these tasks, and how. Second, AI may help us understand the potential and limitations of the computational or representational power of various sorts of hardware, including neurophysiological mechanisms.

These two 'physiological' uses of AI differ from the familiar use of computer technology to model physiological theories, simulate nervous nets, *etc.* In those exercises, the prime intellectual input comes from the

concepts already developed in neurophysiology. But in AI-based studies of
the sort I have in mind, a crucial part — sometimes the larger part — of this
input is supplied by computational research as such.

The first of these physiological advantages has been stressed by David
Marr, whose early theoretical papers on the cerebellum and cerebrum
attracted the interest of brain scientists, and whose more recent
(computational) work on the visual system is deservedly one of the strongest
current influences in AI (Marr, 1982). As Marr put it when embarking on
his computational research-programme:

> The situation in modern neurophysiology is that people are trying to under-
> stand how a particular mechanism performs a computation that they cannot
> even formulate, let alone provide a crisp summary of ways of doing. To rectify
> the situation, we need to invest considerable effort in studying the
> computational background to questions that can be approached in neuro-
> physiological experiments.
>
> Therefore, although [my work] arises from a deep commitment to the goals
> of neurophysiology, the work is not about neurophysiology directly, nor is it
> about simulating neurophysiological mechanisms: it is about studying vision.
> It amounts to a series of computational experiments, inspired in part by some
> findings in visual neurophysiology. The need for them arises because, until
> one tries to process an image or to make an artificial arm thread a needle, one
> has little idea of the problems that really arise in trying to do these things.
>
> *(Marr, 1975, p.31)*

Clearly, these problems arise for the human being trying to thread a needle
or interpret a visual scene, but we are not aware of them because they are
solved at computational levels beyond the reach of consciousness.

Doubtless, our evolutionary heritage has provided us with bodily
mechanisms especially well suited to such tasks, which are the particular
focus of neurophysiology. But to find out just what they are, what
they do, and how they function, we need to understand the task itself.
And since 'task' is essentially a computational concept, albeit one of
everyday English, we must understand the task in computational terms. In
short, Marr's is an anti-reductionist exercise in the sense outlined earlier:
physiology alone, physiology unaided by computational insights, cannot
suffice to explain behaviour or experience.

Marr puts much greater emphasis on neurophysiology, as constraining
the details of computational research, than do some other workers in AI.
For instance, Marr's research on vision considers psychological optics in
some detail and makes a concerted attempt to use (not merely to bear in
mind) knowledge about the anatomy and physiology of the retina and early
stages in the visual pathways. This is not simply because he happens to be
interested in neurophysiology and biological mechanisms whereas others

are not. For some AI-workers on vision consider optics only in very general terms and ignore neurophysiology, on the *principled* ground that physiological (hardware) implementation is theoretically independent of questions about computational mechanisms. This is a widely shared view in AI, and many physiologists are sceptical about AI accordingly.

In principle the view is true, as Marr himself admits (see the quotation above). It is analogous to the fact that the engineer's problem in actually making a governor is theoretically distinct from the cybernetician's problem in identifying the abstract principles of gubernatorial control. But in practice, it may be that we need to study the varying computational powers of distinct (electronic or physiological) hardware if we are to understand how tasks are performed by finite mechanisms in real time. Even if this is agreed, however, doing it is no simple matter. Within AI there are real, and largely unresolved, differences of opinion about the computational potential of relatively low-level, peripheral mechanisms (most of which seem to be biologically specialized to a high degree) as opposed to high-level cerebral mechanisms (which appear rather to be general-purpose in character).

This brings us to the second point noted above, that AI could help us appreciate what computational powers an organism might have *in virtue of its neurophysiology*. This would be useful to a wide range of biologists, not just to 'physiologists' narrowly defined. I have discussed elsewhere some implications germane to a range of ethological problems, concerning both motor action and perception (Boden, 1983a). One relevant example is Hinton's recent research, which is a significant advance in the computational modelling of vision (Hinton, 1981).

Hinton's work is focused on the computation of shape. Most AI vision-workers have assumed that the shapes of three-dimensional objects *have to be* computed by way of high-level, top-down processes. But Hinton has shown that this is not so. (It does not follow, of course, that top-down processes *are not* used by humans or other animals in the perception of shape.) He has designed low-level, dedicated hardware, mechanisms that are capable of cooperative computation, or parallel processing. These systems can perform shape discriminations—such as recognition of an overall Gestalt—commonly believed (even within AI) to require relatively high-level interpretative processes.

Hinton's mechanisms rely on excitatory and inhibitory connections between computational units on various levels that appear to have an analogue in the nervous connectivity of our own visual system. He not only provides an *example* of a mechanism that can compute shape in a surprising fashion, but also gives a *general proof* that many fewer computational units are necessary for the parallel computation of shape than one might initially have supposed. This proof lends some more physiological weight to the model, since the human retina apparently has enough cells to do the job.

Hinton's results suggest also that the way in which an object is represented may be radically different, depending on whether it is perceived as an object in its own right or as a part of some larger whole. This might account for the phenomenological differences between perceptual experiences of which we are reminded by those philosophers (e.g. Dreyfus, 1972) who argue that AI is essentially unfitted to model human minds—or bodies. Hinton believes that his computational model of spatial relations can be used to understand motor control in a way analogous to the mechanisms of muscular control in the human body. His preliminary (unpublished) work suggests an efficient procedure for computing a jointed limb's movements and pathway through space—a problem that can be solved by traditional computing techniques only in a highly inefficient manner.

Many commonly-expressed philosophical criticisms of AI and cognitive psychology, like that of Dreyfus just mentioned, may be invalidated by these recent computational developments. Such criticisms are explicitly anti-reductionist in nature, accusing AI itself of attempting to reduce all psychological phenomena to forms of serial processing. Evidently, to be an 'anti-reductionist' can involve highly varying claims, and one may consistently support one form of anti-reductionism without endorsing another.

If AI is relevant to psychology, to physiology, and to ethology, what of biology more generally conceived? The idea that a computational approach might contribute to theoretical biology is not entirely novel, for various biologists have made comparable suggestions:

Language . . . I suggest may become a paradigm for the theory of General Biology.

(Waddington, 1972, p.289)

The view of the organism as an hypothesis-generating and testing system . . . could transform biology by placing model construction and observation at the centre of the biological process, not at the evolutionary periphery, the phenomenon of Mind.

(Goodwin, 1972, p.267)

The classical cases of pattern regulation whether in development or regeneration . . . are largely dependent on the ability of the cells to change their positional information in an appropriate manner and to be able to interpret this change.

(Wolpert, 1969, p.8)

The problems of biology are all to do with *programs*. A program is a list of things to be done, with due regard to circumstances.

(Longuet-Higgins, 1969, p.229)

All these quotations except the last rely on familiar cognitive terms such as 'language', 'knowledge', 'interpretation', or 'information', rather than the more technical 'program' or 'computation'. But, as I have argued elsewhere (Boden, 1981), their approach to theoretical biology is significantly similar to that of people influenced by AI. And all, of course, are anti-reductionist in the sense defined above.

Piaget is another example of someone who sought to complement the more basic biological concepts by others of a higher theoretical level. His lifelong opposition to reductionism in biology, as well as in psychology, was grounded in his interest in development. Developmental change brings about forms that are in some sense novel. In his account of 'equilibration', Piaget tried to illuminate the way in which new, more differentiated, structures arise out of simpler ones, whether in psychological, embryological, or evolutionary development. He recognized the profound theoretical problem of how it is possible for harmonious structural novelties to develop, along with novel integrative mechanisms whereby the overall regulation of the system is maintained. And he tried to give an account of spontaneous, as opposed to reactive, structural development. However, his concept of equilibration was so vaguely expressed that it provided no clear questions, still less clear answers (Boden, 1982).

Computational studies have not given clear answers either. But they may help give us a better sense of the sorts of questions that need to be addressed in understanding adaptive development, whether in body or mind. From a computational point of view, the generation of new forms, as well as their consequent evaluation, must take place within certain structural constraints (Boden, 1983b). In general, to understand any sort of development or creativity would be to have a theory of the various transformations, at more or less basic levels, by which the relevant structural potential can be selectively explored.

A theory of morphological development, for instance, would explain how it is possible for a gill-slit to be transformed into a thyroid gland, or a normal blastula into a deformed embryo or non-viable monster; also, it would explain why certain fabulous beasts (such as mermaids, but *not* unicorns) could only have been imagined, not created. In biology as in psychology, we need some account of generative processes that can explore the space defined by background constraints, and of more truly creative processes that can transform these constraints themselves.

If an adaptively developing system is to be able to judge the 'interest' of its explorations, it must have some form of evaluation criterion. Lenat's 'automatic mathematician' is an AI program designed to shed light on this matter with respect to the example of mathematical creativity (Lenat, 1977). The program starts with elementary set-theory, generates new concepts, and decides which to explore further. It uses several hundred heuristics (not just a few transformational rules) to explore the space defined by a hundred primitive concepts of set-theory. From this elementary base the program generates and follows up various concepts of number-theory (including one minor theorem that had not previously been defined). Among its heuristics are some which evaluate the (mathematical) interest of newly generated concepts.

In biological, as opposed to psychological, development it is natural selection that functions (*post hoc*) as the evaluative criterion. This fundamental tenet of neo-Darwinism is not in question. But there are various biological phenomena which, on a neo-Darwinist account of evolution, are very puzzling. For example, the fraction of DNA that does not code for the synthesis of specific proteins increases phylogenetically, and species have evolved remarkably quickly. Work in automatic programming has suggested that such facts are not explicable in terms of the neo-Darwinist mutational strategy ('Random-Generate-and-Test'), because its combinatorics are horrendous (Arbib, 1969). Instead, some strategy of 'Plausible-Generate-and-Test' is needed, whereby mutations of a type likely to be adaptive become increasingly probable. The initial heuristics evolve by random mutation and natural selection, but—since they are embodied as DNA and their 'target' for interpretation is itself DNA—they can then develop by modifying each other. They function as heuristics recommending certain 'copying errors' and preventing others, and the transformational processes they influence are gene substitution, insertion, deletion, translocation, inversion, recombination, segregation, and transposition.

These genetic transformations have been discussed in the light of procedures known in AI as 'Production Rules' (Lenat, 1980). Production Rules are IF–THEN rules specifying a certain action, given a particular condition. Lenat suggests that the IF . . . part of the heuristic might be specified by proximity on the DNA molecule, whereas the THEN . . . part could direct gene rearrangement, duplication, placement of mutators and intervening sequences, and so on. For instance, one heuristic might be that gene recombinations should involve neighbour-genes rather than genes at opposite ends of the DNA string: in a creature where genes for morphologically related structures happened to lie next to each other, this heuristic would encourage mutations of both genes together, which would tend toward a structurally integrated evolution. Although these ideas are highly speculative, and their value is not proven, they suggest that concepts drawn from AI might be useful in thinking about biological evolution.

But perhaps AI, in relation to the life sciences, is not mere harmless speculation but rather a dangerous illusion? Certainly, many people assume that AI simply cannot be a humanely anti-reductionist project. Indeed, they scorn AI as an essentially inhuman *reductionist* influence, one that offers an 'obscene' and 'deeply humiliating' image of man (Weizenbaum, 1976).

In general, anti-reductionists commonly complain that reductionist approaches to life and mind somehow rob us of our humanity—to which reductionists are unfortunately apt to retort that theoretical emphasis on 'humanity' is no more than self-indulgent sentimentality. Skinner's attacks on notions like 'freedom' and 'dignity' are notorious, and the molecular biologist Monod has managed to *épater les bourgeois* by such relentlessly reductionist remarks as: 'The cell is a machine. The animal is a machine. Man is a machine'. Psychological approaches based on AI also assume that man is a machine, even daring to compare the mind with the metallic artefacts of the electronics workshop. So how can AI be anything but a form of reductionism? If man is a machine, what room is there for humanity?

If Monod's remark 'Man is a machine' is taken to mean that the bodily processes underlying and generating human behaviour and experience (including moral conduct and insights) are describable by physics or molecular biology, it is—as far as AI is concerned—true. But if it is taken to mean that the concepts of natural sciences such as these suffice to express the nature and regulation of the human mind, then—according to AI—it is false.

Much as the steam-engine governor cannot be described, *qua* governor, in physical terms but only in cybernetic language, so the mind cannot be conceptualized physiologically, but only computationally. Describing a system (whether person or computer) *as* a symbol-manipulating system is conceptually distinct from describing the physical hardware that embodies the computational powers concerned. The former type of description requires computational concepts, whereas the latter employs the terms of physics, chemistry, and physiology. As the poet Blake foresaw, these natural sciences have had a dehumanizing influence, encouraging a 'single vision' that has insidiously undermined people's sense of personal autonomy and responsibility. This is unsurprising, for no science that lacks the concept of *representation* can even acknowledge humanity, still less explain it.

Computational psychology does not support the mechanization of the world-picture brought about by the natural sciences, and by 'scientific' forms of psychology such as behaviourism. For, unlike these, it can distinguish 'subjective' truths (about ideas, aspirations, and beliefs) from 'objective' truths (about brains and other physical things). What is more, it concentrates firmly on the former, admitting the influence on our lives of shared cultural beliefs, of individual ideas, interests, purposes, and choice, and of self-reference and self-knowledge (Hofstadter, 1980).

In sum, AI emphasizes the richness and subtlety of our mental powers, a richness that has often been intuitively glimpsed (at least by literary artists) but never theoretically recognized by scientists. Many humanists reject the reductionist influences of natural science with such passion that they come close to adopting mysticism. There may be good reasons for embracing mysticism, but the undeniable inability of natural science to conceptualize subjectivity is not one of them. Nor is the 'wholistic' integration of biological phenomena such as embryogenesis or regeneration. In providing rigorous hypotheses about the mental processes that underlie humanity and make it possible, AI promises a scientific understanding of the most recent product of evolution, intelligence. Its promise has yet to be fulfilled, but the crucial point is that AI sees mind not as mystical, merely as mysterious.

REFERENCES

Arbib, M. (1969). Self-Producing Automata—Some Implications for Theoretical Biology. In 'Towards a Theoretical Biology' (C. H. Waddington, Ed.), Vol. 2, pp.204–226. Edinburgh University Press, Edinburgh.

Boden, M. A. (1972). 'Purposive Explanation in Psychology'. Harvard University Press, Cambridge, Massachusetts.

Boden, M. A. (1977). 'Artificial Intelligence and Natural Man'. Basic Books, New York.

Boden, M. A. (1981). The Case for a Cognitive Biology. In 'Minds and Mechanisms: Philosophical Psychology and Computational Models' pp.89–112. Cornell University Press, Ithaca, New York.

Boden, M. A. (1982). Is Equilibration Important? Br. J. Psychol. **73**, 165–173.

Boden, M. A. (1983a). Artificial Intelligence and Animal Psychology. New Ideas in Psychology, **1**. (In press)

Boden, M. A. (1983b). Failure is Not the Spur. In 'Adaptive Control in Ill-Defined Systems' (O. Selfridge, M. Arbib and E. Rissland, Eds). Plenum Press, New York. (In press)

Dreyfus, H. L. (1972). 'What Computers Can't Do: A Critique of Artificial Reason'. Harper and Row, New York.

Goodwin, B. C. (1972). Biology and Meaning. In 'Towards a Theoretical Biology' (C. H. Waddington, Ed.), Vol. 4, pp.259–275. Edinburgh University Press, Edinburgh.

Hinton, G. E. (1981). Shape Representation in Parallel Systems. 7th Int. Joint. Conf. A.I. (Vancouver), 1088–1096.

Hofstadter, D. (1980). 'Godel, Escher, Bach: An Eternal Golden Braid'. Penguin, Harmondsworth.

Lakatos, I. (1970). Falsification and the Methodology of Scientific Research Programmes. *In* 'Criticism and the Growth of Knowledge' (I. Lakatos and A. Musgrave, *Eds*), pp.91–196. Cambridge University Press, Cambridge.

Lenat, D. B. (1977). Automated Theory Formation in Mathematics. *Proc. 5th Int. Joint Conf. Art. Int. (Cambridge, Massachusetts)*, 833–842.

Lenat, D. B. (1980). The Heuristics of Nature: The Plausible Mutation of DNA. Stanford University Computer Science Dept., Report HPP-80-27.

Longuet-Higgins, H. C. (1969). What Biology is About. *In* 'Towards a Theoretical Biology' (C. H. Waddington, *Ed.*), Vol. 2, pp.227–236. Edinburgh University Press, Edinburgh.

Marr, D. (1975). Analyzing Natural Images: A Computational Theory of Texture Vision. MIT AI Dept. AI-Memo 334, Cambridge, Massachusetts.

Marr, D. (1982). 'Vision'. MIT Press, Cambridge, Massachusetts.

Rosenblueth, A., Wiener, N. and Bigelow, J. (1943). Behavior, Purpose, and Teleology. *Philosophy of Science* **10**, 18–24.

Waddington, C. H. (1972). Epilogue. *In* 'Towards a Theoretical Biology' (C. H. Waddington, *Ed.*), Vol. 4, pp.283–289. Edinburgh University Press, Edinburgh.

Weizenbaum, J. (1976). 'Computer Power and Human Reason: From Judgment to Calculation'. Freeman, San Francisco.

Wiener, N. (1948). 'Cybernetics, or Control and Communication in the Animal and the Machine'. Wiley, New York.

Wolpert, L. (1969). Positional Information and the Spatial Pattern of Cellular Differentiation. *J. Theor. Biol.* **25**, 1–47.

14

A Socio-Naturalistic Approach to Human Development

CHRIS SINHA

Abstract

The epigenetic approach to development and evolution is contrasted with the dominant paradigms in both evolutionary biology and cognitive science, which are identified respectively as neo-Darwinism and neo-rationalism. An outline and critique of computational approaches is presented, focusing upon the concept of representation. The notion of 'ecological validity' is examined with respect to social action-theoretic and direct perception approaches to cognition, communication and perception. The epigenetic psychobiologies of James Mark Baldwin and Jean Piaget are contrasted. An outline of a socio-naturalistic, epigenetic theory of the relations between behaviour, representation and adaptation in evolution and development is presented.

INTRODUCTION

There is a certain paradox in the inclusion of this chapter in a volume entitled 'Beyond Neo-Darwinism'. The paradox consists in the fact that the critical appraisal I shall offer of contemporary psychology hinges upon the contention that, even today, insufficient attention is paid by psychologists

Some of the arguments in this chapter will be presented more fully in my forthcoming book, 'Language and Representation: a socionaturalistic approach to conceptual development'; to be published by the Harvester Press.

to the importance of embedding theories of human mental processes within the theories and findings of evolutionary biology. At the same time, I wish to argue for an interpretation of evolutionary and human developmental processes which departs significantly from both of the two 'modern syntheses' which currently hold respective sway in the disciplines of evolutionary biology and cognitive psychology.

The first of these modern syntheses, or dominant paradigms, will be familiar to biologists as neo-Darwinism. That is, the so-called 'central dogma' that the only source of evolutionary change is natural (and sexual) selection of randomly-generated, small scale variations in genetic material which is hermetically sealed from environmental perturbation; supplemented by the findings of modern population genetics, and the model-theoretically derived mechanisms advanced by the sociobiologists, such as kin selection, differential reproductive strategies etc. (Wilson, 1975).

The second paradigm will be less familiar, even to psychologists, although most current work in cognitive psychology takes place within it. Though younger, in its modern-synthetic form, than neo-Darwinism, it has a longer ancestry. I propose to call it the *neo-rationalist paradigm*, to indicate its origins in the central debate in the recent history of psychology, which has been between the nativist and rationalist camp, as represented for example by Noam Chomsky and J. A. Fodor, and the environmentalist-associationists in the tradition of J. B. Watson and B. F. Skinner. Psychologists will be familiar with this debate, and with the resultant rout of traditional environmentalism. Biologists may not be so well acquainted with it, so for their benefit I shall shortly attempt a thumbnail sketch of its main features.[1].

There are two positive reasons for denoting the dominant paradigm in psychology (or cognitive science: which latter embraces linguistics, philosophy of mind, neuroscience and artificial intelligence, as well as psychology) as neo-rationalist. The first reason is that a modern synthesis, so closely bound up with the rise of an inter-disciplinary cognitive science as to be virtually identified with it, has in fact emerged and become consolidated in recent years into a coherent account of human behaviour and mental processes. This account, combining elements of the methodology and philosophy of traditional experimental psychology with the nativist and mentalist orientation of traditional rationalism, and enriched by insights yielded by the development of computer science, is what I call neo-rationalism.

The second reason for creating the category of neo-rationalism is to re-position it in relation to an alternative paradigm, also historically long standing, which happens to stand in relation to neo-rationalism in precisely the same way as it does to neo-Darwinism: as the only serious and coherent alternative account. This alternative is *epigenetic naturalism*. Thus, my argument presupposes the paradigmatic opposition shown in Fig. 14·1.

Evolutionary biology	Cognitive science
Neo-Darwinism	Neo-rationalism
Epigenetic naturalism	

FIG. 14·1 Alternative paradigms in evolutionary and cognitive science.

The distinction between 'rationalist' and 'naturalist' psychologies was first employed in recent times by Fodor (1980), and his use of it appears to have commanded broad consent from both camps. The main bone of contention seems to be that 'rationalists' and 'naturalists' alike deny kinship with 'behaviourism'. In fact, I shall suggest that classical learning theory is best seen as part of a general 'reflexological synthesis' in early twentieth century psychobiology, combining elements of both empiricism and naturalism. A consequence of the demise of classical empiricism as an adequate theory of human mental processes has been the break-up of the reflexological synthesis, its replacement by the new synthesis of neo-rationalism, and a widening gulf between psychological and biological frames of explanation.

The term 'neo-rationalism' highlights the historical filiation of modern 'rational psychology' (Fodor's term) with the dominant tendency of Western thought, which has oscillated between 'empiricist' and 'rationalist' solutions to the same *sort* of questions, approached within the same presuppositional framework. As Fodor (1980:64) himself notes, 'there's a long tradition, including both Rationalists and Empiricists, which takes it as axiomatic that one's experiences (and, a fortiori, one's beliefs) might have been just as they are even if the world had been quite different from the way that it is'. This 'contingency assumption' lies at the heart of both empiricist and rationalist theories, and is, I shall argue, enshrined in the modern synthesis in the guise of the competence–performance distinction.

If the intellectual precursors of neo-rationalism include Aristotle, Descartes, Leibniz, Locke and Hume, those of epigenetic naturalism include Heraclitus, Spinoza, Goethe, Hegel and Marx. The 'modern' version of the epigenetic approach has been fashioned largely by such figures as C. S. Pierce, J. M. Baldwin, L. S. Vygotsky, G. Bateson and C. H. Waddington. The most systematic and prolific exponent of (epi)genetic epistemology as a unified science of life and mind has been Jean Piaget; as will become clear, however, certain aspects of Piaget's theory indicate that he, like his greatest mentor, Kant, uncomfortably straddles the rationalist *versus* naturalist divide. Be that as it may, Piaget's theory represents a fundamental reference point for the articulation of an alternative, epigenetic psychobiological paradigm. To conclude this introduction, I shall note that I use, from time to time, the terms 'cognitivist' and 'computational' to denote, respectively, the theoretical and methodological propositions which are programmatically fused in the modern neo-rationalist synthesis. 'Cognitivism', interpreted widely, consists in the proposition that intelligent behaviour can only be explained 'by appeal to internal cognitive processes, that is, rational thought in a very broad sense' (Haugeland, 1978:215). 'Computational' approaches are those which, as the name implies, see cognitive processes as analagous with computer programs.

NEO-RATIONALISM: FROM REFLEXOLOGY TO REPRESENTATION

The nature of cognitive science's 'modern synthesis' can best be explained by outlining the presuppositions and goals of its research programme. This is directed at nothing less than a complete understanding of the nature of human mental processes, and the neural mechanisms underlying the translation of these mental processes into behaviour. Thus, cognitive science has decisively liberated itself from the behaviourist straightjacket, which precluded any explanation which moved beyond the description of observable behaviour. Against classical reflexology, which proposed a direct causal link between brain/nervous system and behaviour, the modern synthesis asserts the necessity of the study of *mind* as an autonomous level, possessing causal efficiency with respect to behaviour, and circumstantially constrained to, rather than epiphenomenal of, lower level neural and biochemical processes.

The modern synthesis, then, is anti-reductionist and mentalistic. It studies human behaviour in order to reveal the workings of mind, the structures and processes permitting the operations of reasoning and the manipulation of symbolic information: its encoding, storage, retrieval, transformation and transmission. The fundamental premise of cognitive science is that human behaviour is *rule governed* and *generative*. That is to say, transformation rules intervene between different stages in coding processes in order to permit goal-directed problem solving.

Representation

A central concept in cognitive science is representation. Information may formally be represented in different ways, in order procedurally to effect various strategies directed to specified goals. The mode of specification of the goals of strategies and procedures is itself representational: an end-state in a chain of transformations which may lawfully be derived from the properties of the formal system constituting the representational 'store' (syntactic rules, semantic memory, visual-spatial imagery etc.). The human (or representational) subject is conceived of as possessing a 'mental model' (Johnson-Laird, 1980) of the world, or world sectors, and cognitive processes consist in operations carried out upon the model(s).

The concept of representation is extended, in the modern synthesis, in two further ways. First, the representations constituting the cognitive subject are assumed themselves to be *represented*, or instantiated, in the functional architecture of the brain and nervous system, and in neurochemical processes. Thus, the brain itself is seen as a *representation* of mind. This inverts the traditional reductionist view, in which 'mind' was seen as a mere trace of *neural* process. In the modern synthesis the brain is rather an embodiment or representation of *mental* process.

Computation

Further, it is no longer assumed that the *particular* instantiation of mental processes embodied in human brains is either necessary, or constitutive of mind. Mental processes may equally be instantiated in artificial physical systems, such as computers. I shall refer to this assumption of the autonomy of cognitive process from neural process as the *contingency assumption*. As I shall argue, the contingency assumption is absolutely basic to the modern neo-rationalist synthesis, and profoundly differentiates it from epigenetic and other naturalisms.

The contingency assumption underlies the well known distinction between (applied) simulation studies, on the one hand, and (pure) 'artificial intelligence', or more generally, computational models, on the other. Simulations are, in principle, constrained to empirical evidence in the psychological or neurological domain which they are intended to 'model' in a very literal sense. Let us refer to simulation as Modelling$_1$. It can then be contrasted with Modelling$_2$, in which the 'model' is constrained, not to neuropsychological evidence, but to (internal) consistency criteria, and (external) correspondence to a formal object, such as a grammar.

Although both Modelling$_1$ and Modelling$_2$ may be implemented in computer programs, I shall restrict the term *computational modelling* for

Modelling$_2$, since the latter, rather than using computation as an adjunct to the description of behaviour, sees computation as itself a fundamental model of cognitive processes. The extension of the concept of representation to include the brain-as-a-representation is closely linked, therefore, to a second extension of representation to include, or be synonymous with, *computation*. An essentially computational model of mental representation is presupposed by much work in cognitive science, in which, to put it simply, representational content is subordinated to computational form. The representational system is seen as being constituted by *rules*, of a formal and computational nature, and the elements upon which the rules operate are symbolic values, or strings of symbolic values, which are themselves defined over the formal rule-system.

Modelling$_2$, though it is bound by formal and theoretical considerations with respect to cognitive processes, is not bound to *behaviour* in the way in which Modelling$_1$ is, since it is assumed that (contingent) processing limitations, inherent in human neurobiology, intervene between mental processing and behavioural outcome. Thus, the contingency assumption also underpins the distinction between *competence* (constraint to a formal object guaranteeing internal consistency), representing the 'ideal' speaker-hearer, or more generally cognitive subject; and *performance* (constraint to 'contingent' limiting conditions on processing), representing the 'actual' human subject.

Language and mind

It is evident from the foregoing that modern cognitive science is deeply indebted to theoretical linguistics. Indeed, if one were to assign a birthdate to cognitive science, as good a one as any would be the publication of Chomsky's (1959) devastating review of Skinner's *Verbal Behaviour*. To return to the arguments advanced in that paper is to gain a proper perspective upon the achievements of cognitive science on its road to the modern synthesis, and an appreciation both of its theoretical richness and rigour, and of the poverty of the radical behaviourism which it has superseded.

Chomsky makes a number of points in this article and elsewhere, not all of which are relevant to the concerns of this chapter. The most important are the following:

(1) Utterances in a natural language are in some sense occasion-independent and non-stimulus bound, from a grammatical viewpoint. Syntax is not a probabalistic phenomenon, either in terms of linguistic context, or in terms of non-linguistic context.

(2) For any natural language, there are an infinite number of possible

utterances (in a non-trivial sense; that is, one that is independent of particular lexical facts such as the infinite series of integers or proper names).

(3) This *generativity* of natural languages is a consequence of their grammars, and thus grammar is definitive of natural language.

(4) There are certain universal properties of the grammars of natural languages.

(5) These universals are not manifest in the actual forms of utterances/ sentences (surface structure), but reside either in the grammatical deep structure or in the generative and transformational rules relating deep and surface structures. Thus, neither grammars nor the universal properties of grammars are phenomenologically accessible to subjects.

(6) To know a language is (definitively) to know its grammar (*see* 3).

(7) Neither grammars nor the universal properties of grammars are phenomenologically accessible to children acquiring language (*see* 5).

(8) Therefore, humans must be innately equipped with a pre-programmed Language Acquisition Device (LAD), specifying the universal constraints on grammars, and permitting the child to transduce the grammar of his/her native language from the 'degenerate' input of the actual utterances of other speakers.

This argument, to which I hope I have done sufficient justice, may be referred to as the prototypical 'rationalist deduction'. It both demonstrates the necessity for, and demarcates the research programme for, a new cognitive science of the mind and its universal laws and properties: a universalistic, nomological and, as I shall try to show, individualistic science of the mind. Since this chapter is about neither linguistics nor psycholinguistics, I do not wish either to endorse or to criticise the deduction at this stage. Rather, I shall try to draw out some of its implications.

Modularization and maturation

The modern synthesis, resting upon the twin pillars of the contingency assumption and the rationalist deduction, is perhaps best typified in terms of its answers to two perennial questions. The first question is: what is the relationship between mind and brain? The second question is: what is the relationship between experience and knowledge (or understanding)? To these two questions, the modern synthesis answers, respectively, *modularization* and *maturation*. Taken together, these answers provide a unified theory of development.

The computational models which have the widest currency within the framework of the modern synthesis are (ideally) constrained to well formulated theories of particular domains. The paradigm for such theories is provided by natural language grammars. The logical next step is to search for quasi-grammatical, generative formalizations of non-linguistic domains. Thus, the programmatic evolution of cognitive science has been dominated by the search for self-consistent, autonomous and generative 'grammars', of action, of vision and so forth. Within this paradigm, goal and method coincide in the imperative of formalization: since computability is presupposed by the paradigm, only theories which constitute formal computational descriptions of the domain under study are acceptable candidates. The paradigm therefore introduces *closure*, at two distinct levels.

Closure$_1$. Only formally coherent (thus closed) systemic descriptions are deemed scientific, and unknown mechanisms defined over unspecified variables are unacceptable. This condition upon theory-building violates a fundamental 'open-ness' presupposition of science, that models should be non-prescriptive of mechanisms that cannot be evidentially supported. The common resort within the paradigm is to vague mechanisms and/or default characterization of values yielded by unknown mechanisms. This is a perfectly acceptable procedure, but it is in principle incompatible with the claim that model and reality coincide. In effect, the neo-rationalist paradigm is trapped within an irresolvable oscillation between correspondence and coherence models of truth.

Closure$_2$. The condition that models should be self-consistent (coherent) *within* domains, together with the aim that such models should be exhaustive (fully explanatory), not only inevitably leads to *ad-hoc* characterization of unknown variables/values, but also leads to non-isomorphism of terms *between* domains. Closure$_2$ is weaker than Closure$_1$, for co-operative *ad-hocing* can yield inter-domain compatibility, if not strict consistency. However, the net effect of Closure$_1$ + Closure$_2$ is to induce a preferred meta-theory in which not only cognition as a whole, but particular cognitive processes within specific domains, are relatively autonomous.

The result is a *modular theory of mind*, which, when given a physicalist interpretation according to the principle of the brain-as-a-representation, assumes the form of a modular theory of brain function. Such a perspective is articulated by, amongst others, Chomsky (1980), who explicitly advances a 'faculty' theory of mind and a modular theory of brain function. Marshall (1980), in his response to Chomsky's proposal, notes the similarity between this and the eighteenth century 'organology' of Franz-Joseph Gall.

There is no *a priori* reason to dismiss the thesis of modularization. Indeed, this thesis, implying a vertical division of brain function, is at least as plausible as quasi-evolutionary theories suggesting a horizontally specified cerebral functional architecture (e.g. MacLean, 1972). Modularization certainly does not imply strict independence (as opposed to relative autonomy) of different faculties. Rather, mind–brain is conceived in terms of co-operatively interacting functions, each specialized to different cognitive domains, and each pre-programmed with its own developmental timetable and computational system. Such a conception does, however, inevitably raise the issue of how the specialist modules communicate with each other, and in what sort of code. Put less formally, this resolves into the question: is there a language of the mind?

Growth and learning

The new organology of neo-rationalism is also a maturationist theory of development, in which there is little room for *learning* in the traditional sense. Chomsky (1980:14) goes so far as to state that, at least as far as *knowledge* is concerned, 'it is rather doubtful, in fact, that there is much in the natural world that falls under learning'. He relegates learning theory, as traditionally understood in psychology, to 'the study of tasks and problems for which we have no special design, which are at the borders of cognitive capacity or beyond, and are thus guaranteed to tell us very little about the nature of the organism'.

I shall return to the problem of evolutionary design which this theory poses. To continue, however, the appropriate model within the theory for cognitive development is growth, rather than learning. This maturationist position appears, at first sight, to fall in the mainstream tradition of American developmental psychology, as represented in the early twentieth century by such figures as Gesell and G. Stanley Hall; a tradition which explicitly repudiated the socially-oriented epigeneticism of Baldwin. Unlike traditional maturationism, however, the new synthesis is anti-naturalistic and anti-behaviourist. Its method is largely formalist, relying heavily upon computational models, and its focus is upon cognition rather than behaviour.

Traditional maturationism, by contrast, was both naturalistic in method (relying upon observation), and easily accommodated to a behaviourist paradigm. The possibility for such an accommodation was given by the universal acceptance, in the first part of this century, of the *reflex arc* as simultaneously the fundamental neuropsychological unit (the unit of association); the paradigm case of organism–environment transaction (the S–R link); and the basic mechanism of learning (through classical or operant conditioning). Such a paradigm, essentially empiricist in nature, can combine classical learning theory with the recognition that certain neural circuits are maturationally established, rather than established strictly through experience. It contrasts with the modern synthesis insofar as the latter sees classical learning as peripheral, is not founded upon the reflex arc, and substitutes internal constraints for external ones. Figure 14·2 illustrates the differences between the old and the new paradigms.

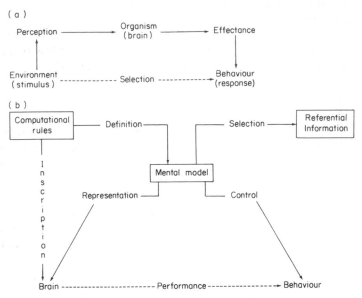

FIG. 14·2 From reflexology to representation. (a) The first synthesis: reflexology; (b) the modern synthesis: cognitivism.

Figure 14·2(a) representing the 'old' associationistic and reflexological synthesis, is simply a slightly elaborated version of the sort of 'stimulus-organism-response' diagram familiar from traditional accounts of learning theory. The important point to note is that, according to the view that it represents, it is the environment which *selects* (or 'elicits') behaviour, through either association-by-contiguity (classical conditioning) or reinforcement and extinction (operant conditioning). Developmental change can occur either through the maturation of pre-established neural circuits, or by means of the establishment of new circuits through learning. Classical learning theory, in fact, provides a perfect correlate, in terms of change within the lifespan of the individual organism, to the neo-Darwinian theory of natural selection of individual organisms in a phylogenetic timespan.

Both theories account for increased adaptation on the part of the organism by ascribing a primary role to the environment. The theories are also complementary insofar as maturational developmental change, whether morphological or behavioural, can be accounted for by means of natural selection—as in the case of 'instincts'—leaving learning theory to account for the residue of ontogenetic change. Further, this 'learned' residue is assumed to be of an extremely plastic nature—it is reversible and completely inconsequential as far as the genes transmitted by the individual are concerned. Classical learning theory is therefore predicated upon a division between instinctual and learned behaviour which precisely parallels Weismann's barrier in neo-Darwinian theory. Finally—and this is important to note—both neo-Darwinism and reflexology assign a central place to *behaviour*. Though much of the subject matter and evidence for Darwinian natural selection concerns morphological characters, it should not be forgotten that the *mechanism* of selection operates upon functioning organisms, upon the behavioural correlates or consequences of structural features, and not upon the structure itself. It is precisely this which distinguishes neo-Darwinian from Lamarckian evolutionary theory.

By contrast, the 'cognitivist' or computational model represented in Fig. 14·2(b) locates the control of behaviour in a 'mental model', constitutively defined by computational rules. This model, in order to represent a particular, local and current 'world-state', *selects* relevant referential information from environmental input. The selection process, however, is also a *construction* process, insofar as the value of referential information is itself defined over the computational rules: in this sense, the only *actual* 'environment' *is* the mental model. The real environment (supposing such to exist: as we shall see, this is not a *necessary* presupposition of the theory) enters into the mental model only as selected referential information (values upon which the computational rules subsequently operate).

A parallel process is assumed to operate in the brain. Where the mental model is defined in terms of the computational rules, the brain is assumed to have these rules pre-programmed or inscribed in it. And the access which the organism has to the environment is solely in terms of the mental model 'represented', at any given moment, in the brain. Further, given the contingency assumption, the systemic unit labelled 'brain' might more generally be labelled 'hardware'—denoting any physical system capable of performing the necessary representational/computational work. *Which* particular physical system is chosen to implement the model is relevant only insofar as the performance characteristics of the system will affect the actual behavioural output.

Radical nativism, mind and nature

The computational rules are therefore in no sense derived from the environment; rather, they are innate algorithms enabling the system to *select* relevant features of input in order to *construct* the environment. The selection and construction of the environment by the system, as opposed to the selection of behaviour by the environment, most fundamentally distinguishes the new from the old synthesis. This selective constructivism applies both to ontogenesis and to mature functioning, in that innate algorithms enable the subject to select from environmental input only those features which are either developmentally or currently relevant to the construction process in hand. A good example of the constructivist approach favoured by contemporary cognitive science is provided by accounts of visual perception which stress the importance of inferential procedures in permitting subjects to build coherent representations of scenes from ambiguous or incomplete visual information (e.g. Minsky, 1975; Oatley, 1978).

Such a conception raises the difficult problem of how it is that the cognitive system comes in the first place to 'know' which features of the 'real' environment are those which are necessary and sufficient for the construction of the representational environment. This question is as much philosophical as psychological—indeed, more so, for the psychological and computational solution is implied by the theory: such 'knowledge' is simply built in. Such a pragmatic solution, however, is unable to escape the charge that, although it explains everything, it thereby also explains nothing.

A more rigorous philosophical justification for such a procedure has been offered by Fodor (1976, 1980), who extends the 'rationalist deduction' a further, and crucial, step. Computational models of mind of necessity require that there exist a specifiable set of (referential) symbolic values. These values are defined over the rule system, which is itself non-derivable from empirically observable reality. The operation of the rules yields representations, and these representations, according to Fodor, *are* mental states. Fodor further argues that there can, in principle, be no adequate means of specifying the 'objects' of mental states in terms of the languages of the special sciences such as physics: thus, it must be assumed that the system-internal values, as well as the formal rule systems, are innate; either as atomic 'primitives', or in the form of further rule systems such as 'meaning postulates' (Carnap, 1956). The computational approach appears therefore to be committed to a radical nativism, predicated upon a dualistic distinction between mental objects and the actual objects of the real world.

This neo-rationalist argument is used by Fodor to defend a general philosophical approach to psychology which he calls 'methodological solipsism' (*see also* Putnam, 1981, for a critical discussion of this concept), consisting essentially in the proposition that the study of cognitive processes is to be seen as completely autonomous from the study either of the perceived environment or of their developmental history.

Such a position is, of course, uncongenial to a great many psychologists, since large chunks of their traditional subject matter are instantly thereby deemed to be either unscientific, or uninteresting, or both. The 'methodological solipsist' stance also highlights another basic feature of the neo-rationalist position: that is, its tendency to cast the problem of knowledge in terms of the mental states of individual subjects, rather than, or in opposition to, knowledge being conceived as a social process and product. To this extent, Fodor's methodological solipsism is simply an extreme variant of a wider tenet of modern cognitivism, which may be referred to as 'methodological individualism'.

As well as a philosophical and psychological aspect, however, the neo-rationalist synthesis also possesses a biological aspect. Thus, Fodor (1976) casts the theory in terms of an argument for an innate, computational 'language of the mind': a hypothesis which neatly ties in with the modularization thesis advanced by Chomsky. This aspect of neo-rationalism returns us to the problem of how such a complex representational system can possibly have emerged from biological evolution.

It will be recalled (page 340) that what I called the 'reflexological synthesis' provided a theory of learning which was, in its determinate features, theoretically interlocked with the neo-Darwinian theory of natural selection. No such theoretical concordance with evolutionary theory can be claimed for the modern, neo-rationalist synthesis, for two fundamental reasons. First, because of the cognitivist and constructivist insistence that the relevant 'environment' be selected by the subject, rather than vice-versa. Second, because the competence–performance distinction, rooted in the contingency assumption, accords to behaviour a secondary role vis-à-vis the computational mechanisms governing and controlling it.

Further, the argument-to-radical-nativism provided by Fodor has equally negative implications for evolutionary theory as it does for developmental theory. In the modern synthesis, 'representation', or computational knowledge, is seen as distinct from knowledge of the physical properties of the universe, which are considered as phenomenologically inaccessible. This conception of representation, which depends upon a strict distinction between mind and nature, is indicative of the positivist and physicalist inheritance of modern neo-rationalism. Despite, or because of, the mentalistic, rather than classically empiricist, orientation of the computational approach, it remains unable to account for the emergence of mind from nature, and is therefore in many respects profoundly anti-evolutionary in its implications. It should come as no surprise, then, that Chomsky, for example, continually stresses the (innate) discontinuity between human mental processes and those of other species, and the biological uniqueness of manifestations of human mentality such as language.

It is my argument, then, that although neo-rationalism does not explicitly disavow evolutionary theory, its deep structure provides no compelling reason for accepting it, and indeed may be read more as an argument in favour of special design. Thus, although its nativistic and maturationist bias appears at first sight to afford an easy accommodation with biology, the fact is that no mechanism is provided to account for the *origins* of representation in naturally evolving life. In this sense, neo-rationalism merely reflects the predominant, but implicit, *biologism* of contemporary psychology, rather than representing a genuine opening to interdisciplinary theorization.

THE 'ECOLOGICAL' ALTERNATIVE

The appeal of cognitivism has never been universal, and even within cognitive theory the computational approach has been subject to criticism in a number of counts. These range from its lack of concern with context and with actual behaviour, to its 'dehumanising' equation of machine with human intelligence (Boden, 1977, Weizenbaum, 1976, de Beaugrande, 1983). In this section, I shall outline alternative approaches which draw upon (mainly) naturalistic evidence and methodologies to construct accounts of human and non-human behaviour in their 'ecologically valid' contexts. I shall also attempt to evaluate the extent to which such approaches meet the implicit challenge of the neo-rationalist synthesis: to develop a convincing alternative account of representation and its development.

Ecological validity (1): context, rules and meanings

As has been noted, a basic feature of the computational approach is its lack of concern with behaviour as this actually occurs in natural situations. Cognitive science does, indeed, sometimes employ arguments based upon laboratory experiments with human subjects. More usually, however, at least in Artificial Intelligence, the 'data' are pre-theorized in a formal sense (in the way in which a grammar might count as data). This lack of concern with behavioural data opens the approach to the fundamental objection that, whatever else it is doing, it simply is *not* producing explanations of how subjects actually perform tasks involving reasoning, inference and so on in natural settings.

This criticism of lack of ecological validity has been levelled at the entire tradition of laboratory-based experimental psychology (Cole, Hood and

McDermott, no date; Neisser, 1976), but it has greatest force in relation to computational models. It is significant, furthermore, that the ecological validity argument, when applied to human psychology and behaviour, has been closely associated with attempts to reinstate the *social* dimension of cognition, particularly in relation to its development (Butterworth and Light, 1982; Flavell and Ross 1981).

There has, in recent years, been a growing recognition by developmental psychologists of the importance of studying the social-ecological matrix within which human development naturally occurs (Bronfenbrenner, 1979), and a renewed emphasis upon the socio-communicative origin and elaboration of cognitive processes (Bruner, 1975; Donaldson, 1978). These developments have led, too, to an appreciation of the nature of the experimental setting itself as a 'socio-dialogic context' (Karmiloff-Smith, 1979; Sinha and Walkerdine, 1978; Freeman, Sinha and Stedmon, 1982). These recent debates in developmental psychology, over the relative weight to accord to intra-individual and inter-individual cognitive processes, echo the Marxist-derived critique Vygotsky offered half a century ago, of Piaget's approach to genetic epistemology (Vygotsky, 1962, 1978).

The basic issue at stake here concerns the extent to which it is either possible or useful to attempt a theorization of individual mental representations and processes, without reference to the representation and evaluation of social actions by interpretive communities. This problem is, of course, fundamental to the human sciences, and strikes at the heart of the claims of the computational approach to scientificity. Contemporary approaches to the psychosocial and philosophical analysis of action (von Cranach and Harré, 1982) increasingly stress the dialectical interdependence between actors' mental states (beliefs, wants, intentions etc.) and interactants' evaluations of act-performance and outcome. This fundamentally communicative epistemology, rooted both in critical social theory (Apel, 1979; Bhaskar, 1979; Habermas, 1971), and in linguistic (particularly pragmatic) theory (Grice, 1975), is distinctly orthogonal to the main stream of cognitive science, although some socio-communicative and pragmatic computational models are perhaps in principle compatible with it (Schank and Abelson, 1977).

In general, it is probably inadvisable to offer a premature judgement on whether computational and socio-communicative approaches represent radically different, or complementary perspectives. However, two points may be noted. First, it is apparent that the methodological individualism inherent in the dominant, neo-rationalist paradigm is by no means the only possible broad perspective for cognitive and communicative theory. This by no means implies that a single, coherent and widely agreed social action-theoretic paradigm currently exists. However, advocates of widely-varying positions, ranging from naturalistic (Bhaskar, 1979) to hermeneutic (Shotter, 1978, 1982); and from Wittgensteinian (Hamlyn, 1978, 1982) and neo-Kantian (Putnam, 1981), to dialectical materialist (Sinha, 1982), all find some measure of common ground in their insistence upon the importance of *intersubjective agreement* as a key attribute of human rational thought in the widest sense. Such a perspective, I shall argue, is wholly compatible with a naturalistic psychology: hence the title of this chapter.

The second important point is that the exclusive focus on formalization of much current cognitive science is unconducive, in practice if not in principle, to systematic investigation of the pragmatic negotiation of behavioural meaning within episodes of social exchange. It is a moot point whether rigorous formalization of pragmatic rules is in principle possible—probably not. But so long as cognitive science relegates these rules to a separate category of 'performance', it will be unable to effect a fruitful rapprochement with many of the most exciting developments in contemporary linguistic theory (Andor, 1980; de Beaugrande and Dressler, 1981).

The possibility, and indeed necessity, for such a rapprochement is indicated by the fact that both computational and social action-theoretic approaches presuppose and theorize the category of *representation*: in this respect, they both differ significantly from the 'direct perception' approach which I discuss below. There remains, however, the problem of accounting in a principled way for the evolution and development of representation without resorting to the radical nativist arguments of neo-rationalism.

Ecological validity (2): direct perception *versus* representation

The ecological validity argument is not restricted to the issue of the interpretation of human behaviour within its social and communicative context. In fact, for most psychologists the phrase is predominantly associated with an approach which derives in a more straightforward fashion from the study of the adaptive circuits linking organisms and environments within ecosystems. The psychobiological premises of what is generally known as the 'ecological approach' are stated by Shaw and Turvey (1980:96) in the following terms:

> logically dependent . . . animal–environment synergy (or reciprocity) . . . identifies (a species of) animal and its environment as reciprocal components comprising an (epistemic) subsystem.

I shall attempt to expand this somewhat opaque statement in my discussion of the concept of behaviour on page 350. For now, the salient feature of the 'ecological approach' upon which I shall concentrate is its radical denial of the necessity of invoking the concept of representation at all in the explanation of behaviour, through its employment of the alternative concept of 'direct perception'.

If there is a single person to whom 'ecological psychology' owes its deepest debt, that person is J. J. Gibson. In this brief exposition, I shall concentrate upon Gibson's later formulations of his theory (Gibson, 1979); for a historical account of Gibson's developing thought, including his gradual break with behaviourism, *see* Costall (1981); for a discussion of varieties of 'ecological psychology', *see* Cutting (1982).

Gibson's life-long research interest was in the perception of *visual space*, as a structured, textured and complex psychobiological unity. His method of approach led him to a radical reformulation of classical psychophysics, in which the concept of 'ambient information' replaces previous terms such as 'stimulation' and 'sensation' which were associated with the 'reflexological synthesis'; and to a decisive break with cognitivist models of perception in which the 'percept' is constructed by the subject from degenerate or inadequate information.

Gibson held that the visual field is *directly specified* by the structure of light, the flow of optical texture. Classical psychophysics, he argued, based upon Newtonian optics, upon atomistic notions of the 'stimulus', and upon untenable philosophical distinctions between 'primary' and 'secondary' qualities, was unable to reconcile the *constancy* of our perception of objects with the apparent ambiguity, in real contexts, of the isolated 'cues' which indicate or 'represent' the objects. Hence the cognitivist argument in favour of an inferential supplementation of psychophysical stimulation. Gibson's alternative solution to this problem was to suggest that a *different* psychophysics, a non-Newtonian or *ecological optics*, was required, which would obviate the need for cognitive supplementation of perception[2].

Ecological optics departs from the geometrical and atomistic abstractions of classical optics by invoking the priority of *background* in specifying 'objectness' within a textured (rather than empty) space. Ecological space is conceived as a *surface*, textured by variate patterns of information, whose objective structure is specified by variables within the ambient light (gradients of shade, scatter, intensity etc.). The structure of light therefore contains all the information necessary for the 'pick-up' of invariant properties, such as those involved in motion perspective and object constancy, without the subject having to do any further representational or mental-inferential work.

The theory of direct perception also views the senses as exploratory, attentional organs, of an organism which is fully, actively engaged with its environment, by means of locomotion and through the variable articulation of body parts and orientation of sense organs. Thus, the organism is itself a *part* of the niche. In this sense, the theory of direct perception depends upon the wider propositions of the ecological approach: that a niche is a negotiated, ordered, spatio-temporally structured *relationship* between organism and habitat, in which behaviours are in part transformative of the very environment to which they are adapted.

In his final work (Gibson, 1979), Gibson introduced a further, and extremely important, concept into the theory of direct perception: this is the notion of *affordance*. As the name implies, 'affordances' consist in the properties of objects by virtue of which these objects serve to sustain or permit the actions of the organism. The concept of affordance, then, fixes the theory of direct perception within the wider propositions of the ecological approach. Affordances, maintained Gibson, are themselves directly specified by ambient optical information, actively 'picked up' by an organism which is, in every sense, *interested* in its environment. Objects exist, for the organism, as meaningful parts of the phenomenal world, precisely because they afford action — they are edible, climb-able, graspable etc. — and these affordances are 'grasped' by the perceptual system of the organism in its direct engagement with the optical array.

As Costall (1981) notes, the theory of affordances is an important new departure in Gibson's work, for it points towards a functionalist theory of perception in relation to adaptive action. As I shall argue, however, the theory of affordances is also a Trojan Horse within the Gibsonian epistemology.

To what extent, then, does the ecological approach offer a solution to the problem of representation? First, it escapes the central problem of rationalist (and empiricist) theories, that of the relationship between mental or intentional objects, and real objects, by denying the existence of 'percepts' as 'copies' of real objects. Objects are directly perceived, and the information picked up in direct perception is neither a mediating construct, nor an isomorphic 'picture' of the world, but a *real structure* which specifies objects by virtue of its internal, higher-order variables:

> the question is not how much [the retinal image] resembles the visual world, but whether it contains enough variations to account for all the features of the visual world.
>
> *(Gibson, 1956:62)*

The variables are defined, not in terms of the special predicates of theoretical physics, but in terms of local, 'human scale' (or organism scale) psychophysics (biophysics). To this extent, the ecological approach offers both a real alternative to neo-rationalism, and a formidable challenge to computational approaches in particular, and cognitive science in general. There is perhaps no reason why formal models should not be developed for direct perception processes; but if cognitive science is to proceed in such a direction, it will necessarily have to leave behind some of its long-cherished assumptions: not least those underlying the neo-rationalist synthesis.

Nevertheless, the ecological approach cannot, by definition, offer a solution to the problem of what have traditionally been called in psychology 'higher thought processes'—those processes operating, not upon the proximal environment, but upon symbolic representations. Granted 'direct perception' of the proximal environment, can we then simply graft a computational theory of mind onto it? This seems obviously unsatisfactory, for it leaves an unbridgeable gap between the world we *live* in, experientially, and the world we think about, remember, make plans and draw conclusions about. Of course, for the 'methodological solipsist' this presents no problem—but this takes us back, in an endless loop, to the beginning of the argument.

A radical Gibsonian might, perhaps, hold that representation is an unnecessary concept *tout court*, adopting a neo-behaviourist position and ruling 'mind' out of the bounds of science. As yet, however, behaviourism has singularly failed to provide an adequate riposte to Chomsky's critique, and such a step would undoubtedly be regressive.

A more interesting step would be to build upon the notion of 'affordance', suggesting that conceptual representations consist of mnemonically-stored 'abstractions' of affordances. In the final part of this chapter, I shall suggest something similar, but as I shall now show, this has serious (I think fatal) consequences for direct perception theory, at least as applied to human beings. Much of the human environment consists of *artefacts*, which indeed afford actions, but many of which resemble natural kind objects not even remotely. In what sense is it plausible to say that the functional affordances of a car, for example, or a telephone, are 'directly specified' in the optical array? A car-seat, or a steering wheel, might be said to 'afford' appropriate actions, in a Gibsonian sense, but these sort of affordances cannot, however you add them up, yield *knowledge* of how to drive a car, or what a car is: that is, of the socio-cultural 'affordances' of cars as complex artefacts within a social system.

It is precisely this sort of specific, 'procedural' knowledge (Winograd, 1975), 'knowing how' rather than, or as well as, 'knowing that', which is of greatest interest to many cognitive psychologists and linguists who wish to adopt a social action-theoretic approach to pragmatic understanding. I shall return to this issue below. For now, let us note that the above argument powerfully demonstrates that 'direct perception' cannot specify *knowledge*, whatever else it might specify (e.g. segregated objects of perception and attention). Even more seriously, it demonstrates the questionable status of 'affordance' as a purely *perceptual* category.

In fact, the argument can be applied to natural kind as well as to artefactual objects. To the untutored eye, a piece of rock might 'afford' merely the actions of kicking, using it as a missile, or an interesting paperweight, etc. To a geologist, however, it might 'afford' a hypothesis, evidence for a theory, an indication of a valuable resource etc. In effect, the whole notion of 'affordance' necessarily involves, at least for adult human subjects, knowledge and representation. Its introduction into ecological psychology means that the theory is no longer restricted to perception, but touches upon cognition too.

Unfortunately, ecological psychology has no theory of representation. Thus, if the concept of affordance is to serve as a starting point for such a theory, that concept will certainly require much critical revision. If neo-rationalism fails to provide a link between representation and behaviour, so, ultimately, does the theory of direct perception. In the final part of this chapter, then, I shall turn directly to the issue of that relationship, and examine it from the point of view of the epigenetic alternative.

BEHAVIOUR, REPRESENTATION AND EVOLUTION

There are a number of possible ways in which behaviour may be described. One way might be to limit one's descriptive terms to those which refer solely to the organism, or to its body parts, or their movements relative to one another. Such a description, however, seems inherently unsatisfactory as a description of *behaviour*, although it is a legitimate enterprise in itself. It seems more appropriate, in such a case, to speak of the physiology of movement, and while this may be an indispensable complement to the description of behaviour, the two are essentially different. For behaviour is *motivated* and *meaningful*, consisting not merely in movements, but in movements in relation to an environment. The movements of even the simplest of life forms are adapted to a particular niche. Even where such adaptations appear merely to consist in a one-to-one causal link between a specific type of environmental event and a specific type of movement, as in the S–R links of learning theory, or the fixed action patterns studied by ethologists, a full *behavioural* description necessarily involves reference to the environmental event structure, as well as the movement structure.

As we move to the consideration of more complex behaviours, the connection between behavioural and physiological descriptions becomes increasingly remote. Flight and predatory pursuit both involve rapid locomotion, but the meaning and 'cause' of the movements is different in the two cases. Ethological studies are often concerned with the detailed specification of particular *combinations* of movement, posture and attentional orientation which constitute a particular behavioural routine; but the degrees of freedom in the co-articulation and sequential organization of micro-behaviours within macro-behavioural units increase rapidly with increased phylogenetic complexity. Indeed, it is the relative freedom of behaviour from 'stimulus', and of behavioural goal or end from the detailed motoric means of attaining that goal, which typifies 'intelligent adaptation'.

At a minimum, then, the description of behaviour necessarily involves reference to an environment to which that behaviour is adapted. This is, of course, the basic tenet of the ecological approach. However, the description of intelligent behaviour requires, additionally, and as we have seen, reference to the *representations* of the environment which the organism utilizes (at the highest level, intentional states such as beliefs and desires), which specify a complex repertoire of behaviours which the environment actually or potentially affords, and which contribute to the maintenance or expansion of adaptive fit.

An essential feature of behaviour is that it is *active*, going beyond the bounds of the organism to affect the environment and change it — even if this change simply consists in the substitution of one spatio-temporal segment of the environment for another, as in locomotion or migration. This active nature of behaviour was stressed by Piaget, who defined behaviour as:

> all action directed by organisms towards the outside world in order to change conditions therein or to change their own situation in relation to these surroundings.
>
> *(Piaget, 1979;ix)*

Piaget makes it clear that automatic self-regulation, such as breathing, is *not* behaviour, although it is quite clearly adaptive. He goes on to state that 'behaviour is teleological action aimed at the utilization or transformation of the environment and the preservation or increase of the organism's capacity to affect this environment', and that, further, 'the ultimate aim of behaviour is nothing less than the expansion of the habitable — and, later, the knowable — environment' (Piaget, 1979, pp. x and xviii). It is in this sense that he saw behaviour as the motor and leading edge of evolution and development.

In what sense is behaviour adaptive?

The straightforward neo-Darwinian answer to this question is: behaviour is adaptive insofar as it contributes to inclusive fitness. But there is another common sense in which we consider behaviour to be adaptive, that is, when it is *appropriate* to a particular situation or local environmental segment. For adaptation in this latter sense, Piaget (1979) frequently uses the expression *adequation*.

The two senses of adaptation are, to be sure, related: the inclusive fitness of an organism which consistently behaved inadequately would, indeed, be poor. But a particular behaviour confers selective advantage only in relation to a given context of occasion, for which its instantiation is appropriate (adequate) in terms of local goals. A fundamental deficiency of neo-Darwinism is its inability to conceptualize this distinction. This deficiency is related in turn to neo-Darwinism's difficulty in explaining the central paradox of evolution: why, if an organism is 'adapted' in the sense of inclusive fitness, should evolution occur at all? Where neo-Darwinism invokes purely exogenous selective 'pressure', an epigenetic approach sees the dynamism of evolution as consisting in an endogenously expanding circuit of adaptation, driven by the exploratory fallibility of behaviour. That is to say, it is the possibility of 'inadequacy' that underlies the leading role of behaviour in evolution and development.

Behaviour is goal-directed, and intelligent behaviour is representationally mediated. It is a characteristic of representations that they can be both *erroneous* and *impoverished*. Behaviour, as Piaget emphasized, is defined foremost in terms of goals, and only secondarily in terms of the degree of coincidence between the goal of the behaviour — its 'intended' outcome — and its *actual* outcome, which are frequently divergent; or between 'representation' and 'reality', which are frequently also divergent. We all do make mistakes.

Not only can behaviours be both situationally inadequate and/or unsuccessful; but also behaviours, and overall behavioural repertoires and strategies, can result in environmental consequences which, though a part of the adaptive circuit linking organism to environment, are not seen as the proximal goal of the behaviour. A 'path' may, for example, be an unintended consequence of repeated locomotion from one place to another, but it is nevertheless a useful one. Ecologists emphasize that species *shape* their niche, as well as being shaped by it, and such environmental shaping is constitutive of, as well as adaptive to, the species' survival strategy.

Further, such shaping, for all species, including our own, can introduce distal consequences — food shortages, erosion, pollution, competition with other species — which are *outside* the initial circuit of adaptation. That useful path may be disadvantageous if a predator cottons on to the idea of lying in wait at certain times of the day or night. Such disadvantageous consequences will require a further set of adaptations if survival is to be assured and extinction averted—or, more generally, unpleasurable difficulties surmounted.

Accommodation and assimilation (1): Baldwin

Such secondary adaptations were referred to by the epigenetic psychobiologist James Mark Baldwin[3] as accommodations, and he saw them as playing a crucial role in both phylogenesis and ontogenesis. Accommodation constitutes a mechanism of active selection on the part of the organism, oriented to enhanced adequation, which prescribes a new, or wider, *range* of dynamic adaptation, rather than conferring a fixed selective advantage. The concept of accommodation sharply contrasts, therefore, with such notions as 'Evolutionary Stable Strategy' favoured by neo-Darwinists: accommodations are active, *organismically* selected responses to induced environmental demands, whereas neo-Darwinian strategies are *environmentally* selected from random behavioural variation.

The cumulative process of accommodation leads to what Baldwin called 'organic selection', which he contrasted with the negative and purely limiting role played by natural selection in evolution. Organic selection is a process independent of natural selection, by which individuals extend their adaptive range by accommodation, thus surviving to 'permit variations oriented in the same direction to develop through subsequent generations, while variations oriented in other directions will disappear without becoming fixed' (Baldwin, 1897: 181). Thus, 'individual modifications or accommodations supplement, protect or screen organic characters and keep them alive until useful congenital variations arise and survive by natural selection' (Baldwin, 1902:173). This effect is still referred to as the 'Baldwin effect'. Baldwin conceived of it as playing an increasingly decisive role in evolution, since he maintained that the range of accomodatory plasticity, and hence the importance of organic selection, increased with phylogenetic complexity. Thus, organic selection, combined with imitation (*see below*), increasingly direct, rather than follow, the course of natural selection. As Piaget (1979) noted, Baldwin's theory of organic selection anticipates the concept of 'genetic assimilation' (Waddington, 1975), and the mechanism of 'phenocopy', whereby a phenotypical adaptation is replaced by a variation in the genotype (Ho, this volume).

The process of accommodation, for Baldwin, also underlies the ontogenetic emergence of higher mental processes. Accommodation permits the integration of behaviour with a complexly-structured and variably-respondent environment, including the behavioural actions of other individuals in response to the actions of the infant. A key example given by Baldwin is the development of imitation, which he sees not as a passive 'copy' of the model, but as an active effort by the infant to overcome the resistance of her own body. The first accommodations are those which 'select' from amongst the infant's habitual actions in order to repeat pleasurable experiences. These 'circular reactions' form the basis for subsequent accommodations to the actions which they provoke in others, giving rise to imitations.

Imitations are subsequently stored as an associative 'net' enabling new events and objects to be *assimilated* to the products of past accommodations. According to Baldwin, the earliest representations are therefore mnemonically 'fixed' accommodations. These early representations, however, are inadequate to the full complexity of reality, which resists assimilation to the infant's primitive intentional accommodations. This leads to what Baldwin calls an 'embarrassment', the awareness of 'an inevitable mismatch between what the infant expects and wants and the behaviour of objects' (Russell, 1978:54). The dialectical motor of development, then, consisted for Baldwin in precisely what I have called 'the exploratory fallibility of behaviour'. Subsequent accommodations are also necessitated by the interventions of adults, who draw the child's attention to the inadequacy of her accommodatory representations and actions, guiding the child towards a system of socially-established and intersubjectively agreed judgments.

Baldwin's genetic epistemology is *functionalist*. The notions of accommodation, assimilation and organic selection designate functional mechanisms for the elaboration and progressive adaptation of behaviour to the environment. This environment *pre-exists* the organism (Baldwin, unlike Piaget, did not adopt a radical constructivist approach to knowledge and representation), but is also respondent to the actions of the organism. It is in the divergence between the intended or desired outcomes of behaviours — guided by what Baldwin calls 'interests' — and their actual outcomes, as either directly perceived through the 'control' exerted by reality, or as socially transmitted through the 'mediated control' exerted by other subjects, that accommodatory representations, and ultimately the linguistic concepts of predication and implication, have their origin. Thus, it is the contradiction between need/desire and reality, and the gap between actual and imagined adequation, which stands at the heart of Baldwin's dialectic.

Baldwin's genetic epistemology is also as much social as it is psychobiological in its orientation. In this respect, and in the importance it accords to language and communication, it has evident affinities with Vygotsky's socio-genetic approach, and is appreciably different in emphasis from Piaget's treatment of the same themes, to which I now turn.

Accommodation and Assimilation (2): Piaget

As Russell (1978:86) points out, although Baldwin did more than merely to prefigure Piaget's theory, Piaget equally did more than to simply extend Baldwin's theory. In fact, the two accounts differ more than is at first suggested by their use of the same terminology. In the first place, Piaget inverted the priority assigned by Baldwin to accommodation over assimilation. Second, he introduced an additional key term, *equilibration*. Third, in contrast to Baldwin's functionalism, Piaget's theory is essentially structuralist; his three functional concepts, of which the most important is equilibration, are best understood as constructs intended to account for the problem of structural stability and transformation. Fourth, Piaget emphasized the *coordination of action within the individual* as the basis of intelligence, and his main theoretical propositions concerned the structural elaboration of such coordinations through the stages of sensori-motor and operative intelligence.

As a consequence of these differences, Piaget's theory is a more *radical* epigeneticism than Baldwin's. Whereas Baldwin implicitly accepted the givenness of the environment, both bio-physical and social, in relation to the subject, Piaget emphasized the literal *construction* of reality by the child. This does not mean that Piaget considered himself to be an 'idealist', for he steadfastly insisted that the constructions of intelligence were *necessary*, though not *a priori*, to use his own (questionable) terminology.

The necessity of these constructions (or coordinations) resides, according to Piaget, in their double aspect as, on the one hand, biological adaptations, and on the other, as epistemological universals. Piaget's concern was with categories or concepts of wide generality: space, time, causality and so forth. He was less interested in the development of specific concepts permitting the classification and linguistic labelling of objects, which Vygotsky, for example, was concerned to investigate, than in the 'logico-mathematical' operations defining different stages of cognitive development. In this respect, Piaget's enterprise, as he frequently acknowledged, consisted in an attempt to biologize and geneticize the categories which Kant saw as being necessarily presupposed by human reason. As he put it, the coordinations are 'the necessary result of psychogenetic constructions yet conform to a timeless and general standard' (Piaget, 1977:25).

The coordinations constituting cognitive systems are viewed as literal extensions of the biological autoregulative mechanisms (Piaget frequently cites the concepts of homeostasis and homeorhesis) enabling the organism to maintain itself in dynamic equilibrium in relation to its environment. The coordinations thus function as *assimilatory schemata*, analogous to the behavioural and physiological systems which enable the organism to assimilate food. Piaget frequently uses the metaphor of 'alimentation' when speaking of cognitive activity. All intelligence presupposes and depends upon assimilation, by which alone the structure of reality can be apprehended, and upon which perception itself depends (contrast this with Gibson's theory of direct perception).

Accommodation, however, is also implied by the very process of assimilation: for every instance of adaptive behaviour or cognition must be precisely fitted to the particular object, event or logico-mathematical structure to which it is directed. In order to grasp an object, the kinematic structure and dynamic configuration of the hand and arm musculature must anticipate the shape and position of the object: and such accommodations extend beyond the sensori-motor to the operations of logico-mathematical reasoning. Adaptation, for Piaget, is brought about by a process of *equilibration* between assimilation and accommodation.

Two points are worthy of note, here. First, accommodation for Piaget is an anticipatory and corrective concept which is dialectically necessitated by every assimilative act. In this conception, he diverges from Baldwin, for whom accommodation is an essentially reactive process necessitated by the failure of existing accommodatory capacities to overcome object-resistances. Second, the notion of accommodation is closely linked for Piaget to what he calls the 'figurative' aspect of intelligence, that which treats of objects in their specificity, rather than to the 'operative' and ultimately logico-mathematical aspect of intelligence, which derives not from objects, but from the subjects actions upon objects. This latter, operative aspect of intelligence is seen by Piaget as both more fundamental, and more closely linked to assimilation.

Thus, assimilation for Piaget plays the more basic role. Insofar as assimilation implies accommodation in every instance, it is both more *active* than it was for Baldwin, who saw it as an essentially passive registration; and more *progressive*: it is the 'disequilibrations' brought about by the inadequation of assimilation, and *not* the failure of accommodation, that necessitate and provoke re-equilibration at a higher level. Piaget, in his later writings, was very clear about this (Piaget, 1977, p.39):

> We must distinguish two important categories of disturbances. The first includes those which are opposed to accommodations: resistances of objects, obstacles to reciprocal assimilations of schemes or subsystems, etc. In short, these are the reasons for failures or errors of which the subject becomes more or less aware; the corresponding regulations include negative feedback. The second category of disturbance, *the source of nonbalance* (my italics), consists of gaps which leave requirements unfulfilled and are expressed by the

insufficiency of a scheme . . . The gap, functioning as a disturbance, is therefore always defined by an already activated scheme of assimilation, and the
corresponding regulation then includes a positive feedback which prolongs the
activity of the scheme [pp.18–19] . . . The object not yet assimilated and not
immediately capable of being assimilated constitutes an obstacle . . . and a new
accommodation is then necessary. But as assimilation and accommodation
constitute two poles, and not two distinct behaviours, it is clear that the new
assimilation plays the construction role, and the new accommodation that of
compensation.

In summary, Piaget's constructivist epigeneticism is radically endogenous
and subject-oriented, but its criteria of objectivity are timeless formal
abstractions. In Baldwin's functionalist epigeneticism, on the other hand,
overall systemic development is driven by endogenous processes awakened
by exogenous changes (themselves brought about in part by the subject),
and the criteria of objectivity are those of socially-negotiated, intersubjective agreement.

Adaptation, affordance and representation

It is tempting to suggest that the concept of affordance, as proposed by
Gibson, constitutes a reciprocal construct, in terms of the environmental
niche, to that of assimilation, as proposed by Piaget, in terms of organismic
adaptation. For it is in virtue of its affordances that an object offers itself to
organismic assimilation; and to the extent that the object's affordances fail
to correspond to the assimilatory schemes of the organism—that is, the
object *resists* assimilation—then the organism must perforce accommodate
its schemes and actions to the 'new' properties afforded—that is, to extend
its assimilatory capacities.

Such an interpretation receives support from the extension of Gibson's
theory by Shaw, Turvey and Mace (1982), who propose an organismic
counterpart of environmental affordance in the concept of 'effectivity'. The
'effectivities' of the organism, on this account, would correspond roughly
to those assimilatory schemes adapted to the affordances of the niche.

The human environment, however, as I have noted, has always-and-
already been *intentionally shaped* by previous generations of human agents
into a material culture, consisting of artefacts and symbols embedded
within social practices and institutions. Thus, the physical environment of
the human infant is *meaningful in its material structure*, and *represents*
human consciousness and intentionality. Representation, in the epigenetic
viewpoint, is no longer to be seen as 'merely' mental: consciousness and
knowledge are 'inscribed', not just in brains and nervous systems, but also
in artefacts, institutions, practices, symbols, utterances and languages.

Representation, like behaviour, extends beyond the boundary of the individual organism. The human infant, in development, is engaged in an *accommodatory effort after meaning*, whereby culture and representation is assimilated anew by every generation.

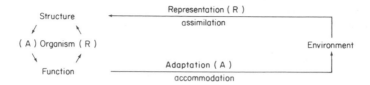

FIG. 14·3 An epigenetic approach to representation and adaptation.

Representation, from this viewpoint, is the structural realization of adaptive action, and adaptation (adequation) is the active engagement of the subject in the use and acquisition of representational systems in pragmatic contexts. Thus, the circuit of assimilation and accommodation is also the circuit of representation and adaptation (*see* Fig. 14·3).

In Fig. 14·3, the directionality of the arrows designates the directionality of *shaping processes:* through adaptive functioning, the organism shapes the environment, including its representational structure; the environment also shapes the organism through being represented in the structure of the organism. Equally, the structure and function of the organism itself mutually shape each other. Just as the environment is structurally represented in the organism, so too does the adaptive functioning of the organism in the environment acquire a material representation in organismic structure; and just as organismic function is adapted to the environment, so too is organismic structure adapted to organismic function.

Representation and adaptation are regarded here as 'equilibrated' or more or less stable *products* of a circuit of accommodation and assimilation, which is the dynamic moment of structural representation/ functional adaptation. Figure 14·3 can therefore be seen as an attempt to reconcile Baldwin's functionalism with Piaget's structuralism.

No distinction is drawn between 'cognitive' and 'morphological' structure: it is assumed that the organism progressively 're-represents' neurologically systems of representation existing in the environment. Equally, behavioural and mental function are assumed to be aspects of a continuous, asymptotic adaptive accommodatory engagement with a reality which is representational in its material structure. No assumptions are made about the relative priorities to accord to assimilation and accommodation. As a hypothesis, one might suppose that functional accommodations to augmented environmental (including social) complexity, when re-represented by other subjects, resist structural assimilation and thus provoke the kind of internal disequilibration that Piaget proposed. This proposal is, however, purely speculative, and by no means excludes the possibility that multiple developmental mechanisms operate in parallel or in alternation.

CONCLUSION

Throughout this chapter, two themes have been stressed. One is the central importance of the concept of representation in contemporary cognitive theory, and the other is the key role of interpretive communities and social interaction in human behavioural and cognitive development. I have argued that neo-rationalist approaches to cognition are misleading, both because they have no adequate theory of social-interactive, pragmatic understanding, and because the radical nativist view they adopt towards development is unsatisfactory from a psychological and a biological perspective. I have also argued that mainstream 'ecological' psychology is handicapped by its inability to theorize the concept of representation, which in turn is related to its refusal to treat of human culture and knowledge in its specificity. In outlining an epigenetic and socio-naturalistic alternative, I do not wish to suggest that general theoretical frameworks are a substitute for the empirical study of human developmental processes. The value of a theoretical analysis lies more in its delimitation of the sort of questions that are most fruitfully submitted to empirical reality.

To conclude optimistically, then, there are many signs that psychologists are turning increasingly to the study of the actual deployment of communicative and cognitive abilities by children in familiar settings, as both an alternative, and a complement, to the long established habit of studying children's performance in unfamiliar settings, using unfamiliar tasks. Irrespective of the particular theoretical views that I have advanced in this chapter, it is probably true to say that a socio-naturalistic *practice* of developmental psychology now enjoys wider support than at any previous time.

Notes

(1). *Rationalism* is traditionally held to be the doctrine that ideas, or aspects of knowledge, are innate; and has historically been opposed to *empiricism*, the doctrine that knowledge is acquired by experience.
(2) It is interesting to note the similarities between Gibson's proposals, and those of

Goethe, first published in 1840, regarding the perception of colour. Goethe, too, opposed Newtonian optics with 'naturalistic' optics:

> The eye may be said to owe its existence to light, which calls forth, as it were, a sense that is akin to itself; the eye, in short, is formed with reference to light, to be fit for the action of light; the light it contains corresponding with the light without . . . [the aim of Goethe's theory being] to rescue the attractive subject of the doctrine of colours from the atomistic restriction and isolation in which it has been banished, in order to restore it to the general dynamic flow of life and action which the present age loves to recognise in nature.
>
> *Goethe, 1970: liiliv*

(3) The most extensive discussion of Baldwin's theory, and its relation to that of Piaget, is to be found in Russell (1978), upon which this account is partly based.

REFERENCES

Andor, J. (1980). Some remarks on the notion of competence. *Behav. Brain Sci.* **3**, 15–16.

Apel, K. O. (1979) The common presuppositions of hermeneutics and ethics: types of rationality beyond science and technology. *Res. Phenomenol.* **9**, 36.

Baldwin, J. M. (1897). 'Le Développement mental chez l'enfant et dans la race'. Cited in Piaget (1979:22).

Beaugrande, R. de (1983). Freudian psychoanalysis and information processing: towards a new synthesis. Technical Report NL-22, University of Florida, Gainesville.

Beaugrande, R. de, and Dressler, W. (1981). 'Introduction to Text Linguistics'. Longman, London and New York.

Bhaskar, R. (1979). 'The Possibility of Naturalism'. Harvester, Brighton.

Boden, M. (1977). 'Artificial Intelligence and Natural Man'. Hassocks, Harvester.

Bronfenbrenner, U. (1979). 'The Ecology of Human Development'. Harvard University Press, Cambridge, Massachusetts and London.

Bruner, J. S. (1975). From communication to language: a psychological perspective. *Cognition* **3**, 225–287.

Butterworth, G. and Light, P. (*Eds*) (1982). 'Social Cognition: studies in the development of understanding'. Harvester, Brighton.

Carnap, R. (1956). 'Meaning and Necessity'. University of Chicago Press, Chicago.

Chomsky, N. (1959). Review of Skinner's 'Verbal Behaviour'. *Language* **35**, 26–58.

Chomsky, N. (1980). Rules and representations. *Behav. Brain Sci.* **3**, 1–15.

Cole, M., Hood, L. and McDermott, R. (no date). Ecological niche-picking: ecological invalidity as an axiom of experimental cognitive psychology. Unpublished manuscript from the Rockefeller University.

Costall, A. (1981). On how so much information controls so much behaviour. *In* 'Infancy and Epistemology' (G. Butterworth, *Ed.*). Harvester, Brighton.

Cranach, M. von and Harre, R. (*Eds*) (1982). 'The Analysis of Action'. Cambridge University Press, Cambridge.

Cutting, J. (1982). Two Ecological Perspectives: Gibson *versus* Shaw and Turvey. *Am. J. Psychol.* **95**, 199–222.

Donaldson, M. (1978). 'Children's Minds'. Fontana/Open Books, Glasgow.

Flavell, J. and Ross, L. (*Eds*) (1981). 'Social Cognitive Development'. Cambridge University Press, Cambridge.

Freeman, N., Sinha, C. and Stedmon, J. (1982). All the cars—which cars? From word meaning to discourse analysis. *In* 'Children Thinking through Language' (M. Beveridge, *Ed.*). Edward Arnold, London.

Fodor, J. (1976). 'The Language of Thought'. Harvester, Hassocks.

Fodor, J. (1980). Methodological solipsism considered as a research strategy in cognitive science. *The Behavioural and Brain Sciences* **3**, 63–73.

Goethe, J. W. von (1970). 'Theory of Colours' (translated by C. L. Eastlake). MIT Press, Cambridge, Massachusetts.

Grice, H. P. (1975). Logic and conversation. *In* 'Syntax and Semantics: 3, Speech Acts' (P. Cole and J. Morgan, *Eds*). Academic Press, New York.

Habermas, J. (1971). 'Towards a Rational Society'. Heinemann, London.

Hamlyn, D. (1978). 'Experience and the Growth of Understanding'. Routledge and Kegan Paul, London.

Hamlyn, D. (1982). What exactly is social about the origins of understanding? *In* 'Social Cognition: studies in the development of understanding' (G. Butterworth and P. Light, *Eds*). Harvester, Brighton.

Haugeland, J. (1978). The nature and plausibility of cognitivism. *Behav. Brain Sci.* **1**, 215–226.

Johnson-Laird, P. N. (1980). Mental models in cognitive science. *Cognitive Science* **4**, 71–115.

Karmiloff-Smith, A. (1979). 'A Functional Approach to Child Language'. Cambridge University Press, Cambridge.

MacLean, P. D. (1972). Cerebral evolution and emotional processes. *Ann. NY Acad. Sci.* **193**, 137–149.

Marshall, J. C. (1980). The new organology. *Behav. Brain Sci.* **3**, 23–25.

Minsky, M. (1975). A framework for representing knowledge. *In* 'The Psychology of Computer Vision' (P. Winston, *Ed.*). McGraw Hill, New York.

Neisser, U. (1976). 'Cognition and Reality'. Freeman, San Francisco.

Oatley, K. (1978). 'Perceptions and Representations'. Methuen, London.

Piaget, J. (1977). 'The Development of Thought: equilibration of cognitive structures'. Blackwell, Oxford.

Piaget, J. (1979). 'Behaviour and Evolution'. Routledge and Kegan Paul, London.

Putnam, H. (1981). 'Reason, Truth and History'. Cambridge University Press, Cambridge.

Russell, J. (1978). 'The Acquisition of Knowledge'. Macmillan, London.

Schank, R. C. and Abelson, R. (1977). 'Scripts, Plans, Goals and Understanding'. Academic Press, New York.

Shaw, R. and Turvey, M. T. (1980). Methodological realism. *Behav. Brain Sci.* **3**, 94–97.

Shaw, R., Turvey, M. T. and Mace, W. (1982). Ecological psychology: the consequences of a commitment to realism. *In* 'Cognition and the Symbolic Processes' (W. Weimer and D. Palermo, *Eds*), Vol. 2 Lawrence Earlbaum, Hillsdale, New Jersey.

Shotter, J. (1978). The cultural context of communication studies: methodological and theoretical issues. *In* 'Action, Gesture and Symbol: the emergence of language' (A. Lock, *Ed.*). Academic Press, London and New York.

Shotter, J. and Newson, J. (1982). An ecological approach to cognitive development: implicate orders, joint action and intentionality. *In* 'Social Cognition: studies in the development of understanding (G. Butterworth and P. Light, *Eds*). Harvester, Brighton.

Sinha, C. (1982). Negotiating boundaries: psychology, biology and society. *In* 'Towards a Liberatory Biology'. Allison and Busby, London.

Sinha, C. and Walkerdine, V. (1978). Children, logic and learning. *New Soc.* **43**, 62–64.

Vygotsky, L. S. (1962). 'Thought and Language' MIT Press, Cambridge, Massachusetts.

Vygotsky, L. S. (1978). 'Mind in Society: the development of higher psychological processes' (M. Cole, V. John-Steiner, S. Scribner, E. Souberman *Eds*). Harvard University Press, Cambridge, Massachusetts and London.

Waddington, C. H. (1975). 'The Evolution of an Evolutionist'. Edinburgh University Press, Edinburgh.

Weizenbaum, J. (1976). 'Computer Power and Human Reason'. Freeman, San Francisco.

Wilson, E. O. (1975). 'Sociobiology: the new synthesis'. MIT Press, Cambridge, Massachusetts.

Winograd, T. (1975). Frame representations and the declarative-procedural controversy. *In* 'Representation and Understanding' (D. G. Bobrow and A. Collins, *Eds*). Academic Press, London and New York.

Author Index

Subject Index